DEEP-SEA FISHES

This is Volume 16 in the
FISH PHYSIOLOGY series
Edited by William S. Hoar, David J. Randall, and Anthony P. Farrell

A complete list of books in this series appears at the end of the volume.

DEEP-SEA FISHES

Edited by

DAVID J. RANDALL
Department of Biology
University of British Columbia
Vancouver, British Columbia
Canada

ANTHONY P. FARRELL
Department of Biological Sciences
Simon Fraser University
Burnaby, British Columbia
Canada

ACADEMIC PRESS
San Diego   London   Boston   New York   Sydney   Tokyo   Toronto

Front cover photograph: *Anoplogaster cornuta*. © by Norbert Wu/Mo Yung Productions

This book is printed on acid-free paper. ∞

Copyright © 1997 by ACADEMIC PRESS

All Rights Reserved.
No part of this publication may be reproduced or transmitted in any form or by any means, electronic or mechanical, including photocopy, recording, or any information storage and retrieval system, without permission in writing from the publisher.

Academic Press
a division of Harcourt Brace & Company
525 B Street, Suite 1900, San Diego, California 92101-4495, USA
http://www.apnet.com

Academic Press Limited
24-28 Oval Road, London NW1 7DX, UK
http://www.hbuk.co.uk/ap/

Library of Congress Catalog Card Number: 76-84233

International Standard Book Number: 0-12-350440-6

PRINTED IN THE UNITED STATES OF AMERICA
97 98 99 00 01 02 QW 9 8 7 6 5 4 3 2 1

CONTENTS

Contributors ix
Preface xi
Norman Bertram "Freddy" Marshall xiii

1. What Is the Deep Sea?
Martin V. Angel

  I. The Nature of Water 2
 II. Ocean Gradients 10
III. Morphology of Ocean Basins 18
 IV. Biophysics and Oceanic Food Webs 30
    References 37

2. Systematics of Deep-Sea Fishes
Stanley H. Weitzman

  I. Introduction 43
 II. A Classification of Living Fishes Occurring near or below 500 to 600 m, with an Annotated List of Deep-Sea Fish Orders and Families 46
    References 74

3. Distribution and Population Ecology
Richard L. Haedrich

  I. How Many Deep-Sea Species Are There? 79
 II. Pelagic Habitats 82
III. Demersal Fauna: Shelf, Slope, and Rise 83
 IV. Distribution Patterns 83

V. Feeding Relationships 99
VI. Age Determination 102
VII. Reproductive Strategies 103
References 106

4. Feeding at Depth

   John V. Gartner, Jr., Roy E. Crabtree, and Kenneth J. Sulak

   I. Introduction 115
   II. Feeding Habits of Deep-Sea Fishes 118
   III. Patterns in the Diets of Deep-Sea Fishes 128
   IV. Sources of Food in the Deep Sea 172
   V. Deep-Sea Energetics Related to Feeding 176
   VI. Future Directions in Deep-Sea Fish Research 180
   References 182

5. Buoyancy at Depth

   Bernd Pelster

   I. Introduction 195
   II. The Problem of Buoyancy 196
   III. Swim Bladder Function 201
   IV. Lipid Accumulation 214
   V. Watery Tissues 223
   VI. Hydrodynamic Lift 227
   VII. Conclusions 229
   References 230

6. Biochemistry at Depth

   Allen G. Gibbs

   I. Introduction 239
   II. Effects of Pressure on Biochemical Systems: Protein Interactions and Enzyme Kinetics 241
   III. Tolerance Adaptations: Maintenance of Biochemical Function in the Deep Sea 244
   IV. Capacity Adaptation: Biochemical Correlates of Organismal Metabolism 263
   V. Future Directions: Phylogenetic and Molecular Approaches 268
   References 271

CONTENTS

7. Pressure Effects on Shallow-Water Fishes
   Philippe Sébert

I.	Introduction	279
II.	The Fish as a Model	280
III.	Methods	282
IV.	Effects of Short-Term Pressure Exposure	283
V.	Acclimatization of Fish to Hydrostatic Pressure	299
VI.	Comparison of Shallow-Water Fishes and Deep-Water Fishes	307
VII.	Conclusion	313
	References	314

8. Sensory Physiology
   John Montgomery and Ned Pankhurst

I.	Introduction	325
II.	Olfaction/Chemoreception	326
III.	Vision	328
IV.	Touch	333
V.	Octavolateralis Systems	333
VI.	General Comments	342
	References	346

9. Laboratory and *in Situ* Methods for Studying Deep-Sea Fishes
   Kenneth L. Smith, Jr., and Roberta J. Baldwin

I.	Introduction	351
II.	Laboratory Studies	352
III.	*In Situ* Studies	359
IV.	Future Directions	373
	References	375

INDEX	379
OTHER VOLUMES IN THE FISH PHYSIOLOGY SERIES	387

CONTRIBUTORS

Numbers in parentheses indicate the pages on which the authors' contributions begin.

MARTIN V. ANGEL *(1), Southampton Oceanography Centre, Southampton SO14 3ZH, England*

ROBERTA J. BALDWIN *(351), Marine Biology Research Division, Scripps Institution of Oceanography, University of California, San Diego, La Jolla, California 92093*

ROY E. CRABTREE *(115), Florida Marine Research Institute, Department of Environmental Protection, St. Petersburg, Florida 33701*

JOHN V. GARTNER, JR. *(115), Department of Natural Science, St. Petersburg Junior College, St. Petersburg, Florida 33711*

ALLEN G. GIBBS *(239), Department of Ecology and Evolutionary Biology, University of California, Irvine, Irvine, California 92697*

RICHARD L. HAEDRICH *(79), Department of Biology, Memorial University, St. Johns, Newfoundland A1B 5S7, Canada*

JOHN MONTGOMERY *(325), School of Biological Sciences, University of Auckland, Auckland, New Zealand*

NED PANKHURST *(325), Department of Aquaculture, University of Tasmania, Launceston, Tasmania 7250, Australia*

BERND PELSTER *(195), Institut für Physiologie und Limnologie, Universität Innsbruck, A-6020 Innsbruck, Austria*

PHILIPPE SÉBERT *(279), Laboratoire de Physiologie, UFR Médecine, 29285 Brest, France*

KENNETH L. SMITH, JR. *(351), Marine Biology Research Division, Scripps Institution of Oceanography, University of California, San Diego, La Jolla, California 92093*

KENNETH J. SULAK *(115), Florida-Caribbean Science Center, Biological Resources Division, U. S. Geological Survey, Gainesville, Florida 32653*

STANLEY H. WEITZMAN *(43), Division of Fishes, National Museum of Natural History, Smithsonian Institution, Washington, District of Columbia 20560*

PREFACE

The oceans constitute three-quarters of the earth's surface, and the major portion of the volume of the oceans can be classed as deep sea. The now-classic text *Aspects of Deep Sea Biology* by N. B. Marshall (Hutchinson's Scientific and Technical Publications, London, 1954), was written when gaining access to this very large region, representing a significant proportion of the biosphere, was quite difficult. Deep-sea research remains technically challenging and often very expensive, but technical developments have eased the situation somewhat. The result is a growing knowledge base, as reflected in *Deep-Sea Biology,* edited by J. D. Gage and P. A. Tyler (Cambridge University Press, 1991). However, this knowledge base remains fragmentary, based on foci of information rather than on a broad foundation. We hope that this volume on the physiology of deep-sea fishes will help strengthen these foundations.

This book not only brings together what we know of the physiology of deep-sea fishes, but attempts to describe in general terms the biotic and abiotic environments and the techniques used to investigate deep-sea fishes. As such, the book serves as both a general and a specific source of information about the fishes of the deep sea. Finally, we hope that this book will also convey something of the fascination of this little-known environment and stimulate others to enter the field.

We dedicate this book to the memory of Professor Norman Bertram "Freddy" Marshall, FRS.

WILLIAM HOAR
DAVID RANDALL
ANTHONY FARRELL

NORMAN BERTRAM "FREDDY" MARSHALL

Professor Norman Bertram Marshall, F.R.S., the distinguished ichthyologist and student of the deep sea, died in February 1996, almost exactly half a century after he began his work on fish as a deputy keeper at the British Museum (Natural History) in London (now known as The Natural History Museum). Although his papers and influential books were signed "N. B. Marshall," he was universally and affectionately known as Freddy, and it was a measure of his kindliness and how much he was appreciated and respected that Anton Bruun once remarked "Marshall? Ah yes, he is the man whenever you hear his name anywhere in the world everyone smiles." In person, Freddy was of average height and strong build. In later life he grew a white beard which gave him a cherubic appearance that went well with his smile and frequent chuckle in conversation. This kindly and cheerful persona concealed an active and questing intellect; Freddy never sought to display his interests and (unlike most scientists) disdained any competitiveness in conversation. Occasionally though, in discussions of some scientific topic, the conversation would turn to a modern American poet like Wallace Stevens or to some abstruse point in comparative philosophy and religion. He read very widely and was knowledgeable about music, but he rarely disclosed his expertise. Readers of his books, however, will perhaps have noticed his familiarity with the French structuralists.

Freddy described himself as a marine biologist with a special interest in fishes, particularly those that live in the deep sea. He not only was the leading world authority on deep-sea fishes like the macrourids and on deep sea biology as a whole, but also had prodigious knowledge of many branches of marine biology. This made the books he wrote extremely influential, beginning with *Aspects of Deep Sea Biology* in 1954 (Hutchinson's Scientific and Technical Publications, London), which became the standard text on the subject. A reviewer of the French translation in 1968 justly remarked that "Au total l'ouvrage qui nous est offert est probablement le meilleur ouvrage qui ait jamais été écrit sur la question." Like several of his other books, it was beautifully illustrated by his wife Olga.

Freddy had a straightforward writing style, often using apt quotations, and wore his knowledge lightly, which made *Aspects of Deep Sea Biology* a very readable, yet remarkably complete survey of the subject from its history to marine biogeography, and within a few years the book became much used by undergraduates. Several workers on deep-sea fish have commented that it was this book that determined the course of their later careers as ichthyologists. The chapter on counteracting gravity contains a few illustrations of the swimbladders of deep-sea fishes and shows that he was already working on his most significant scientific paper, the classic 122-page *Discovery Report* of a few years later (1960). Freddy had long been interested in the teleost swimbladder, having earlier (1953) written a *Biological Review* (with Harden Jones) on the various functions of the teleost swimbladder. Even before this, in his second paper from the museum, he had suggested and given cogent evidence in support of the view that the deep scattering layer was the result of reflection from the swimbladders of bathypelagic fish. As an undergraduate at Cambridge (where he got a double first), Freddy had been interested not only in fish, but also in experimental embryology and in biochemistry, and when he came to the Natural History Museum, this breadth of interest made his papers unusual for their holistic approach. He himself felt he owed a debt to Dr. Albert Parr, the Director of the American Museum of Natural History, who came to visit the London museum soon after Freddy had taken up work there. Parr, who had come to work on the deep-sea fish in the collections, discussed fish classification with Freddy and suggested that the best way to know fishes was to make a general study of one of their organ systems. Freddy made the fortunate choice to study the varieties of swimbladders in deep-sea fish, and soon found that this offered a valuable field for research. His various papers on the teleost swimbladder are interesting not only for the new discoveries made, but also because they reveal the way in which Freddy brought order into what had been a rather confused field, using different kinds of evidence, in particular comparative studies of swimbladder distribution in relation to depth. His choice of the swimbladder as an organ system to study led to his remarkable *Discovery Report* devoted to the swimbladder structure of deep-sea fishes in relation to their systematics and biology, to the use of swimbladder structure in classification, to the buoyancy mechanisms of fishes without swimbladders, and to his ideas about the economy of construction and simplicity in bathypelagic fishes such as the *Cyclothone* group of species. This formed the basis of several papers and of much of the book he was working on when he died.

Freddy began his academic career at Hull, where he joined the oceanography department under A. C. Hardy in 1937, working on the plankton collected by Hardy's plankton recorders. He published three papers on zooplankton distribution, being particularly interested in *Sagitta elegans* and *S. setosa,* on which F. S. Russell was then working. However, he had always been interested in fish and even as an undergraduate at Cambridge had made five vacation trips to the Arctic on commercial trawlers seeing such rat-tail species as *Macrourus berglax* and *Coryphaenoides rupestris,* on which he was many years later to write a monograph for the Sears Foundation. As his professor, Stanley Gardner wrote that remaining cheerful and friendly on such trips was as good a test of character as could well be found, and in later years, on *Discovery* cruises, Freddy was a welcome shipmate. When trawls were brought up, he was able to identify almost all the specimens, not only the fish, and at other times enlivened the company with a remarkable store of ditties. As is easy to imagine for anyone who knew him, he took a most active part in late night rugger scrums.

Not long after he had joined Hardy at Hull, the war interrupted his career, and he first was drafted as a radar officer to a gun site on the Thames estuary (an experience that perhaps led to his interest in deep scattering layers) and then, like many biologists, joined operational research. He volunteered in 1944 for "Operation Tabarin," which was a project for occupying Antarctic bases, and after collecting 25 huskies from Hebron in northern Labrador, spent two years at the base in Grahamland on the Weddell Sea. Here he made collections of marine animals, catching enough notothenioids to feed 13, and on a depot-laying trip found that seals fed on benthic notothenioid eggs during the Austral summer.

His service in the Antarctic was recognized later by the award of the Polar Medal in 1953 and by a mountain peak named after him by the Royal Geographic Society in 1966. Freddy characteristically remarked that most of his colleagues had only had glaciers named after them, which would merely end up as lumps of ice fit only for Martinis. As he sorted and catalogued his collections at the museum on his return from the Antarctic, he saw the wonderful fish collections in the New Spirit building and determined to apply to work there. Thus began his remarkable 25-year sojourn at the museum, working on marine fishes.

Freddy was not at all a typical museum fish man, pursuing taxonomic niceties with all the relish that the reorganization of higher categories seems to evoke in its devotees. He was later to poke gentle fun at cladist colleagues by speaking of those Irish taxonomists, the O'Morphy family (apomorphy, synapomorphy, and their relatives). At the time he began at the museum in 1947, where he took over "fish" after J. R. Norman's death, teleost classification was in a real state of uncertainty, and there were no clear taxonomic rules for systematists to operate by. Freddy had a liberal supply of common sense, and a good deal later, after Hennig had become widely known, often remarked with some truth that his common sense had given systematic results similar to the application of cladistic methods! Although he wrote systematic papers, often with colleagues like Bertelsen, Krefft, and Iwamoto, his main interest was in the way fish functioned, how they were adapted to life in the depths of the sea. Much of this was apparent in his excellent little book of 1971 *Explorations in the Life of Fishes* (Harvard University Press, 1971), an expanded version of lectures he had given at Harvard. His next book, in 1979, *Developments in Deep-Sea Biology* (Blandford Press, Poole, England, 1979), came after his retirement from Queen Mary College, where he had moved in 1972 from the Natural History Museum, and was a masterly synthesis of the advances in our understanding of the deep sea in the 25 years following his first book.

I first got to know Freddy while visiting the Natural History Museum, where he used to take visitors to legendary lunches at a nearby pub. Overseas ichthyologists who studied with Freddy found it astonishing that he managed to write so many significant papers yet seemed to spend so little time at the museum during the day; but he was very well organized and worked on the train when commuting from Saffron Walden each day. Later, I got to know him better when he visited Plymouth regularly. He often stayed to work with Eric Denton on buoyancy, and we worked together on the Mauthner fibers of deep-sea fishes and on a little book on fish biology. It was while I was working on this book with Freddy that he astonished me by the extent of his knowledge and his kindness in criticism, and he delighted me with his often terrible jokes. At this time I served on various grant committees and it was striking that when Freddy gave a reference for someone, he was simply unable to say anything unkind; the committee members soon realized with affection that Freddy's references were invariably excellent.

He was an exceptional man who lived his life by high standards and was fortunate to receive what I suppose all scientists prize most, the respect and admiration of his fellows. Elected to the Royal Society in 1970, in the next year he was the second recipient of the Rosenstiel Gold Medal for services to oceanography. Few marine biologists have had such an influence in so wide a field for so long, and I can think of none who was regarded with such affection. Freddy died in the village near Cambridge to which he had retired, some six miles from the village in which he had been born 81 years earlier.

QUENTIN BONE

1

WHAT IS THE DEEP SEA?

MARTIN V. ANGEL

I. The Nature of Water
   A. Chemical Constituents of the Oceans
   B. The Density of Seawater
   C. Dissolution of Gases
   D. Light in Water
   E. Sound in Water
II. Ocean Gradients
   A. Bathymetric Profiles
III. Morphology of Ocean Basins
   A. Large-Scale Ocean Circulation
   B. Upwelling
IV. Biophysics and Oceanic Food Webs
   A. Patterns of Productivity and Biogeography
   B. Comparisons with Coastal Zones
   References

The oceans constitute the largest habitat on Earth. Seawater covers 71% of its surface to an average depth of 3800 m. The hypsographic curve (Fig. 1) shows that continental shelves (0–200 m deep) cover approximately 5% of the Earth's surface, slopes (200–3000 m) cover 13%, abyssal depths of 3000–6000 m cover 51%, and hadal depths >6000 m cover <2%. The total volume of the oceans is $1.368 \cdot 10^9$ km^3, providing living space that Cohen (1994) estimates to be 168 times that offered by terrestrial habitats. This immense volume of seawater contributes about 0.24% of the total mass of the Earth and has a major influence on its climate. The deep ocean is characterized by being permanently unlit by sunlight and occupying depths >1000 m. It is a biome encompassing about 75% of the biosphere, in which most of the abiotic and biotic factors, whose variability generates so much of the ecological diversity that is familiar to us on land and in shallow waters, show relatively little variation in both time and space. But do the deep oceans really offer an almost invariant environment that is

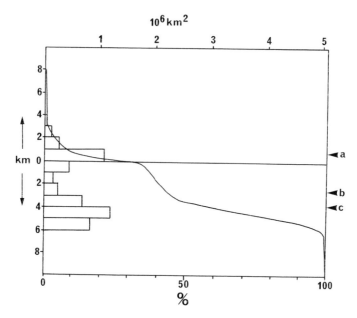

Fig. 1. Hypsographic curve of the world, showing areas in 10^6 km^2 (upper axis), and percentages of the Earth's surface covered by land [with a range of elevations (km)] and by oceans [with a range of various bathymetric depths (lower axis)]. (a) The mean elevation of the land (840 m); (b) the mean sphere depth (2440 m); (c) the mean depth of the ocean (3800 m).

physiologically challenging only because of resource limitations? Before this question can be addressed and the oceans explored as a habitat, it is important to appreciate the chemical characteristics of the oceans' major component—water.

I. THE NATURE OF WATER

Water is an exceptional substance. It has abnormally high boiling and freezing points, compared with oxides of the other elements close to it in the periodic table. This characteristic results from its molecules associating in liquid-phase water. Ice, the solid phase of water, is just as exceptional in that its density is lower than liquid-phase density and hence it floats; this lower density results from the "looser" packing of the molecules within the solid crystal structure. Water has a very high heat capacity, which serves to dampen the impact of variations in heat input and loss. It also results in ocean currents playing a major role in modifying global climate by transferring copious amounts of heat from low latitudes to high latitudes.

1. WHAT IS THE DEEP SEA?

The latent heats of ice formation (80 cal/g) and evaporation (537 cal/g) are the highest for any liquid. Hence the melting of 1 g of ice lowers 80 g of liquid water by 1°C. Similarly the evaporation of water from the sea surface lowers the skin temperature considerably, and such processes are important in determining water mass characteristics (see following discussion).

Water is amphoteric (i.e., it can function as either a base or an acid). Thus, although pure water has a neutral pH, through its electrolytic dissociation it can be a provider of either hydrogen or hydroxyl ions. Hence it is a good solvent for many inorganic and organic compounds that contain radicals that can dissociate, such as hydroxyl bonds. However, organic compounds lacking such radicals are mostly insoluble in water, including those compounds that are important structural components of cells and membranes. Because water is the major chemical constituent of the internal and external milieu of fishes at all levels of organization, from cells, to tissues, to whole bodies, its physicochemical characteristics are fundamental to understanding fish physiology (Dorsey, 1940).

A. Chemical Constituents of the Oceans

Most of the 92 natural elements have been detected dissolved in seawater, although the majority occur only in trace concentrations (Open University, 1989). *Salinity,* the term used to quantify the total quantity of the dissolved salts in seawater, generally ranges between 33 and 37 and averages about 35 (*note:* salinity, once described in parts per thousand, is now defined as a dimensionless ratio and so does not have units). Most constituents of seawater are conservative (unreactive or inert), so their concentrations vary in direct ratio to salinity. This is particularly true for the major constituents (i.e., those occurring at concentrations >1 ppm), which contribute to 99.9% of the salinity value (Table I). Most are in steady state (i.e., their chemical budgets are in balance), although their turnover rates (i.e., total mass of substance in oceans/rate of supply or removal) can vary consideraly. The mean ocean mixing rate is of the order of 500 years (Stuiver *et al.,* 1983), so those substances with turnover rates of about 10^5 years will be uniformly distributed, except very close to major point sources. However, there are many substances whose concentrations fluctuate widely within the oceans, notably those that play a role in biological systems. Fish regulate their internal ionic concentrations of many of these elements; for example, potassium and calcium are concentrated, whereas others—particularly sodium—are excreted. Some of the rarer elements are essential in trace concentrations but become toxic if present in greater concentrations. Emissions from hydrothermal vents (Parson *et al.,* 1995) result in high local concentrations of heavy metals such as manganese, cobalt, and mercury,

Table I
Average Abundances of the Major Constituents of Seawater

Element	Abundance (mg liter^{-1})	Total amount in 10^{12} tonnes
Chlorine	$1.95 \cdot 10^4$	$2.57 \cdot 10^4$
Sodium	$1.08 \cdot 10^4$	$1.42 \cdot 10^4$
Magnesium	$1.29 \cdot 10^3$	$1.71 \cdot 10^3$
Sulfur	$9.05 \cdot 10^2$	$1.20 \cdot 10^3$
Calcium	$4.12 \cdot 10^2$	$5.45 \cdot 10^2$
Potassium	$3.80 \cdot 10^2$	$5.02 \cdot 10^2$
Bromine	$6.7 \cdot 10^1$	$8.86 \cdot 10^1$
Carbon	$2.8 \cdot 10^1$	$3.70 \cdot 10^1$
Nitrogen[a]	$1.15 \cdot 10^1$	$1.50 \cdot 10^1$
Strontium	8	$1.06 \cdot 10^1$
Oxygen[a]	6	7.93
Boron	4.4	5.82
Silicon	2	2.64
Fluorine	1.3	1.72

[a] Elemental oxygen and nitrogen are usually not considered to be major constituents because they are dissolved gases.

which the fish must either avoid, tolerate, or control, but these vents may fulfill a significant evolutionary function by creating variability in an otherwise chemically monotonous environment.

B. The Density of Seawater

The density of seawater plays a key role in ecological processes through determining the stability of oceanic water columns and contributing to the patterns of ocean circulation. Seawater density is determined by three factors: hydrostatic pressure, temperature, and salinity. Away from the turbulent wind-mixed layer of the upper few tens of meters, and near the seafloor where the frictional forces and the effects of rough bottom topography combine to create a well-mixed benthic boundary layer, even quite small differences in densities of contiguous layers of water can prevent mixing. Thus ocean waters tend to be highly structured vertically. In the deep-sea environment where there are few sensory indicators, chemoreception can be expected to play an important role in intraspecific and interspecific communication. Hence in deep-sea fishes, the elaboration of chemical receptors can be expected to be associated with behavioral adaptations linked to the preferential lateral spread of chemical cues along isopycnal

1. WHAT IS THE DEEP SEA?

surfaces (i.e., surfaces of constant density) (Marshall, 1971; Bone *et al.*, 1995).

Hydrostatic pressure is more or less a function of depth. The pressure (p) at any depth (z) is a function of the weight ($g\rho$) of the overlying water per unit area (where g is the gravitational constant and ρ is the density of the seawater). The variations in atmospheric pressure at the surface can play an important role in large-scale physical processes, such as El Niño Southern Ocean (ENSO) events (see following discussion). Water is only very slightly compressible, so there are only slight increases in its *in situ* density with increasing depth. Surface seawater with a density 1028.1 kg m^{-3} and a temperature of 0°C will increase in density to 1028.6 kg m^{-3} if it is lowered to a depth of 100 m without allowing its temperature to change. If lowered still further to 1000 m, its density will rise to 1032.8 kg m^{-3} and to 1046.4 kg m^{-3} at a depth of 4000 m. If its temperature is raised and maintained at 30°C without changing its salinity, its density at the surface will be 1021.7 kg m^{-3}; at 100, 1000, and 4000 m, its density will be 1022.2, 1026.0, and 1038.1 kg m^{-3}, respectively. A seawater sample collected at depth and brought back to the surface fully insulated will undergo adiabatic cooling as a result of its slight volumetric expansion.

Variations in the density of surface seawater result mainly from imbalances between the quantity of water lost by evaporation from the sea surface and the input quantity of fresh water from rainfall. Where rainfall is the higher value, the buoyancy of the water in the upper wind-mixed layer increases and thus stabilizes the upper water column. Where the quantity of water lost by evaporation is the higher value, the salinity and hence the density of the surface waters increase. This denser surface water will then sink beneath neighboring lighter water masses at convergences, sliding down layers of equal density, or pycnoclines (McCartney, 1992). Because within the body of the ocean there is limited mixing between water of different densities, these sinking "water masses" retain characteristic properties of temperature and salinity and can be tracked over extensive distances moving within the deep circulation patterns of the ocean (Dickson *et al.*, 1988). Even finer details of the large-scale circulation can be followed using natural and anthropogenic chemical tracers such as chlorofluorocarbons (CFCs) (Smethie, 1993) and radioactive isotopes (Schlosser *et al.*, 1995). In a few areas large outflows of fresh riverine water play a significant role in reducing the density of surface seawaters, most notably in the Arctic, where the outflows of the large Russian rivers stabilize the upper water column, reducing the fertility of Arctic waters.

Unlike fresh water, which has a density maximum at 4°C above its freezing point, the density of seawater continues to increase until it reaches its freezing point at about −1.9°C (note that the higher the salinity, the lower the freezing point). When seawater freezes, the ice that is formed is virtually free of salt, so the remaining liquid water is saltier and denser. This process underlies the formation of Bottom Waters in the Weddell Sea and off Greenland, which ensure that the deep ocean is cold and well-ventilated with oxygen. Because fish blood is isotonically equivalent to 50% seawater, where there is active deep-water formation, fish require adaptations for supercooling if they are to survive.

C. Dissolution of Gases

Gases that have low reactivity with water generally have low solubilities (e.g., nitrogen, oxygen, CFCs), whereas those that react chemically with water have high solubilities (e.g., sulfur dioxide, nitrogen dioxide, ammonia). Carbon dioxide, which reacts relatively slowly with water to form carbonic acid, is usually considered to be of low solubility. However, its solubility is controlled by the chemical equilibria governing the reactions of the aqueous carbonate–bicarbonate system that plays a major role in buffering seawater against substantial changes in alkalinity, so that the natural range of pH in seawater is 7.7–8.2 (Brewer *et al.,* 1995). Gaseous exchanges between the surface waters and the atmosphere across the sea-surface interface result in rapid equilibration of the partial pressure of the gases in solution in the surface waters and the atmosphere (Thorpe, 1995). As partial pressures of gases increase with temperature, so their solubilities decrease. (This is opposite to the effect of temperature on the solubility of salts, which increases as the temperature rises.) Thus more oxygen and carbon dioxide can dissolve in the surface waters of cold polar seas than in warm tropical seas. So Bottom-Water formation not only supplies dissolved oxygen to the interior of the ocean but also removes carbon dioxide from the upper ocean. There is evidence that anthropogenic emissions of carbon dioxide to the atmosphere have already resulted in the reduction of pH in the deep ocean by 0.1 (Sarmiento *et al.,* 1992), so further reductions can be expected within the next few decades.

In the North Atlantic, water that sinks along the polar front is of similar density to North Atlantic Bottom Water, so it freely mixes with it to form North Atlantic Deep Water. This water mass pervades the deep waters of all the major oceans via the "great conveyor" (Broecker, 1992). Thus in most of the deep ocean, there is enough oxygen dissolved in the water of the ocean's interior to support aerobic respiration. In those regions of the ocean where there is a deep chlorophyll maxima, more oxygen may be produced by photosynthesis than is being utilized for respiration, so the partial pressure of oxygen may even exceed that of the atmosphere (i.e., the water is supersaturated in oxygen). Respiration and dissolution of calcium

1. WHAT IS THE DEEP SEA? 7

carbonate in deep water can result in the substantial elevation of carbon dioxide partial pressures, so that upwelled waters actually vent carbon dioxide back into the atmosphere. But when the rate of primary production is high, for example, during the spring bloom at temperate latitudes in the North Atlantic, the partial pressure of carbon dioxide in the surface waters can be lowered so much that the ocean absorbs carbon dioxide from the atmosphere.

As hydrostatic pressure increases, the partial pressures of the dissolved gases decrease and their solubilities increase. Thus the energy required to extract oxygen for respiration, carbon dioxide for incorporation into skeletal calcium carbonate, and other gases to inflate swim bladders increases substantially with depth. Moreover, because gases are so much more compressible than water, their densities increase rapidly with increasing hydrostatic pressure, thus reducing their functional value for regulating buoyancy. However, their acoustic characteristics are maintained, so there are examples of gas bubbles being retained, even at abyssal depths, for the detection of sound.

In the eastern Tropical Pacific and in the northwestern Indian Ocean, the oxygen demand created by the very high sedimentary input of organic material from the highly productive surface waters, combined with the fact that the source waters are "old" and thus already depleted in dissolved oxygen, results in the development of strong oxygen minima wherein oxygen concentrations become so low they are almost undetectable. Within sediments, oxygen concentration profiles show declining quantities of free oxygen with depth until reaching a redox boundary, at which free oxygen disappears completely. Associated with the redox boundary, both in the sediments and in the water column, are major changes in the chemistry of compounds that are redox sensitive; for example, ferric ions become reduced to ferrous ions. Ecologically, the most important shift is in the oxidative metabolism undertaken by microorganisms. Denitrification reduces nitrates to nitrites and eventually to nitrogen. Further reductions in the redox potential result in sulfate being reduced to sulfide, which is highly toxic to aerobes, and eventually sulfide production is succeeded by methanogenic activity.

D. Light in Water

Light is a major ecological influence in the upper 1 km or so of the ocean, but in the deep ocean its influence wanes. In the euphotic zone it is a key factor regulating the rate of photosynthesis. However, water is translucent but not transparent, and it selectively absorbs and scatters light of different wavelengths. Red wavelengths are most rapidly absorbed (except where turbidity is very high), and blue-green wavelengths penetrate to the greatest depths, ~1 km, in the clearest oceanic waters. Even pure water scatters light, but the scattering is greatly enhanced by suspended particles; therefore, the higher the particle loading, the more rapidly the light is attenuated. Profiles of light intensity and the proportional changes in its composition play an important role in determining the zonation of the communities in the upper ocean through adaptations to counter visual predation. Most of the fish species that inhabit the near-surface waters of the ocean by day are countershaded. However, at depths >250 m, the pattern of light intensity becomes symmetrical (Fig. 2), with the brightest light coming from vertically overhead and the dimmest being backscattered

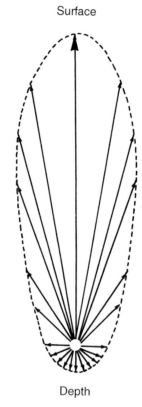

Fig. 2. Distribution of relative light intensity in the water column at midwater depths of 250–750 m. The length of the arrows indicates the relative intensity of the light arriving from each direction at the axial point. Thus the brightest light comes from directly overhead, and there is a symmetrical decrease (three-dimensional) in intensity as the angle of observation is rotated vertically, so that the dimmest light is backscattered from the deep water directly below. Redrawn from Denton (1970).

1. WHAT IS THE DEEP SEA?

from the depths below (Denton, 1970). At these depths the dominant types of fishes (e.g., myctophids) have black backs, mirror sides, and lines of ventral photophores. The dominant decapod crustaceans are half-red and half-transparent; the red pigment is a carotenoid obtained from the animals' diet and is functionally black because all the red wavelengths of daylight have been absorbed. The pigment also has the maximum absorbence of the blue-green wavelengths of most bioluminescence. At depths of 700–1000 m, the mirror-sided fishes disappear and are replaced by other species that are mostly uniformly dark, but some still have ventral photophores. Most of the decapod crustaceans are uniformly red, which may explain why a few fishes have evolved cheek light organs that emit light in the far-red portion of the spectrum and also have a retinal pigment that, unusually, can detect far-red light. However, except for light produced by the organisms, the majority of the deep ocean (i.e., depths at greater than about 1000 m) is permanently dark.

At all depths, bioluminescence, with the light being produced by either the organism's own luciferin/luciferase system or the bacteria they "culture," is an almost universal feature of most oceanic species. Not surprisingly, in the absence of daylight, color ceases to play a major role in communication and/or camouflage, and so bioluminescence takes over color's functions for interspecific and intraspecific signaling. The physiological characteristics of each individual species determines the environmental range (or niche) within which it can survive and compete successfully. However, physiological characteristics that enhance competitive fitness within the specific range of environmental conditions experienced within a certain depth range may well limit an organism's ability to compete elsewhere. Thus a species of *Argyropelecus,* with its mirror sides, elaborate ventral photophores, (Denton, 1970), and highly modified swimming behavior (Janssen *et al*, 1986), is superbly adapted to life at daytime depths of 250–600 m, but if displaced either higher or lower in the water column, these adaptations would render it very susceptible to visual predation. Such limitations resulting from a high degree of specialization have evolutionary implications. Studies of the geological records of invertebrates, notably mollusks, imply that specialist species have higher speciation and higher extinction rates than do generalist species, but the unchanging physical characteristics of the oceanic water column over evolutionary time appear to have favored the evolution of highly specialized morphological types, which have then been able to outcompete new, less well-adapted immigrants.

E. Sound in Water

Water is much more "transparent" to sound than to light. But as frequencies increase (and wavelengths shorten), attenuation increases, especially as the wavelengths approach the acoustic diameter of suspended particles. Particles (or bodies) whose size exceeds the wavelength of the sound will tend to backscatter the sound, and dense concentrations of finer particles attenuate the sound through Tyndall scattering. Thus sound with a frequency of 10 kHz can generally penetrate to full ocean depths and is backscattered only by the larger fishes, whereas 150-kHz sound will penetrate only to a maximum of 400 m and is backscattered by high concentrations of zooplankton (Urick, 1975). So, not only is sound used extensively by oceanographers for underwater investigation and communication, but it is also quite widely used by fish and marine mammals as a means of intraspecific communication and echolocation.

II. OCEAN GRADIENTS

In the open ocean, horizontal gradients are far weaker than vertical gradients and are often confused by turbulent eddies. The strongest and most predictable gradients are vertical, so distributional patterns of species and their morphological and physiological characteristics are often closely linked and adapted to these vertical gradients. Moreover, many fundamental ecological processes are strongly influenced by the vertical structure of the water column. Many of the ecologically important gradients have a complex relationship with depth.

Many vertical gradients are related to the density structure of the upper water column. At low latitudes this structure is predominantly a function of its thermal characteristics, with the depth of the strong temperature gradient (the thermocline) being particularly important. But at high latitudes, particularly in the northern hemisphere, the density structure is dominated by the salinity structure (the halocline). In the wind-mixed layer—the upper few tens of meters of the water column—the water is turbulently mixed and so it is relatively uniform in temperature, salinity, and nutrient content and even in the distribution of phytoplankton. If the stratification of the water column becomes stable enough to prevent the wind-mixed layer from eroding the thermocline, the phytoplankton are retained suspended in sunlit waters. There they can flourish until either all the available nutrients (nitrate, phosphate, and silicate) are used up or some other limiting factor, such as the availability of iron, inhibits plant growth (De Baar, 1994). The zone that is illuminated by enough sunlight to support photosynthesis is termed the *euphotic zone*. If the lower boundary of the euphotic zone lies deeper than the lower limit of the wind-mixed layer (i.e., the thermocline and its associated nutricline), supplies of nutrients are sufficient to maintain high levels of primary production despite

1. WHAT IS THE DEEP SEA?

losses through sedimentation and the diel vertical migrations of grazers (Longhurst and Harrison, 1988). However, more often than not the lower boundary of the euphotic zone lies at or above the thermocline, so that as the available nutrients are used up as a result of photosynthesis and are removed from the euphotic zone by sedimentation, plant growth is supported only by regenerated nutrients. The component of primary production supported by recycled nutrients is often termed *old production* and that supported by nutrients supplied by vertical mixing is termed *new production;* the ratio between new and old production is described as the f ratio (Eppley and Peterson, 1979). Thus the basic biological process of photosynthesis is closely regulated by the vertical distributions of nutrients and micronutrients and the gradients of light and temperature.

The major nutrients (nitrate, nitrite, ammonia, phosphate, and silicate), together with some of the essential trace compounds (e.g., iron) and even some toxic metals (e.g., cadmium), become depleted in the euphotic zone during periods of persistent stratification, which inhibits vertical mixing. In deep water, nutrients are regenerated through chemical and microbial breakdown of sedimenting organic and detrital materials. Resupply of nutrients in the euphotic zone has a major influence on the key biological processes. At low latitudes the resupply from subthermocline depths either is a result of upwelling (Summerhayes *et al.*, 1995) or occurs by the very slow process of vertical diffusion; most production is supported by nutrients within the euphotic zone. During wintertime at latitudes >40°, the upper water column is cooled until the upper part of the water column becomes isothermal and uniform in density, so storms result in convective mixing.

In the European sector of the northeastern Atlantic, the water column is less stable as a result of the influence of the Mediterranean Outflow Water, so the convective overturn in wintertime extends to >500 m in the region of the Bay of Biscay (Parsons, 1988). In contrast, in the North Pacific the low salinity of the near-surface waters stabilizes the density profile so that the convective overturn extends only to depths of 150 m. However, the rate at which nutrients are resupplied depends on not only the upwelling and vertical mixing processes but also the dissolved nutrient content of the deep source waters, and this status is a function of the water's age (i.e., the time that has elapsed since the water was last at the surface). In the Atlantic, where deep-water formation is most active, the bottom and deep waters are "young"; that is, they have been at the surface relatively recently and thus have a high oxygen content (Mantyla and Reid, 1983). However, because they have not been enriched by remineralization processes, their nutrient content is low. In contrast, the "older" deep waters of the Pacific and Indian oceans have a relatively low oxygen content of 3–4 ml O_2 liter^{-1}, and are enriched with nutrients (Levitus *et al.*, 1993) (Fig. 3).

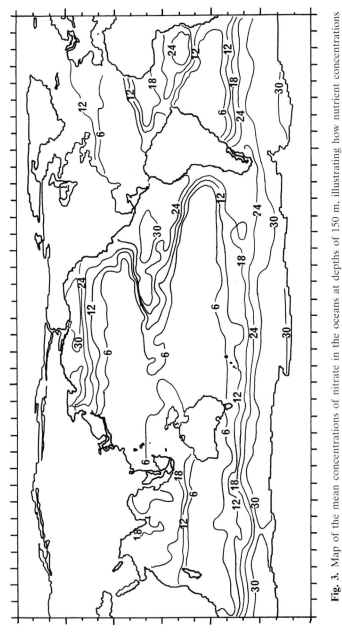

Fig. 3. Map of the mean concentrations of nitrate in the oceans at depths of 150 m, illustrating how nutrient concentrations match the gross patterns of the thermohaline circulation. Note how the highest concentrations of nitrate at these depths occur in the North Pacific, the eastern Tropical Pacific, and the Southern Ocean, all localities where it is postulated that the availability of iron is limiting primary production. From Levitus *et al.* (1993).

1. WHAT IS THE DEEP SEA?

Water masses generated in different regions of the oceans contain widely varying but characteristic concentrations of nutrients. As a result, there are marked differences in the potential productivity in the ecological provinces of the ocean, which are then transmitted to the deep ocean. For example, although the sources of water upwelled off the coasts of California and northwest Africa are both from depths of 150 m, the water off California contains more nutrients and thus stimulates higher productivity. Even within provinces there are differences. For example, the productivity of water upwelled off the coast of northwest Africa is higher to the south of Cap Blanc because the source water for the upwelling there is South Atlantic Central Water, which is "older" and therefore richer in nutrients than North Atlantic Central Water, which is the source for upwelling to the north (Gardner, 1977).

The distributions of some metals are also controlled biologically to some extent (e.g., barium, cadmium). Other substances have localized inputs (via river, atmosphere, continental margins, or hydrothermal vents) or may be scavenged by detrital fluxes. There are some interesting contrasts between some related metals; for example lead-210 and polonium are both members of the same radioactive decay series. In continental rocks, uranium-238 decays to form radon. Radon is a gas that is released into the atmosphere, where it decays into lead-210 and is washed into the ocean in rainfall. There lead-210 behaves conservatively (inert), so its behavior and distribution in the ocean water column can be accurately predicted. Lead-210 decays into polonium, which, in contrast to its parent element, behaves like a nutrient and is biologically scavenged in the upper water column. Consequently ratios between lead-210 and polonium can be used as indicators of organic flux rates, and many detritivores tend to have high concentrations of polonium in their guts. Ratios between other radioactive isotopes have been used to estimate the ages of deep-sea fish [e.g., the changing ratios of lead-210:radium-226 in the otoliths of orange roughy *Hoplostethus atlanticus* (Fenton *et al.*, 1991)].

A. Bathymetric Profiles

The strength and interaction of the vertical gradients lead to strong vertical structuring in bathymetric distributions of the pelagic and benthic assemblages; these distributions are often described as being zoned (Fig. 4). The interfaces between the zones reflect a spectrum of biological responses by the changing assemblage of individuals and populations and tend to be steep clines extending over tens of meters, rather than sharp discontinuities, and also tend to fluctuate in time and space.

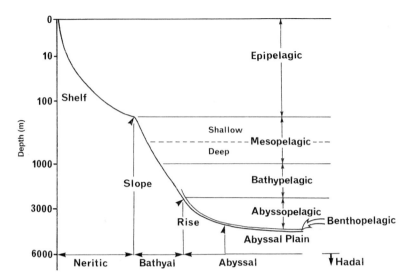

Fig. 4. Schematic representation of the pelagic and benthic zonation in the oceans. Note that the depth scale is plotted logarithmically, and that the interfaces between the zones are clinal rather than clearly defined boundaries. The depths of the interfaces also show local and seasonal variations and may be obscured by dominant hydrographic features such as the edge of the Gulf Stream on the eastern seaboard of North America.

The epipelagic zone includes the euphotic zone and the seasonal pycnocline (where and when it occurs) and is usually considered to extend to depths of 200–250 m. During winter at high latitudes when the wind-mixed layer may extend far deeper into the water column, even the clearest of these pelagic boundaries becomes quite indistinct. Many of the planktonic inhabitants of the epipelagic zone are transparent or translucent. The mesopelagic zone underlies the epipelagic zone and extends down to about 1000 m, the upper limit we have used to define the deep ocean. The mesopelagic zone can often be subdivided at about 600 to 700 m into shallow and deep zones on the basis of the predominance of mirror-sided fish and half-red–half-transparent decapod crustaceans (shallow zone) in the upper region and the predominance of nonreflective fishes and totally red decapods (deep zone) in the lower region. Most diel migrants from the daytime shallow mesopelagic zone readily cross the seasonal pycnocline up into the wind-mixed layer at night, whereas most migrants (mostly micronekton) from the deep mesopelagic zone halt just below the thermocline. The majority of macroplankton inhabiting the deep mesopelagic zone are nonmigrants.

The boundary between the mesopelagic and bathypelagic zones is generally at about 1000 m, the depth at which daylight apparently ceases to play a significant role in organism behavior and distributions. At temperate latitudes it is also the lower limit for diel-migration micronekton and can coincide with the deep oxygen minimum and the base of the permanent thermocline. In addition, it coincides with the maximum in species richness of both pelagic assemblages and megabenthos (Angel, 1993).

There is another less well-defined change in the pelagic assemblages at 2500–2700 m, which at 42°N, 17°W Angel (1983) noted coincided with a sharp decline in fish abundances so that they ceased to be a dominant component of the micronekton. It may also coincide with a depth that is critical physiologically. In some preliminary field experiments Menzies and Wilson (1961) compared the survival of benthic specimens brought up to the surface with that of littoral species lowered to a range of depths. In both groups of species, depths of 2500 to 2700 m proved to be lethal and quite sharply defined the limit. This zone, termed the *abyssopelagic zone,* extends down to close to the bottom to within 100 m of the seafloor, where there is the benthopelagic zone, which is encompassed within a layer of isothermal and isohaline water described by hydrographers as the benthic boundary layer (BBL). In regions where there is high mesoscale eddy activity, benthic storms extend the BBL upward to as much as 1000 m above the bottom (Weatherly and Kelley, 1985). Wishner (1980) was the first to note that planktonic standing crops more than double within the benthopelagic zone. Many of the species occurring there are novel (Angel, 1990).

For benthic communities, the shelf break at the edge of the continental margin, usually at a depth of about 200 m, marks the edge of the open ocean. Around the margin of Antarctica the shelf break is deeper, at 500 m, because the heavy ice loading has depressed the Earth's crust. Beyond the shelf break, the continental slope then falls steeply away into deep water. Along passive continental margins, the base of the continental slope is marked by a change in the gradient that marks the upper edge of the continental rise (often but not invariably at about 3000 m) and of true abyssal depths. The gradient of the rise slackens with depth and almost imperceptibly merges with the edge of the abyssal plain. Along active margins the continental slope typically plummets down to hadal depths of a trench system.

Gage and Tyler (1991) discuss the range of zonation schemes for benthos suggested by a variety of authors. These frameworks have been devised based on either subjective interpretations of succession downslope changes or statistical analyses such as cluster or factor analyses that classify sample data on the basis of a similarity coefficient. (It is important to note that changes in the coefficients used can shift the boundaries of the zones quite extensively.) The fauna inhabiting continental slopes down to depths of around 3000 m is often described as being "bathyal," and the animal life of the continental rise and abyssal plains is described as being "abyssal." The boundary between bathyal and abyssal faunas is often poorly defined faunistically, with little lateral consistency. However, such lack of a clear-cut pattern may prove to be an artifact arising from the logistical difficulty of accumulating an adequate data base (Koslow, 1993). Determination of whether there is any ecological or physiological link between the gradient seen in the water column between the bathypelagic and the abyssopelagic zones and the shift between bathyal and abyssal faunas in the benthos awaits investigation. There may, however, be a distinction that is significant evolutionarily between abyssal and bathyal faunas. The former has no barriers to lateral spread, whereas bathyal species that cannot survive at abyssal depths can only spread along-slope. Thus the zoogeographical scheme for the distribution of abyssal species produced by Vinogradova (1979) shows abyssal provinces as immense areas bounded by ocean ridges and continental margins, whereas bathyal (and hadal) assemblages are restricted to ribbons lying parallel to the continental margins.

Because primary production is concentrated in the upper sunlit layers of the ocean, except for the small isolated pockets where chemosynthesis occurs, all life in the ocean is supported by the downward transfer of organic material from the euphotic zone. Availability of organic matter decreases and the standing crops of the communities decrease exponentially with increasing depth. Organic matter is transferred down mostly via sedimentation, but a significant, albeit small, proportion is transported actively by vertically migrating animals (Longhurst and Harrison, 1988). The majority of diel vertical migrants feed mostly during the shallow phase of their migrations, where and when food is more available. Thus when they migrate down, their stomachs are fuller, and consequently they transport organic material as gut contents. There is also a death flux, if they die or get eaten in deep water. In addition, the migrants carry down nutrients and carbon dioxide, which are excreted at depth. Diel migrations are mostly restricted to the upper 1 km of the water column (i.e., to the deep mesopelagic zone), but in the central oligotrophic gyres, migrations by pelagic decapods extend to depths of 1200 m (Domanski, 1986) and migrations by some myctophid fish such as *Ceratoscopelus warmingeri* extend to depths of 1600–1700 m (Angel, 1989). Consequently daytime concentration profiles of biomass often show a subsurface maximum at depths of 500–700 m, but otherwise show concentrations decline with depth (Fig. 5). Standing crop at 1000 m is 10% of that in the euphotic zone and declines to 1% at 4000 m (Angel and Baker, 1982); these data reflect the proportions of primary production that sedimentary fluxes supply to the deep ocean. There is also a shift in the average size of the assemblages so that the ratio of planktonic to

1. WHAT IS THE DEEP SEA?

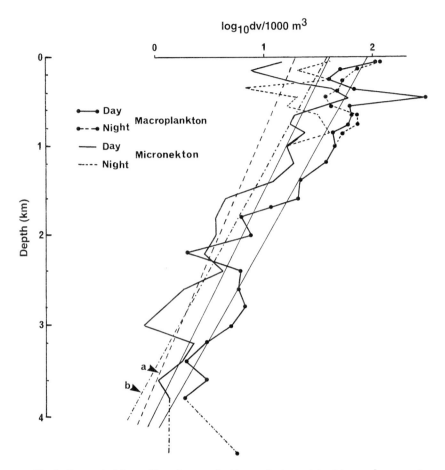

Fig. 5. Day and night profiles of macroplankton and micronekton biomass [expressed as $\log_{10}$ ml displacement volume (dv) per 1000 m^3] 42° N, 17°W with their straight-line regressions superimposed. Also superimposed are straight-line regressions for micronekton from two other stations in the northeastern Atlantic: at 20°N 21°W (a) and at 49°40 N, 14°W (b). The slopes of all the regressions lie between 0.004 and 0.005. Modified from Angel and Baker (1982).

micronektonic biomass is usually >10:1 in the euphotic zone but declines to parity at depths around 1000 m (e.g., Angel, 1989).

The biomass of benthic communities shows a similar bathymetric decline (Rowe, 1983; Lampitt et al., 1986). A possible exception to this trend is shown by the benthopelagic scavengers, which rely on "large packages" or corpses large enough to reach the seabed intact. Their biomasses appear to be less affected by depth and more closely aligned with the productivity of the upper waters.

Species diversity shows some consistent changes with depth. Species richness (the numbers of species that can be caught) increases with depth and usually shows a maximum at 1000–2000 m. Species evenness also increases with depth, and this increase continues to even greater depths. Thus species richness based on rarefaction curves generally reach maxima at depths of 2–3 km (Rex, 1983), although the numbers of species actually identified often decline below depths of 1–2 km; this is exemplified by data for polychaetes in the Rockall Trough reported by Paterson and Lambshead (1995), who identified maximum numbers of species at depths of 1000 m, but their richness curves based on rarefaction showed a maximum at depths of ~2000 m, where the numbers of species they actually identified were substantially lower.

III. MORPHOLOGY OF OCEAN BASINS

Our knowledge of how the morphology of ocean basins and the circulation patterns in the global ocean have changed over geological time is constantly improving (Parish and Curtis, 1982) (Fig. 6). Imprints of past ocean circulations, altered as a result of the changing gross distribution of continents over geological time, have been identified in present-day distributions (Van der Spoel *et al.,* 1990; White, 1994). One-off geophysical (or vicariance) events, such as the opening and closing of the Panama Isthmus, the Messinian salinity crisis in the Mediterranean, and to a lesser extent the results of fluctuations in sea levels during the glacial cycles, have both created and broken down barriers to distributions. Biodiversity studies have tended to focus on the processes that are presently maintaining diversity and community structure and have paid surprisingly little attention to the evolutionary origins of present biogeographical distributions (White, 1994; Angel, 1997).

The distributions of the continents on the Earth's surface are asymmetrical, contributing to the disparity between the characteristics of the various oceans and basins. Ocean covers 60.7% of the northern hemisphere, compared with 80.9% of the southern hemisphere. One result of this difference is seen in the more extreme seasonal ranges of sea-surface temperature at temperate latitudes in the northern hemisphere compared with those in the southern hemisphere (Fig. 7). The boundaries of four of the major oceans are largely determined by the distribution of the continental land masses.

The largest and oldest ocean by far is the Pacific, which has a total area of around $165.38 \cdot 10^6$ km^2, a mean depth of 4200 m, and a maximum depth of 11,524 m in the Mindanao Trench. It is fully open to the Southern Ocean but has only a shallow connection with the Arctic Ocean via the Bering Strait.

1. WHAT IS THE DEEP SEA?

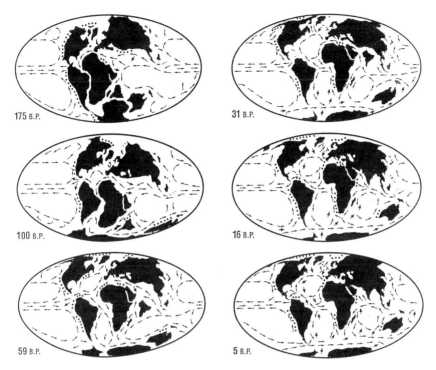

Fig. 6. The redistribution of the continental land masses as a result of continental drift at various intervals following the beginning of the fragmentation of the supercontinent Pangea about 200 million years B.P. The arrows indicate the likely patterns of surface currents generated by the winds and influenced by the Earth's rotational effects. Redrawn from *Palaeogeogr. Palaeoclimatol. Palaeoecol.* **40,** J. Parish and R. L. Curtis. Atmospheric circulation, upwelling and organic-rich rocks in the Mesozoic and Cenozoic eras, 31–66. Copyright 1982 with kind permission of Elsevier Science-NL, Sara Burgerhartstraat 25, 1055 KV Amsterdam, The Netherlands.

At present it is connected to the Atlantic only via the Drake Passage to the south of Cape Horn. However, as recently as 5 million years ago there was a shallow-water connection through the Panama Isthmus, so now, although there is close similarity between the shallow tropical water faunas of the two oceans, there are marked differences between the deep-living faunas. Exchanges would also have been possible between mesopelagic species whose life-histories include a shallow-living larval phase of sufficient duration to be advected through the connecting channel. The Pacific's connection with the Indian Ocean is constricted to the north of Australasia through the islands of the Indonesian Archipelago, but it is open to the south.

The second largest ocean is the Atlantic, which has an area of $82.22 \cdot 10^6$ km^2 (half that of the Pacific). The North Atlantic began to open up

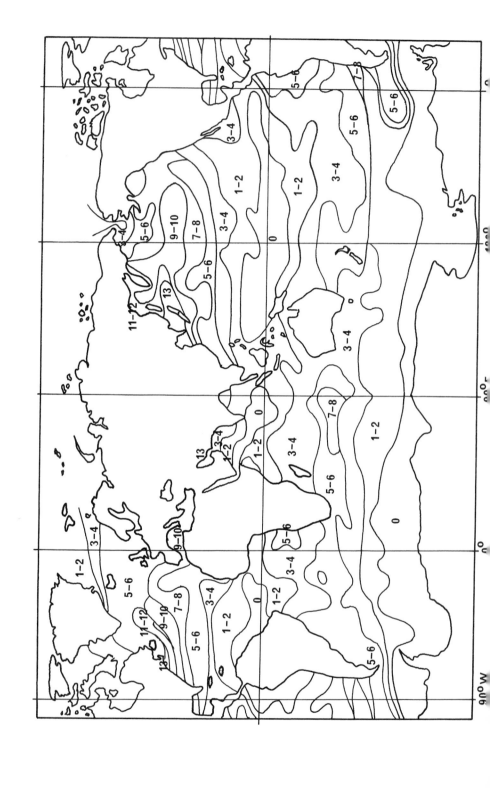

at the beginning of the Jurassic era about 200 million years ago as the supercontinent of Pangea began to fragment and what was to become the continental land mass of the Americas began to separate from the Afro-Eurasian continent. The South Atlantic began opening much later, around 100 million years ago. The Atlantic has an average depth of 3600 m, reflecting its relatively young age. Its maximum depth of 9560 m is in the Puerto Rico Trench. It is the only ocean with a major connection to the Arctic Ocean to the north. Its hydrography is greatly affected by outflows from the Mediterranean and Caribbean. It also receives inflow from the Indian Ocean around the south of the Cape of Good Hope, and further to the south it is bounded hydrographically by the Southern Ocean. These inflows are balanced by outflows of North Atlantic deep water (NADW), which supplies deep water to all the other major oceans via the "great conveyor" (Broecker, 1992) and may well provide the mechanism for gene flow between widely separated populations of deep-living species; for example, populations of the fish *Hoplostethus atlanticus* from the North Atlantic to the southwest of Ireland and from south of Australia have such similar molecular biology that active gene flow must be occurring (Elliot *et al.*, 1994). It is argued that critical evaluation of many of the deep-ocean species now considered to have cosmopolitan ranges will reveal that they are divided into discrete and isolated (geographically and genetically) populations of species that show little if any morphological separation (Wilson and Hessler, 1987).

The Indian Ocean ($73.48 \cdot 10^6$ km^2) is unusual in that it is connected only to one polar ocean, the Southern Ocean. In the northern hemisphere it is closed off by continental Asia, and as a result it is strongly influenced by the atmosphere's interactions with the land masses to the north. These interactions generate seasonal cycles of reversing monsoon winds, which also dramatically reverse the surface currents at tropical and subtropical latitudes. Thus in the northwest Arabian Sea and to a lesser extent in the Bay of Bengal, surface conditions oscillate between being highly productive during the southwest monsoon and highly oligotrophic during the northeast monsoon. This variation creates unique deep-sea conditions, with oxygen concentrations in subthermocline waters down to depths of 1000 m fluctuating broadly.

The Arctic Ocean is not only a truly polar ocean but is also a Mediterranean-type sea, being almost entirely enclosed by land. Nearly

Fig. 7. Map of the seasonal ranges in sea-surface temperature °C in the global ocean. Ranges tend to be greater in the northern hemisphere, which has the greater area of land. Redrawn from Van der Spoel and Heyman (1983).

half of its total area of $14.06 \cdot 10^6$ km^2 consists of broad areas of shallow continental-shelf seas, which are particularly extensive to the north of the Eurasian continental land mass. Its greatest depth of 4400 m occurs in the Fram Basin not far from the geographical North Pole. It has a narrow and shallow connection with the North Pacific via the Bering Strait, through which there are only limited exchanges of water. In contrast, its connection with the North Atlantic is broad and deep, allowing much freer exchanges of water, which greatly influence the hydrography of both oceans. A large inflow of relatively warm Atlantic water enters from the Norwegian Sea feeding the Spitsbergen Current. This pushes the southern boundary of the winter pack ice far to the north. The major outflow is via the East Greenland Current, which carries cold water, pack ice, and icebergs carved from the Greenland glaciers well south along the eastern seaboard of Canada. Much of the Arctic Ocean remains covered throughout the year with multiyear (up to 5 years old) pack ice that ranges in thickness from 1.5 to 4 m. During summer, the areal coverage of pack ice shrinks by only about 10%. Voluminous outflows of fresh water from the great Russian rivers create a stable haline stratification, which keeps productivity relatively low throughout much of the Arctic.

The Southern Ocean is a very different ocean. It is bounded poleward by the continent of Antarctica. To the north not only is it open to exchanges with the other major oceans, but it also is not readily separable from them on the basis of geographical features. Hydrographically its northern limit is defined by the Antarctic Convergence (Foster, 1984), which is where Antarctic intermediate water forms, sinks, and spreads equatorward at its quasi-equilibrium depth of 1000 m below the subtropical water mass. The precise location of the Antarctic Convergence fluctuates both seasonally and interannually. The major feature of the circulation in the Southern Ocean is the circumglobal current, the West Wind Drift, which developed when circumpolar deep-water connections were established about 35 million year ago and appears to have initiated the onset of cooling of bottom waters throughout the global ocean. The flow of this current is constrained by the narrowness of the Drake Passage between the tip of South America and the Antarctic Peninsula, and this has a profound influence on the general oceanic circulation. The areal extent of pack ice fluctuates from about 20 million km^2 in the austral winter to 5 million km^2 in summer (Gloersen *et al.*, 1992). So, unlike the Arctic, extensive areas of multiyear pack ice occur only in the Weddell Sea. It is also in the Weddell Sea where bottom-water formation is the most active, supplying deep water to all the world's oceans via the "conveyor belt" system of currents. Large tabular icebergs are spawned from the broad ice-shelves of the Ross and Weddell

Seas where the shelf depths are unusually deep (400 to 500 m), isostatically depressed by the weight of the ice.

Surface waters of the Southern Ocean to the south of the Antarctic Convergence are constantly rich in nutrients (Levitus *et al.*, 1993). Primary production never exhausts the available nitrate in the surface waters in the Southern Ocean (along with the North Pacific and the eastern Tropical Pacific) because, it is postulated, the production is limited by a lack of iron (Martin *et al.*, 1990). Another notable feature of the Southern Ocean is the exceptionally high sedimentation of silicate that occurs beneath the Antarctic Convergence (Shimmield *et al.*, 1994).

The geological morphology of some of the oceanic basins can be important in determining the ecological characteristics of the oceans. For example, in the southeastern Atlantic, the Walvis Ridge between South Africa and the mid-Atlantic Ridge blocks the northward spread of Antarctic bottom water and so modifies the hydrography of the whole Atlantic. However, probably the most important geological processes affecting deep-ocean ecology are those associated with hydrothermal vents (Parson *et al.*, 1995). Along the central ridges of each ocean basin are the spreading centers where the active formation of new ocean crust is taking place. The formation of new crust seems to be the driving force that pushes the main tectonic plates apart. The central rift valleys are underlaid by magma chambers from which liquid basalt is extruded episodically, creating the new crust. As the liquid basalt cools and solidifies, it cracks. These cracks provide the conduits for the development of deep convective circulations of seawater within the new, still hot, crust. When the water is vented, it not only is often superheated to temperatures as high as 350°C but also is greatly enriched with metallic sulfides as a result of chemical interactions (diagenesis) at high pressures and temperatures between the water and the crustal rocks (Tunnicliffe, 1991). As the vent fluids discharge, they mix with the cold ambient seawater and the sulfides are precipitated as dense black plumes (hence the term *black-smokers*). The sulfides provide the basis for chemosynthetic activity in the immediate vicinity of the vents: bacteria oxidize the sulfides to sulfate, providing the basis for the high biomass and unique assemblages of species that cluster around the vents. Although chemosynthesis makes a very tiny contribution ($\sim 0.03\%$) to global productivity, it still provides about 3% of the organic carbon available at abyssal depths (Jannasch, 1994). So the vents may have a major local influence on the ecology of the deep ocean.

As the seafloor continues to spread outward, the underlying crust cools and shrinks, so that crustal depths increase toward the continental margins. Although the ocean floor is also covered by a progressively thicker drape of sediment as it ages, creating the vast areas of abyssal plains that dominate many ocean basins, water depths deepen away from the midocean ridges.

The oceanic margins impinging on the continental land masses may be either active or passive (Fig. 8). Around the Pacific the margins are active. The crust is buckling down (subducting) beneath the continental land masses to form deep trenches. These trenches are usually bounded on the landward side by a chain of active volcanoes—the so-called Ring of Fire in the Pacific. As the crustal rocks age, they become stiffer, and thus the deepest trenches occur where the subducting crust is oldest. Hadal depths (>6000 m) occupy <2% of the Earth's surface and occur as linear features, often isolated by long distances (Vinogradova, 1979). Thus each system of trenches tends to have a highly endemic fauna and highly contrasting ecologies, depending on the local sedimentation regimes.

The presence of trenches along the margins of the continents has an important effect on sedimentation regimes and hence the ecology of the abyssal plains. The trenches trap any sediment transport and turbidity flows that result from mass-wasting events and slope failures triggered by the heightened seismic activity along the continental margins. Therefore the deep abyssal plains of the Pacific remain unaffected by major turbidite flows. In contrast, along passive margins where there is no active subduction and the continental slope is bounded by the continental rise, any mass-wasting event results in massive and catastrophic sediment and turbidity flows. These flows have an unimpeded path to spread right across the adjacent abyssal plains to the outliers of the midocean ridge. Recent ocean drilling has identified extensive turbidite deposits throughout many of the deep basins in the northeastern Atlantic (Weaver *et al.*, 1995). Throughout the Holocene, turbidity flows have occurred repeatedly, usually at times the sea level was changing rapidly during the switch from interglacial to glacial periods and vice versa. In the Norwegian Sea there is a massive feature known as the Storegga Flow, which appears to have been the result of about three successive failures of the Norwegian slope. It consists of about 500 km^3 of debris (Bugge *et al.*, 1988), which covers about a third of the Norwegian basin. The most recent failure occurred about 7000 before present time (B.P.) and probably caused a tsunami that devastated North Atlantic coastal ecosystems. These debris flows likely eradicated benthic communities over immense areas of the ocean floor and may well have created a mosaic of isolated habitats within which speciation of taxa with limited dispersive ability could have occurred, creating a high regional diversity. It is now recognized that meiobenthic and macrobenthic organisms are unexpectedly rich in species (Grassle and Maciolek, 1992) and are much more speciose than megabenthic species (including fish). (Megabenthos is defined as fauna large enough to appear in photographs of the

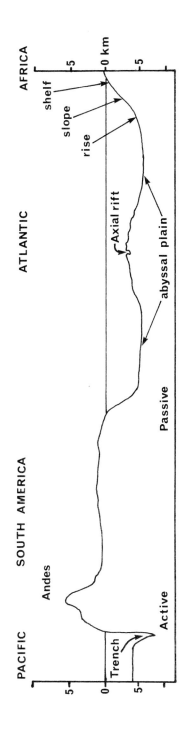

Fig. 8. Schematic representation of a section across the South Atlantic and into the Pacific showing the differences in the morphologic features of active and passive continental margins. Note that there is a vertical exaggeration of ×100.

seabed, macrobenthos is sediment fauna retained on 1- to 0.5-mm meshes, and meiobenthos is fauna retained on 32-μm mesh; however, different investigators use different sieve sizes.) Maybe the impacts of these repeated turbidity flows have created sufficient isolation for this high diversity of small species to have evolved. If so, the macrofaunas and meiofaunas of the Pacific may not have the same overall rich diversity as in the Atlantic, although locally they appear to be just as diverse.

A number of smaller seas separated from the main ocean basins have very different hydrological and ecological characteristics. For example, the Red Sea and the Mediterranean have anomalously warm, deep waters of high salinity, and they appear to lack a typical bathypelagic fauna; instead, a few components of the mesopelagic fauna occur at unusually great depths. In contrast, the Baltic is largely estuarine in character and during the last glaciation was a freshwater lake. Around the East Indies there are a series of deep basins that may have played an important role in speciation of shallow- and deep-water species. During periods of low sea level, land barriers emerged, isolating some of the deep-water basins. Some inshore pelagic species were isolated for long enough for speciation to have occurred (Fleminger, 1986), and some of the faunistic boundaries in the region may have their origins in the changing faunistic linkages. The deep-water faunas have not been carefully studied to see if they, too, show evidence of similar speciation events. Note that when terrestrial faunas were isolated, barriers to the spread of marine species were removed, and vice versa.

A. Large-Scale Ocean Circulation

Ocean circulation is driven by latitudinal variations in solar radiation (and hence variations of heating and cooling), precipitation and evaporation, transfer of frictional energy across the ocean surface by winds, and planetary forcing resulting from rotation of the Earth. The pattern of trade winds is determined by the development of Hadley cells in the atmosphere and the development of polar high-pressure systems (Fig. 9), and the influence of the Intertropical Convergence Zone (ITCZ). Longitudinal instabilities in the ITCZ in the western Pacific play an important role in the generation of El Niño Southern Oscillation (ENSO) events (Donguy, 1994), which

---

Fig. 9. Schematic representation illustrating how shifts in the position of the Intertropical Convergence Zone (ITCZ) between the western and central Pacific play a major role by tilting and depressing the thermocline (A) and, through locally lowering surface salinity (35.0 isohaline) via the effects of the heavy rainfall associated with the ITCZ, influencing ENSO events in the Pacific (B). Modified from Donguy (1994).

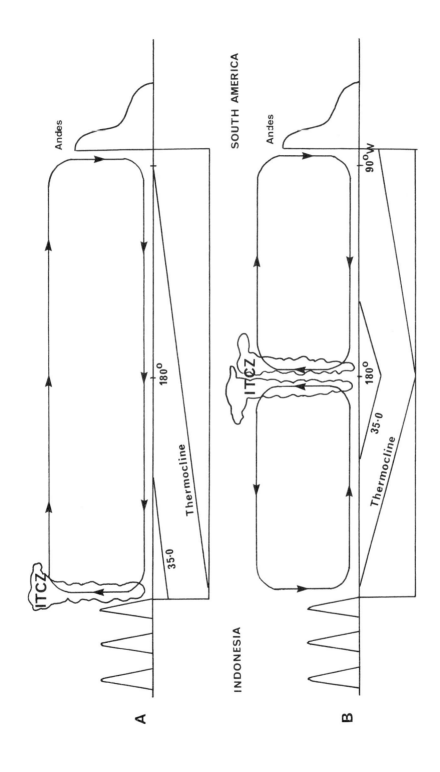

cause substantial fluctuations in sea-surface temperatures, ocean productivity, and weather patterns that are transmitted via planetary waves eastward along the equator and then poleward along the western margins of the continents. Under normal conditions the trade winds generate major gyral circulation features bounded by the major frontal systems, such as the polar fronts and the subtropical convergences. These fronts not only coincide with the boundaries of water masses but also are often major biogeographical boundaries for pelagic communities. However, relatively few species have geographical ranges that coincide exactly with these fronts. Changes in environmental conditions across the fronts are subtle in comparison with the physiological tolerances of the individual species, so the species can survive being advected across them, albeit with reduced viability.

At smaller scales (10 to 100 km), the major source of variability in the pelagic ecosystem are mesoscales, eddies, and rings, both warm core and cold core (Joyce and Wiebe, 1992). Eddies of these scales are almost ubiquitous throughout the ocean and are akin to the weather systems in the atmosphere. But whereas a cyclonic feature in the atmosphere typically has a lateral dimension of 1000 km and a height of 10 km, typical dimensions of oceanic eddies are 100 and 5 km, respectively. Atmospheric eddies seldom persist for more than a week or so, but oceanic eddies can persist for 1 to 2 years, although many disappear by coalescing into their source waters. This occurs because, although oceanic eddies are much smaller than atmospheric eddies, the higher density of the liquid medium means that oceanic eddies contain about 1000 times more dynamic energy.

Some of the most striking eddy features are to be seen in remotely sensed images of either sea-surface temperatures or ocean color (chlorophyll) of eastern boundary current regions. For example, along the margins of the Gulf Stream, meanders often pinch off, forming ring structures. Along its inshore margin, the Gulf Stream forms warm-core anticyclonic rings that contain a body of warm Sargasso Sea water wrapped around with Gulf Stream water, moving across the shelf where it is surrounded by much cooler Shelf water (Joyce and Wiebe, 1992). Conversely, along the offshore boundary of the Gulf Stream, cold-core cyclonic rings are formed that contain a central core of relatively cold Shelf water ringed by Gulf Stream water; these rings advect at speeds of 5–10 km/day through the warmer waters of the Sargasso Sea. Around each ring is a meandering jet current. Where the meandering jet current is turning clockwise, potential vorticity effects result in there being divergence (upwelling); where this current is turning anticlockwise, there is convergence (downwelling). These localized effects influence nutrient supplies, locally enhancing primary production where there is upwelling and depressing it where there is downwelling. The resultant patchiness in primary production and in the phytoplankton standing crop influences the zooplankton and its consumers. Grazer populations increase in divergences through either reproduction or immigration and in turn attract micronektonic predators and larger predators such as whales and the large pelagic fishes.

The fate of species originally entrapped within an eddy is largely determined by their migratory behavior. Species that do not undertake diel vertical migrations tend to persist within an eddy and are passively advected within it. They can show signs of malnourishment and physiological stress if the ambient conditions deteriorate for the species (Wiebe and Boyd, 1978). In contrast, migrating species that were originally entrapped in the eddy when it formed tend to get spun out of it relatively quickly, because as they undertake their daily vertical excursions, they traverse the differential shears within the water column. So within rings and eddies, the assemblages of species change more rapidly than might otherwise be expected. This generates chaotic heterogeneity in the distributions of pelagic species with fractal characteristics similar to that of the eddy structure of the water. Behavioral, feeding, and reproductive strategies in pelagic species can be expected to be adapted to this heterogeneity in the biotic and abiotic environment. The effects of mesoscale features influence deep-sea environments. Many eddies extend all the way from the surface to the bottom and generate "benthic storms" in deep water (Weatherly and Kelley, 1985; Kontar and Sokov, 1994). They may also advect the early planktonic stages of benthic species far beyond their normal distributional ranges, as has been observed for planktonic foraminifers (Fairbanks *et al.*, 1980).

B. Upwelling

At latitudes >40°, the main mechanism resupplying nutrients to the euphotic zone is the seasonal mixing that occurs when winter cooling breaks down the stratification. In the subtropics and tropics where the stratification persists throughout the year, the resupply of nutrients via vertical mixing is limited except where there is upwelling. As discussed earlier, there is some localized upwelling along divergent fronts around eddies and bordering some of the major oceanographic features. Much more significant upwelling occurs in western boundary coastal regions where trade winds blow equatorward, causing the surface waters to be pushed offshore and replaced by cooler subthermocline waters. There are five major coastal upwelling regions: (1) along the Peru/Chile coast, (2) in the California Current regions, (3) off the coast of Mauritania (northwest Africa), (4) in the Benguela Current region off the coasts of Namibia and southwest Africa, (5) and in the northwest Arabian Sea. Upwelling is usually both seasonal and episodic (Summerhayes *et al.*, 1995); even so, the high productivity of these regions makes them important centers for fisheries. There are also important open-ocean upwelling regions, notably along the equator in the central and eastern Pacific and the eastern Atlantic and also offshore in the Arabian Sea as a result of the Findlater jet. In general, the very high biomasses of zooplankton in these regions are dominated by a relatively few species, some of which have life-history characteristics involving extensive ontogenetic migrations into deep water, which appear to be adaptations to maintaining the population within the upwelling system (Smith, 1984). Seasonal peaks in export of organic matter to the neighboring deep ocean generate a seasonality in the deep-sea communities.

IV. BIOPHYSICS AND OCEANIC FOOD WEBS

Primary production, the fixation of carbon dioxide by green plants to form organic molecules using energy from sunlight, requires not only sunlight but also the availability of essential nutrients. In shallow water where the seabed is illuminated and the substrate is stable enough to allow fixed plants to grow, a substantial proportion of the primary production is by macroalagae or a small number of higher plants (e.g., mangroves and sea grasses). The size and/or concentration of plant biomass enable grazers and browsers to be quite large in size, so fish can be herbivorous. The larger plants create finely structured three-dimensional habitats analogous to those of terrestrial environments. Variations in the local geology and differences in exposure to waves and currents, suspended sediments, and differing tidal regimes and runoff from land create much finer scaled mosaics of habitats in littoral and sublittoral habitats in which different species and communities coexist. Similarly, in tropical waters, corals containing symbiotic photosynthetic algae create an even more complex fine-scaled diversity of microhabitat supportive of a greater diversity of species.

Offshore over deeper water, where sunlight penetration to the seabed is insufficient for photosynthesis to occur there, suspended phytoplankton is solely responsible for all primary production apart from the 0.03% produced by chemosynthesis at hydrothermal vents. Phytoplankton cells are small and their turnover is rapid, so the standing crop of plant biomass is small and dilute. Moreover, there appear to be only some 5000 species of phytoplankton in the oceans (Tett and Barton, 1995), compared with an estimated 250,000 species of green plants on land. Oceanic herbivores either are suspension feeders or feed on individual particles and so functionally they, too, have to be very small relative to terrestrial herbivores. These grazers are mostly small zooplankton, except in those regions (or seasons) where large diatoms are the dominant primary producers and larger species are able to sieve the cells out of suspension. Thus the anchovetta in the upwelling region off Peru/Chile is able to graze diatoms, directly sieving them out of suspension on its gill rakers. Even so, in the North Pacific the abundant populations of large copepods, which formerly were considered to be herbivorous, have now been shown not to exploit the phytoplankton directly but to be mainly detrital feeders (Dagg, 1993). In oligotrophic regions, >50% of primary production is by picoplankton—cells <2 μm in diameter. Consumers of picoplankton are, perforce, mostly very small and constitute the microbial food web from which relatively little of the primary production eventually flows into the food chain exploited by fishes. Food webs tend to be longer, and more carbon is recycled before it reached end-consumers such as fish.

The rapid removal of autotrophs by grazing or sedimentation processes results in the sizes of the standing stocks of grazers and detritivores being relatively much smaller in the ocean than in terrestrial ecosystems. Cohen (1994) points out that it is reasonable to assume that the residence time of carbon is roughly proportional to the mean generation time. Thus because generation times scale allometrically with body size, the mean sizes of oceanic and terrestrial biota can be compared. Cohen deduces that the mean adult body size and length of oceanic organisms are smaller by a factor of $3.8 \cdot 10^8$ and $1.9 \cdot 10^4$, respectively.

Elton (1935) argued that there are globally uniform principles governing the functioning of ecosystems, with body size being a fundamental characteristic:

> Animals form food chains in which the species become progressively larger in size or, in the case of parasites, smaller in size. A little consideration will show that size is the main reason underlying the existence of these food chains. . . . We have very little information as to the exact relative sizes of enemies and prey, but future work will no doubt show that the relation is fairly regular throughout all animal communities.

The size spectrum and spatial distribution of primary producers in open-ocean ecosystems are strikingly different from those in terrestrial and even most shallow-water ecosystems. Standing crops of plants can often be much smaller than annual primary production. Turnover rates are high, and residence times of organic carbon in oceanic biomass have been estimated to be 0.08 years compared with 11.2 years in terrestrial ecosystems (Harte, 1988)—a 140-fold difference, which is large enough to be real even if the data are imprecise (Table II).

The small size and unpredictable occurrence of phytoplankton in the oceans appear to have inhibited the evolution of specific associations between animal and plant species, associations that are a notable feature of terrestrial ecosystems. Away from shallow coastal waters, plants seldom

Table II
Comparison of Biophysics of Oceans and Continents[a]

Parameter	Continents	Oceans
Surface area (10^8 km^2)	3.6	1.5
Surface area as percentage Earth's surface (%)	71	29
Mean depth of life zone (km)	3.8	0.05
Volume of life zone (10^9 km^3)	1.37	0.0075
Volume percentage of total (%)	99.5	0.5
Standing crop of plants (10^{27} kg C)[b]	~2	560
Biomass per unit area (10^3 kg C km^{-2})	5.6	3700
Biomass per unit volume (10^3 kg C km^{-3})	1.5	75,000
Dead matter (10^{15} kg C)	~2	1.5
Dead organic matter per unit area (10^6 kg C km^2)	5.5	10
NPP y^{-1c}	25–44	~50[b]
NPP per unit area (10^3 kg C km^{-2} yr^{-1})	69	330
NPP per unit volume (10^3 kg C km^{-3} yr^{-1})	18	6700
Carbon residence time in living biomass (yr)[a]	0.08	11.2
Ratios—ocean:land		
Mean adult body	1:1.4 × 10^7	
Mean adult body length	1:240	
Mean adult body mass	1:3.8 × 10^8	

[a] From Cohen (1994).
[b] Based on Harte (1988).
[c] NPP, Net primary productivity.

provide a physical substrate for the herbivores, *Sargassum* weed being the obvious exception. In addition, autotrophs are almost entirely restricted to the upper sunlit depths, which constitute a very small fraction (~2.5%) of the total living space within the oceans. This limits the distributional ranges of herbivorous grazers to the upper waters and also, because their food is so tiny, causes many of these grazers to be physiologically constrained and quite small in size. Their small size then limits their ability to regulate their vertical ranges. Even those that are large enough to be capable of diel vertical migration are still limited to as little as 10–15% of the total ocean volume. Thus herbivores are absent from most oceanic volume; in this way detritivores become the basis for food chains in most deep-ocean scenarios. Platt *et al.* (1981) found that oceanic food webs can be modeled more closely on the basis of size spectra rather than functional relationships. Perhaps another significant consequence of detrital feeding is that only in exceptional conditions do large quantities of detrital organic material accumulate in the deep ocean.

An important implication of the differences in food-web structure is that there are few opportunities for specialization. This may account for the low global species richness of open-ocean fishes compared with freshwater

1. WHAT IS THE DEEP SEA?

species; for example, the >690 species (with 84% endemism) reported from the Zaire River together with the >600 species (with 96% endemism) in Lake Malawi, of which 92.5% are cichlids (Ribbink, 1994), probably nearly equal the total numbers of fish species in the deep ocean.

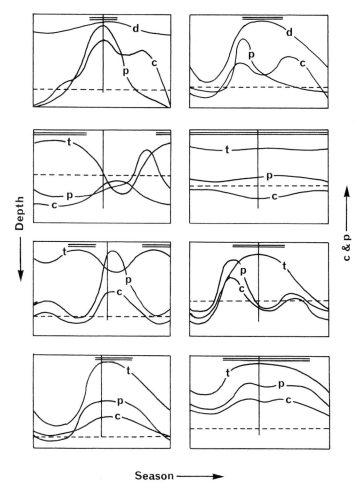

Fig. 10. Schematic illustrations of the general features of Longhurst's eight basic types of annual production cycle. The relative depth below the surface (the upper line of each illustration) of the mixed layer is determined by either temperature (t) or salinity (d). Fluctuations in relative chlorophyll concentrations (c) and rates of primary productivity (p) are shown (with zero being the base of each illustration) about the time of midsummer (vertical line). The seasons when the pycnocline is illuminated are shown by the double horizontal lines. The relative proportion of the annual production that occurs above and within the deep chlorophyll maximum is indicated by the dashed line. Redrawn from Longhurst (1995).

A. Patterns of Productivity and Biogeography

In the open ocean the annual quantity and seasonal cycling of primary production is determined by vertical stratification, the light cycle, and the persistence of nutrient supplies. Longhurst (1995) interpreted surface chlorophyll data from satellite imagery in the context of information on mixing processes. He identified just eight basic types of production cycle in the global ocean (Fig. 10). He identified three basic production domains in the open ocean—polar, temperate, and tropical—which differ fundamentally in their seasonal cycles of water column stability, nutrient supply, and illumination. He considered a fourth category, coastal domains, which are fragmented into very much smaller scale regions. The classical latitudinal patterns in biodiversity appear to be related to or even determined by these differences (Angel, 1993). Longhurst subdivided these basic domains into 56 biogeographical provinces, using climatological Coastal Zone Color Scanner (CZCS) chlorophyll and sea-surface temperature (SST) data from the Nimbus-7, together with data on mixed-layer depths and nutrient climatology (Levitus *et al.*, 1993). These provinces are delimited by recurrent features in ocean currents, fronts, topography, and sea-surface chlorophyll distributions (Fig. 11). These boundaries

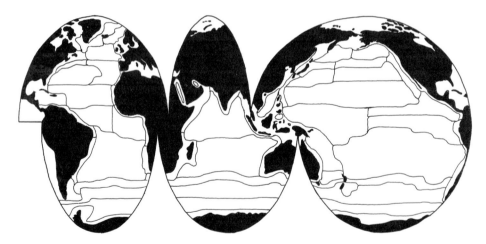

Fig. 11. The distribution of the biogeochemical provinces based on the productivity cycles illustrated in Fig. 10 and the approximate climatological positions of major oceanographic discontinuities identified from a combination of remotely sensed and hydrographic data. These provinces correlate closely with classical biogeographic provinces identified in each ocean, suggesting that these large-scale distribution patterns are determined by bottom-up processes. There are indications that these provinces are also mirrored in the deep benthic communities. Redrawn from Longhurst (1995).

match many of the classical biogeographical boundaries identified empirically (but often without precision) through the analysis of distributional data (Backus, 1986).

Within each province the structure of pelagic food webs, and hence of the communities of pelagic fishes, is likely to be relatively consistent. Moreover, because the quantities and dynamics of export production (i.e., the amounts of organic carbon exported through the base of the euphotic zone) are also likely to be directly influenced by the production cycles and the community structure, the differing seasonal patterns of input of organic carbon to the bottom-living communities are likely to affect changes in their structure and dynamics. Initial evidence of there being some coherence between Longhurst's provinces and the zoogeographical distributions of benthic abyssal species has been presented by Rex *et al.* (1993).

Merrett (1987) had already suggested that benthic abyssal fishes show a clear faunal boundary in species richness and dominance at around 40°N in the northeastern Atlantic, coincident with the boundary between the markedly pulsed seasonal inputs and the far less variable sedimentary inputs of subtropical and tropical domains. At temperate latitudes, sediment trap records show that the sedimentary fluxes vary by over two orders of magnitude (Wefer, 1989) and there is heavy seasonal deposition of phytodetritus on the seafloor (Billett *et al.*, 1983; Rice *et al.*, 1994), whereas at lower latitudes, such as off the coast of Bermuda (Deuser, 1987), the sediment trap fluxes vary by about an order of magnitude throughout the year and there have been no reports of deposition of phytodetritus. Thurston *et al.* (1994) have further shown that whereas the size spectra of the meiobenthic and macrobenthic faunas from these domains look very similar, there are substantial changes in the megafaunal components (Fig. 12). Basically, the large-deposit feeders that dominate the megafaunal component at high latitudes are almost totally missing from the subtropical communities. There are also marked changed in the necrophage communities that are reliant on "large lumps" (Stockton and Delaca, 1982). Deployments of baited cameras by Thurston *et al.* (1995) showed that at 21°N and 31°N the necrophage community consists almost entirely of the decapod prawn *Plesiopenaeus armatus*, whereas at 48°N two fish (*Coryphaenoides armatus* and *Pachycara bulbiceps*) and decapod crustaceans (*Munidopsis* spp.) dominate the community. Curiously enough *P. armatus* was still abundant in trawl catches at the temperate locality but did not feed at the baits. Haedrich and Merrett (1992) report that in the Porcupine Seabight, 35% of the demersal fish species feed purely on pelagic prey and 52% feed on a mixed diet of pelagic and benthic organisms.

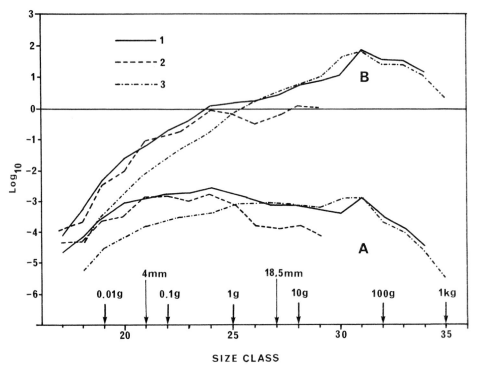

Fig. 12. Size spectra of abyssal megabenthos, macrobenthos, and meiobenthos from two sites either side of the divide between temperate and subtropical conditions. (A) The $\log_{10}$ abundances per square meter; (B) the $\log_{10}$ grams wet weight per square meter. Lines 1 represent epibenthic sledge samples from 4850 m on the Porcupine Abyssal Plain at 48°50'N, 16°30'W, a temperate, highly seasonal locality; lines 2 represent epibenthic sledge samples from 4940 m on the Madeiran Abyssal Plain at 31°05'N, 21°10'W, an oligotrophic subtropical gyre locality; lines 3 represent otter trawl samples from the Porcupine Abyssal Plain showing only how the spectra are extended using a larger sized (and meshed) trawl. Note how the size spectra are extremely similar over a broad range of the smaller sized organisms sampled by the sledge, but at subtropical latitudes the megafaunal size classes (>30) almost completely disappear. From Thurston et al. (1995).

B. Comparisons with Coastal Zones

Coastal waters show large fluctuations in space and time as a result of the great influence of tides and interactions with bottom topography. The physical and chemical environment is buffered by a large volume of water, so the ranges of most environmental parameters encountered are more extreme. There are only a few examples of consistently sampled transects that have extended offshore from continental shelf waters, out across the shelf break, and into oceanic water, whereby reliable comparisons of the faunas can be made. However, Hopkins *et al.* (1981) sampled pelagic species along such a transect in the Gulf of Mexico and found that the numbers of pelagic species increased sharply over the continental slope and began to decrease again further offshore. These differences in local species richness (diversity) are reversed at global scales, with the numbers of shelf species greatly exceeding those of oceanic species (Angel, 1997). This finding presumably reflects the finer scaling, both in time and in space, of coastal ecosystems together with the greater restrictions in genetic exchange between assemblages. For example, coastal (and bathyal) environments are ribbons, with most exchanges being possible only along shore. Such boundaries are now being breached by long-distance transportation of species in ballast waters. Over geological time there have been much greater variations; for example, a mere 7000 years ago the southern region of the North Sea was still dry land.

REFERENCES

Angel, M. V. (1983). A vertical profile of planktonic ostracods at 42°N 17°W from depths of 1500–3900 m. "Applications of Ostracoda" (R. F. Maddocks, ed.), pp. 529–548. University of Houston, Geoscience Department, Houston, Texas.

Angel, M. V. (1989). Vertical profiles of pelagic communities in the vicinity of the Azores Front and their implications to deep ocean ecology. *Prog. Oceanogr.* **22,** 1–46.

Angel, M. V. (1990). Life in the benthic boundary layer: Connections to the mid-water and sea floor. *Philos. Trans. R. Soc. London, A* **331,** 15–28.

Angel, M. V. (1993). Biodiversity of the pelagic ocean. *Biol. Conserv.* **7,** 760–772.

Angel, M. V. (1997). Pelagic biodiversity. *In* "Marine Biodiversity: Patterns and Processes" (R. F. G. Ormond, J. Gage, and M. V. Angel, eds.), in press. Cambridge Univ. Press, Cambridge.

Angel, M. V., and Baker, A. de C. (1982). Vertical standing crop of plankton and micronekton at three stations in the North-east Atlantic. *Biol. Oceanogr.* **2,** 1–30.

Backus, R. H. (1986). Biogeographical boundaries in the open ocean. *UNESCO Tech. Papers Mar. Sci.* **49,** 9–13.

Billett, D. S. M., Lampitt, R. S., Rice, A. L., and Mantoura, R. F. C. (1983). Seasonal sedimentation of phytoplankton to the deep-sea benthos. *Nature (London)* **302,** 520–522.

Bone, Q., Marshall, N. B., and Blaxter, J. H. S. (1995). "Biology of Fishes," 2nd Ed. Blackie, Glasgow and London.

Brewer, P. G., Glover, D. M., Goyet, C., and Shafer, D. K. (1995). The pH of the North Atlantic Ocean: Improvements to the global model for sound absorption in sea-water. *J. Geophys. Res.* **100**(C5), 8761–8776.

Broecker, W. S. (1992). The great ocean conveyor. *In* "Global Warming: Physics and Facts" (B. G. Livi, D. Afemeister, and R. Scribner, eds.), pp. 129–161. American Institute of Physics, New York.

Bugge, T., Belderson, R. H., and Kenyon, N. H. (1988). The Storrega slide. *Philos. Trans. R. Soc. London, A* **325,** 357–388.

Cohen, J. E. (1994). Marine and continental food webs: Three paradoxes? *Philos. Trans. R. Soc. London B* **343**, 57–69.

Dagg, M. J. (1993). Sinking particles as a possible source of nutrition for the large calanoid copepod *Neocalanus cristatus* in the subarctic Pacific Ocean. *Deep-Sea Res., Part 1* **40**, 1431–1445.

De Baar, H. J. W. (1994). Von Liebig's Law of Minimum and plankton ecology (1899–1991). *Progr. Oceanogr.* **33**, 347–386.

Denton, E. J. (1970). On the organization of reflecting surfaces in some marine animals. *Philos. Trans. R. Soc. London, B* **258**, 285–313.

Deuser, W. G. (1987). Variability and hydrography and particle flux: Transient and long-term relations. *In* "Particle Flux in the Oceans" (E. T. Degens, E. Izdal, and S. Honjo, eds.), *Mitteilungen aus dem Geologisch-Paläontologischen Institut der Universität Hamburg* **62**, 179–193.

Dickson, R. R., Meincke, J., Malmberg, S.-A., and Lee, A. J. (1988). The "Great Salinity Anomaly" in the northern North Atlantic 1968–1982. *Progr. Oceanogr.* **20**, 103–151.

Domanski, P. A. (1986). The Azores Front: A zoogeographical boundary? *UNESCO Tech. Papers Mar. Sci.* **49**, 73–83.

Donguy, J.-R. (1994). Surface and subsurface salinity in the tropical Pacific Ocean, relations with climate. *Progr. Oceanogr.* **34**, 45–78.

Dorsey, N. E. (1940). Properties of ordinary water-substance in all its phases: Water vapour, water and all the ices. *Am. Chem. Soc. Monogr.* **81**.

Elliot, N. G., Smolenski, A. J., and Ward, R. D. (1994). Allozyme and mitochondrial DNA variation in orange roughy, *Hoplostethus atlanticus* (Teleostei: Trachichthyidae). *Mar. Biol.* **119**, 621–627.

Elton, C. (1935). "Animal Ecology." Macmillan, New York.

Eppley, R. W., and Peterson, B. J. (1979). Particulate organic matter flux and planktonic new production in the deep ocean. *Nature (London)* **282**, 677–680.

Fairbanks, R. G., Wiebe, P. H., and Bé, A. W. H. (1980). Vertical distribution and isotopic composition of living planktonic Foraminifera in the western North Atlantic. *Science* **207**, 61–63.

Fenton, G. E., Short, S. A., and Ritz, D. A. (1991). Age determination of orange roughy *Hoplostethus atlanticus* (Pisces: Trachichthyidae) using ^{210}Pb:^{226}Ra disequilibria. *Mar. Biol.* **109**, 197–202.

Fleminger, A. (1986). The Pleistocene equatorial barrier between the Indian and Pacific Oceans and a likely cause for Wallace's line. *UNESCO Tech. Papers Mar. Sci.* **49**, 84–97.

Foster, T. D. (1984). The marine environment. In "Antarctic Ecology" (R. M. Laws, ed.), Vol. 2, pp. 345–372. Academic Press, London.

Gage, J. D., and Tyler, P. A. (1991). "Deep-sea Biology: A Natural History of Organisms at the Deep-sea Floor." Cambridge, Univ. Press, Cambridge.

Gardner, D. (1977). Nutrients as tracers of water mass structure in the coastal upwelling off northwest Africa. *In* "A Voyage of Discovery: George Deacon 70th Anniversary Volume" (M. V. Angel, ed.), pp. 327–340. Pergamon, Oxford.

Gloersen, P., Campbell, W. J., Cavalieri, D. J., Comiso, J. C., Parkinson, C. L., and Zwally, H. J. (1992). "Arctic and Antarctic Sea Ice, 1978–1987: Satellite Passive-Microwave Observations and Analysis." NASA Scientific and Technical Information Service, Washington, D.C.

Grassle, J. F., and Maciolek, N. J. (1992). Deep-sea richness: Regional and local diversity estimates from quantitative bottom samples. *Am. Nat.* **139**, 313–341.

Haedrich, R. L., and Merrett, N. R. (1992). Production/biomass ratios, size frequencies, and biomass spectra in deep-sea demersal fishes. *In* "Deep-sea Food Chains and the Global

Carbon Cycle" (G. T. Rowe and V. Pariente, eds.), pp. 157–182. Kluwer Academic Publishers, Dordrecht, The Netherlands.

Harte, J. (1988). "Consider a Spherical Cow: A Course in Environmental Problem Solving." University Science Books, Mill Valley, California.

Hopkins, T. L., Milliken, D. M., Bell, L. M., McMichael, E. J., Hefferman, J. J., and Cano, R. V. (1981). The landward distribution of oceanic plankton and micronekton over the west Florida continental shelf as related to their vertical distribution. *J. Plankton Res.* **3,** 645–659.

Jannasch, H. W. (1994). The microbial turnover of carbon in the deep-sea environment. *Global and Planetary Change* **9,** 289–295.

Janssen, J., Harbison, G. R., and Craddock, J. (1986). Hatchetfishes hold horizontal attitudes during diagonal descents. *J. Mar. Biol. Assoc. UK* **66,** 825–833.

Joyce, T. M., and Wiebe, P. H., eds. (1992). Warm core rings: Interdisciplinary studies of Kuroshio and Gulf Stream rings. *Deep-Sea Res.* **39**(Suppl.), S1–S417.

Kontar, E. A., and Sokov, A. V. (1994). A benthic storm in the northeastern tropical Pacific over the fields of manganese nodules. *Deep-Sea Res. 1* **41,** 1069–1089.

Koslow, J. A. (1993). Community structure in North Atlantic deep-sea fishes. *Prog. Oceanogr.* **31,** 321–338.

Lampitt, R. S., Billett, D. S. M., and Rice, A. L. (1986). Biomass of the invertebrate megabenthos from 500 to 4100 m in the Northeast Atlantic Ocean. *Mar. Biol.* **93,** 69–81.

Levitus, S., Conkright, M. E., Right, J. L., Najjar, R. G., and Mantyla, A. (1993). Distribution of nitrate, phosphate and silicate in the world oceans. *Prog. Oceanogr.* **31,** 245–274.

Longhurst, A. R. (1995). Seasonal cycles of pelagic production and consumption. *Prog. Oceanogr.* **36,** 77–168.

Longhurst, A. R., and Harrison, W. G. (1988). Vertical nitrogen flux from the oceanic photic zone by diel migrant zooplankton and nekton. *Deep-Sea Res.* **35,** 881–889.

McCartney, M. S. (1992). Recirculating components to the deep boundary current of the northern North Atlantic. *Prog. Oceanogr.* **29,** 283–383.

Mantyla, A. W., and Reid, J. L. (1983). Abyssal characteristics of the world ocean waters. *Deep-Sea Res.* **30A,** 805–833.

Marshall, N. B. (1971). "Explorations in the Life of Fishes." Cambridge Univ. Press, Cambridge.

Martin, J. H., Fitzwater, S. E., and Gordon, R. M. (1990). Iron deficiency limits phytoplankton growth in Antarctic waters. *Global Biogeochem. Cycles* **4,** 5–12.

Menzies, R. J., and Wilson, J. B. (1961). Preliminary field experiments on the relative importance of pressure and temperature on the penetration of marine invertebrates into the deep sea. *Oikos* **12,** 302–309.

Merrett, N. R. (1987). A zone of faunal change in the eastern Atlantic: A response to seasonality in production? *Biol. Oceanogr.* **5,** 137–151.

Open University (1989). "Ocean Chemistry and Deep-sea Sediments." The Open University, in association with Pergamon, Milton Keynes, U.K.

Parish, J., and Curtis, R. L. (1982). Atmospheric circulation, upwelling and organic-rich rocks in the Mesozoic and Cenozoic eras. *Palaeogeogr. Palaeoclimatol. Palaeoecol.* **40,** 31–66.

Parson, L. M., Walker, C. L., and Dixon, D. R., eds. (1995). "Hydrothermal Vents and Processes." Geological Society Special Publication 87. Royal Geological Society, London.

Parsons, T. R. (1988). Comparative oceanic ecology of the plankton communities of the subarctic Atlantic and Pacific oceans. *Oceanogr. Mar. Biol. Annu. Rev.* **26,** 317–359.

Paterson, G. L. J., and Lambshead, P. J. D. (1995). Bathymetric patterns of polychaete diversity in the Rockall Trough, northeast Atlantic. *Deep-sea Res.* **42,** 1199–1214.

Platt, T., Mann, K. H., and Ulanowicz, R. E., eds. (1981). "Mathematical Models in Biological Oceanography." The UNESCO Press, Paris.

Rex, M. A. (1983). Geographic patterns of species diversity in deep-sea benthos. In "Deep-Sea Biology, The Sea" (G. T. Rowe, ed.), Vol. 8, pp. 453–472. Wiley (Interscience), New York.

Rex, M. A., Stuart, C. T., Hessler, R. R., Allen, J. A., Sanders, H. L., and Wilson, G. D. F. (1993). Global-scale latitudinal patterns of species diversity in the deep-sea benthos. *Nature (London)* **365,** 636–639.

Ribbink, A. J. (1994). Biodiversity and speciation of freshwater fishes with particular reference to African cichlids. In "Aquatic Ecology: Scale, Pattern and Process" (P. S. Giller, A. G. Hildrew, and D. G. Raffaelli, eds.), pp. 261–288. Blackwell, Oxford.

Rice, A. L., Thurston, M. H., and Bett, B. J. (1994). The IOSDL DEEPSEAS programme: Introduction and photographic evidence for the presence and absence of a seasonal input of phytodetritus at contrasting abyssal sites in the northeastern Atlantic. *Deep-Sea Res.* **41,** 1305–1320.

Rowe, G. T. (1983). Biomass and production of the deep-sea macrobenthos. In "Deep-Sea Biology, The Sea" (G. T. Rowe, ed.), Vol. 8, pp. 97–122. Wiley (Interscience), New York.

Sarmiento, J. L., Orr, J. C., and Siegenthaler, U. (1992). A perturbation simulation of CO_2 uptake in an ocean general circulation model. *J. Geophys. Res.* **97**(C3), 2621–2645.

Schlosser, P., Bonisch, G., Kromer, B., Loosli, H. H., Buhler, R., Bayer, R., Bonani, G., and Koltermann, K. P. (1995). Mid-1980's distribution of tritium, ^{3}He, ^{14}C and ^{39}Ar in the Greenland/Norwegian Seas and the Nansen Basin of the Arctic Ocean. *Prog. Oceanogr.* **35,** 1–28.

Shimmield, G. B., Derrick, S., Mackensen, A., Grobe, H., and Pusey, C. (1994). The history of barium, biogenic silica and organic carbon accumulation in the Weddell Sea and Antarctic Ocean over the last 150,000 years. Carbon Cycling in the Glacial Ocean: Constraints on the Ocean's Role in Global Change. *NATO ASI Ser. I: Global Environmental Change* **17,** 555–574.

Smethie, W. M., Jr. (1993). Tracing the thermohaline circulation in the western North Atlantic. *Prog. Oceanogr.* **31,** 51–99.

Smith, S. L. (1984). Biological interactions of active upwelling in the northwestern Indian Ocean in 1964 and 1979, and a comparison with Peru and northwest Africa. *Deep-Sea Res.* **31,** 951–967.

Stockton, W. L., and Delaca, T. E. (1982). Food falls in the deep sea: Occurrence, quality and significance. *Deep-Sea Res.* **29,** 157–169.

Stuiver, M., Quay, P. D., and Ostlund, H. G. (1983). Abyssal water carbon-14 distribution and age of the World ocean. *Science* **220,** 849–851.

Summerhayes, C. P., Emeis, K.-C., Angel, M. V., Smith, R. L., and Zeitschel, B., eds. (1995). "Upwelling in the Ocean: Modern Processes and Ancient Records." Dahlem Workshop Reports, Environmental Sciences Research Reports 18. Wiley, New York.

Tett, P., and Barton, E. D. (1995). Why are there about 5000 species of phytoplankton in the sea? *J. Plankton Res.* **17,** 1693–1704.

Thorpe, S. A. (1995). Dynamical processes of transfer at the sea surface. *Prog. Oceanogr.* **35,** 315–352.

Thurston, M. H., Bett, B. J., Rice, A. L., and Jackson, P. A. B. (1994). Variations in the invertebrate abyssal megafauna in the North Atlantic Ocean. *Deep-Sea Res. I* **41,** 1321–1348.

Thurston, M. H., Bett, B. J., and Rice, A. L. (1995). Abyssal megafaunal necrophages: Latitudinal differences in the eastern North Atlantic Ocean. *Int. Rev. Ges. Hydrobiol.* **80,** 267–286.

Tunnicliffe, V. (1991). The biology of hydrothermal vents: Ecology and evolution. *Oceanogr. Mar. Biol. Annu. Rev.* **29,** 319–407.

Urick, R. J. (1975). "Principles of Underwater Sound," 2nd Ed. McGraw-Hill, New York.

Van der Spoel, S., and Heyman, R. P. (1983). "A Comparative Atlas of Zooplankton: Biological Patterns in the Oceans." Bunge, Utrecht.

Van der Spoel, S., Pierrot-Bults, A. C., and Schalk, P. H. (1990). Probable Mesozoic vicariance in the biogeography of Euphausiacea. *Bijdragen tot de Dierkunde* **60,** 155–162.

Vinogradova, N. G. (1979). The geographical distribution of the abyssal and hadal (ultra-abyssal) fauna in relation to the vertical zonation of the ocean. *Sarsia* **64,** 41–50.

Weatherly, G. L., and Kelley, E. A. (1985). Storms and flow reversals of the HEBBLE site. *Mar. Geol.* **66,** 205–218.

Weaver, P. P. E., Masson, D. G., Gunn, D. E., Kidd, R. B., Rothwell, R. G., and Maddison, D. A. (1995). Sediment mass wasting in the Canary Basin. *In* "Atlas of Deep Water Environments: Architectural Style in Turbidite Sediments" (K. T. Pickering, R. N. Hiscott, and N. H. Kenyon;, eds.), pp. 287–296. Chapman & Hall, London.

Wefer, G. (1989). Particle flux in the ocean: Effects of episodic production. *In* "Productivity of the Ocean: Present and Past" (W. H. Berger, V. S. Smetacek, and G. Wefer, eds.), pp. 139–154. Wiley, New York.

White, B. (1994). Vicariance biogeography of the open-ocean Pacific. *Prog. Oceanogr.* **34,** 257–282.

Wiebe, P. H., and Boyd, S. H. (1978). Limits of *Nemtoscelis megalops* in the northwestern Atlantic in relation to Gulf Stream core rings. 1. Horizontal and vertical distributions. *J. Mar. Res.* **36,** 119–142.

Wilson, G. D. F., and Hessler, R. R. (1987). Speciation in the deep sea. *Annu. Rev. Ecol. Syst.* **18,** 185–207.

Wishner, K. F. (1980). The biomass of deep-sea benthopelagic plankton. *Deep-Sea Res.* **27,** 203–216.

2

SYSTEMATICS OF DEEP-SEA FISHES

STANLEY H. WEITZMAN

I. Introduction
II. A Classification of Living Fishes Occurring near or below 500–600 m, with an Annotated List of Deep-Sea Fish Orders and Families
   A. Class Chondrichthyes—Cartilaginous Fishes
   B. Class Actinopterygii—Ray-Finned Fishes
   References

I. INTRODUCTION

Fortunately, during approximately the past 40 years, since the publication of the classic "Aspects of Deep-Sea Biology" (Marshall, 1954) and the years after its subsequent revision (Marshall, 1979), there have been numerous deep-sea exploratory expeditions supported by governments and private organizations of many nations. These cruises not only increased our knowledge of oceanography, but also greatly contributed to museum collections of deep-sea organisms including fishes. This increase in the number of specimens available for study and the consequent augmentation in the number of different and new deep-sea fish taxa have allowed numerous reinvestigations and completely new inquiries about evolutionary relationships. These inquiries have focused on the 157 (about 33%) extant fish families now known to include some deep-sea fishes, or at least species that occasionally penetrate to the deeper portions of the mesopelagic zone, below 500 to 600 m. These investigations have explored the interrelationships among deep-sea fish taxa as well as their relationships with shallower water fishes. Evolutionary adaptations to deep-sea life have apparently occurred independently many times in at least some of the 22 orders of fish discussed herein. Strikingly, the evolution and adaptation of bony fishes to the deep seas, especially to its pelagic environment, have apparently occurred more extensively among taxa that are derived from the relatively primitive groups of teleost fishes. Fish of the more derived teleost orders, such as the large order Perciformes, including nearly 150 of the well over 400 families of teleosts, have been successful in occupying relatively shallow waters, but have comparatively few deep-sea representatives, especially at the species level. A common assumption among ichthyologists suggests that the reason for this pattern is that the nonspiny rayed fish groups are older and therefore have had more time to evolve into various regions of the deep-sea environment. However, many more data concerning the phylogeny of fishes are needed for confirmation of this rather all too inclusive assumption. An excellent descriptive overview of the nature of many deep-sea fish groups is available (Marshall, 1979).

The large increase in the number of investigations of evolutionary relationships of deep-sea fishes has been greatly stimulated over the past 25 years by improvements in the methods and theories of the study of phylogenetic relationships. During the twentieth century a series of attempts to produce relatively inclusive, up-to-date fish classifications incorporating the most recent available data, interpreted through the use of the latest evolutionary concepts, has brought about major advances in our knowledge of fish evolution. Examination of the more important twentieth century fish classifications testifies to this (Regan, 1929; Berg, 1955; Greenwood et al., 1966; Lauder and Liem, 1983; Nelson, 1994). Future acquisition of data, and application of steadily improving phylogenetic (cladistic) systematic concepts, methods, and procedures to those data, should yield fish classifications that better reflect evolutionary history. Throughout this century attempts to summarize known data, collect additional information, and publish inclusive summaries of knowledge of fish evolution have progressed from almost single-person research programs (Regan, 1929) to multiauthored projects (Greenwood et al., 1966), and finally to attempts to incorporate and summarize a multitude of large and small research reports by a multiplicity of authors (Nelson, 1994). Also, within the last quarter of the twentieth century, a series of multiauthored symposiums addressing the phylogeny of various fish groups were published periodically and continue to be produced (Greenwood et al., 1973; Moser et al., 1984; Cohen, 1989; Johnson and Anderson, 1993; Stiassny et al., 1996).

Phylogenetically informative classifications should provide stable names for phylogenetically arranged fossil and living natural entities or taxa such as orders, families, genera, and species, each of which has its own evolutionary history and a clearly discernible existence in time. Phylogenetic classification, a product of systematics, should be an accurate reflection of the history of genetic, anatomical, and functional diversity as well as taxon biodiversity. Few classifications, even limited ones including only a few taxa, achieve this ideal. If an inclusive ideal classification of animals was accomplished, biologists studying physiology, genetics, anatomy, molecular biology, and many other biological disciplines would have a reliable framework for comparative evolutionary interpretations of their data. New data from many biological fields can help confirm or challenge the stability of existing classifications based on hypotheses of phylogeny derived from comparative morphology. Thus, biologists who are not primarily systematists can help improve current classifications. Unfortunately, at the present, fish classifications having much confirmatory evidence cladistically analyzed, and therefore a relatively higher degree of phylogenetic stability, are rather rare and mostly confined to small monophyletic groups. The general fish classification of Nelson (1994), on which the classification given herein is primarily based, remains far from perfect in providing consistently defended phylogenetic interpretations of the history of the approximately 57 orders, 484 families, 4260 genera, and over 24,000 species currently suggested to be a representation of the biodiversity of fishes.

At present most ichthyologists agree that phylogenetic systematics, or cladistics, is the most logical approach to evaluating data possibly supportive of hypotheses of relationships based on genealogy (i.e., common ancestry). Hypothetical relationships supported by data having genealogical (i.e., phylogenetic) significance are considered to be more informative for comparative discussions of functional and anatomical similarities and dissimilarities of organ systems, organs, and tissues among putatively related taxa than are relationships using taxa based only on or even partly on overall similarity, as in phenetic systematics.

The classification given herein should be viewed as a guide to possible relationships among fishes. Only some sections of the classification have received substantial phylogenetic improvement in recent years. Even in parts of its overall organization it is not phylogenetic. For example, the term *fishes* as used here is a common name, including five classes of aquatic animals that together do not represent a phylogenetic or historically natural group. In terms of genealogy the group "fishes" lacks evolutionary significance because many of its tetrapod descendants are excluded.

A fuller understanding of any given biological system, organ, or other structure in any particular taxon, whether it is a single species or a group of related species or higher taxon, is best accomplished in the context of the history of its evolutionary diversity. This is especially true currently, when humans are becoming increasingly concerned with comparative similarities, dissimilarities, and relationships of physiological processes of humans and other organisms—for example, in immunology. Biologists contribute data useful for an increase of phylogenetic knowledge, but most of these persons lack experience or knowledge about how to place these data into hypothetically significant phylogenetic contexts. On the other hand, many systematists lack experience and knowledge of how to address and utilize data that are not more or less anatomical. There are extensive and fruitful areas for cooperation in phylogenetic studies between systematists and biologists of all disciplines.

II. A CLASSIFICATION OF LIVING FISHES OCCURRING NEAR OR BELOW 500 TO 600 m, WITH AN ANNOTATED LIST OF DEEP-SEA FISH ORDERS AND FAMILIES

Although deep-sea fishes are often considered as those living below 1000 m, the classification herein also includes many families and genera that have species occurring between 500 and 1000 m. This is done because many fishes from these deeper mesopelagic depths have unique and interesting adaptations for deep-water living, and some may ultimately be found to occur at levels deeper than so far recorded. It may be assumed that all references given here to deep-sea taxa refer to marine fishes unless it is clearly stated they come from fresh water—for example, Lake Baikal in Russia. This chapter was greatly aided by reference to a list of fishes distributed below 2000 m (Grey, 1956), but hundreds of new records for deep-sea fishes have been added since that publication.

Only classes and subclasses of fishes and their taxa known to include extant deep-sea fishes are listed. Thus the class Sarcopterygii, including lobefinned fishes and tetrapods, is left out, as is the subclass Chondrostei of the class Actinopterygii (Nelson, 1994). All orders of the two classes listed (Chondrichthyes and Actinopterygii) are provided, but only those family names of deep-sea fishes as previously defined are given. Those orders listed without family names contain no known deep-water mesopelagic, bathypelagic, abyssopelagic, benthopelagic, bathyal, abyssal, or hadal fishes. Orders of extant elasmobranchs and teleostean fishes are listed with common names of many of the included deep-sea and non-deep-sea family-level groups in order to provide the reader with a sense of the proportion and placement of deep-sea fishes in the classification. When a taxon has no common name, an adjectival form of the scientific name is provided.

The classification herein is primarily based on that most recently published (Nelson, 1994), but important alterations have been made based on subsequent studies (Mooi and Gill, 1995; Patterson and Johnson, 1995; Johnson and Patterson, 1996; Baldwin and Johnson, 1996; Harold and Weitzman, 1996). The English common names for fish families and orders are, for the most part, those that have become more or less standard in recent years, especially in North America (Robins et al., 1991a,b).

2. SYSTEMATICS OF DEEP-SEA FISHES 47

The deep-sea families for each order are briefly and selectively discussed. In most cases the genera having deep-sea species are listed along with some information about the depths frequented by those taxa. In cases where a given family has numerous genera with deep-sea species, examples are given and references are provided that will lead to information on the bathymetric distribution of these fishes. References to publications listing depths for species, genera, and families are scattered and exceedingly numerous. Space limitations prevent citing most of these literature sources. Space limitations also preclude listing distinguishing characters of all the included deep-sea taxa, and only a few of the more remarkable characteristics are sometimes mentioned. The literature citations about the phylogeny of the included taxa mostly exclude older literature. In most instances, references to the literature about the identification and relationships of the included fish families can be found in the most recent general fish classification (Nelson, 1994).

Although depths of capture are given for many families and genera, these do not necessarily well represent their bathymetric ranges. Some of the depth ranges given here were derived from tows made with open nets that captured fishes while the net was descending and ascending. Such bathymetric records are tentative. Also, in the early years of oceanography, the only "depth" information given is the length of cable or wire used in the tow and sometimes together with an estimation of the angle of the wire. Depth estimates derived from such information are unreliable. For some taxa the only depth estimates in the literature are given as mesopelagic or bathypelagic, and these statements are repeated for what they may be worth, but even these broad bathymetric estimates are not always reliable. Fortunately many modern collections were taken using closing nets and depth-measuring devices that allow relatively accurate estimates of capture depths. Throughout the text the word meters is abbreviated to m.

Many pelagic deep-sea fish species make diurnal migrations, ascending at night, in some cases to the surface, and descending during the day. Although well studied for some species, reliable information of this kind is not available for many deep-sea fishes. Such information is mentioned for only a few species.

The classification format used is shown in Table I. Standard endings for categories above the ordinal level are not available for fish, but such endings are used for orders and subcategories. These suffixes are also shown in Table I.

A. Class Chondrichthyes—Cartilaginous Fishes

This class consists of two monophyletic subclasses, the Holocephali and the Elasmobranchii, thought to have a common ancestor (Nelson, 1994).

Table I
Classification Format

Category	Standard endings
Class	
Division	
Subdivision	
Infradivision	
Group	
Subgroup	
Infragroup	
Section	
Subsection	
Infrasection	
Order	iformes
Suborder	oidei
Superfamily	oidea
Family	idae
Subfamily	inae

Both subclasses have species living below 500 to 600 m. These fishes have a cartilaginous skeleton that is often partly calcified, but almost never ossified; typical bone has been histologically reported in only one species of catshark, family Scyliorhinidae (Hall, 1982). Distinguishing features common to living taxa of this class are many and have been enumerated (Maisey, 1986; Jamieson, 1991).

A(1).* Subclass Holocephali—Chimaeras

These fishes are characterized by having a gill cover superficial to four gill openings, among many other anatomical details. The classification of this group is provisional and much research and elucidation of the structures of especially the eight fossil orders remain to be accomplished (Nelson, 1994).

1. ORDER CHIMAERIFORMES — CHIMAERAS

This order of mostly benthopelagic slope-dwelling chondrichthians contains three extant families, of which two, Chimaeridae and Rhinochimaeridae, are known to have some species living at depths below 1000 m. The

---

* Entry headings have numbers or letters in straightforward outline format, except where subclassifications are extensive, in which case entry headings have numbers or letters in parentheses following a single letter or number. That single letter or number is that of the main outline entry to which the subclassifications are subordinate.

2. SYSTEMATICS OF DEEP-SEA FISHES

chimaeras are internally fertilized, some with the aid of modified head claspers that are used in courtship as well as pelvic claspers that are used in courtship and copulation. They also have a prominent cephalic laterosensory system. All have a prominent, apparently venomous spine at the anterior border of the first dorsal fin.

a. Family Chimaeridae. Chimaeras. There are about 27 species in two genera, *Chimaera* and *Hydrolagus.* Some reach a depth of 2600 m.

b. Family Rhinochimaeridae. Longnose chimaeras. This family has about six species distributed in three genera: *Harriotta,* perhaps to 2600 m, *Neoharriotta,* and *Rhinochimaera.*

A(2). Subclass Elasmobranchii—Sharks and Rays

Of 43 families of elasmobranchs, 19 have species with a distribution to or below 500 to 1000 m. In addition to net hauls, actual sightings or accurate records of depths reached by sharks are few (Clark and Kristof, 1990). A relatively recent list of sharks (Springer and Gold, 1989) has been updated by Nakaya (Springer and Gold, 1992). Many species are known but remain undescribed.

1. ORDER HETERODONTIFORMES

Port Jackson, horn, or bullhead sharks (one family).

2. ORDER ORECTOLOBIFORMES

Carpet sharks, nurse sharks, whale sharks, and relatives (seven families).

3. ORDER CARCHARHINIFORMES

Ground sharks, requiem sharks, and relatives (eight families). The phylogeny of this order as well as that of its families is in need of further research (Naylor, 1992). One monotypic genus has one spineless dorsal fin, but species in all other genera have two spineless dorsal fins. Five gill slits are present and gill rakers are absent. Species are either oviparous, ovoviviparous, or viviparous. Of the eight families, comprising perhaps 47 or 48 genera and a little over 200 species, only one family, the catsharks, family Scyliorhinidae, has benthic species extending to depths below 1000 m.

a. Family Scyliorhinidae. Catsharks. There are about 86 species arranged in 17 genera. All are marine, and most live in cool waters of upper continental slopes, worldwide (Springer, 1979). All species for which there is information are oviparous. The genus *Apristurus* has over 30 species of bottom-dwelling small sharks, apparently frequently taken between 700 and 1460 m and possibly reaching 1840 m. Other genera with deep-sea species are *Galeus* (to 640 m), *Cephaloscyllum* (to 700 m), and *Halaelurus* (to 900 m).

    b. Family Pseudotriakidae. False catshark. The single species, *Pseudotriakis microdon*, is known to occur as deep as 580 m.

    c. Family Triakidae. Smoothhounds or hound sharks. With nine genera and 39 species, *Galeorhinus galeus* is known down to 550 m; species of *Mustellus* are known down to 752 m.

    d. Family Carcharinidae. Requiem sharks or whaler sharks. There are 13 genera and about 60 species; *Carcharinus albimarginatus* is known as deep as 800 m.

4. ORDER LAMNIFORMES

Mackerel sharks, sand tigers, megamouth shark, thresher sharks, basking sharks, mackerel sharks, and relatives (seven families).

    a. Family Mitsukurinidae. Goblin shark. The goblin shark, *Mitsukurina owstoni*, is known to occur as deep as 1200 m.

    b. Family Pseudocarchariidae. Crocodile shark. The crocodile shark, *Pseudocharcharias kamoharai*, is known to occur as deep as 590 m.

    c. Family Megachasmidae. Megamouth shark. The megamouth shark, *Megachasma pelagios*, is known to occur between 150 and 1000 m deep. This is a filter-feeding shark.

    d. Family Alopidae. Thresher sharks. With one genus and more than three species, one, the bigeye thresher, *Alopias superciliosus*, is known at least down to 500 m.

    e. Family Lamnidae. Mackerel sharks. The white shark, *Carcharodon carcharias*, is recorded from the surface at least down to 1280 m. The longfin mako shark *Isurus pacus* is recorded as deep as 1150 m.

5. ORDER HEXANCHIFORMES

Frill and cow sharks, including the six- and seven-gill sharks (two families). Most of these sharks are confined to relatively deep waters of the continental shelves of tropical and temperate oceans. They have a single

2. SYSTEMATICS OF DEEP-SEA FISHES   51

dorsal fin (rather than two), six or seven gill arches and slits (rather than five, as in other sharks). The additional one or two gill arches compared to other sharks was at one time thought primitive for the elasmobranch orders. It is now considered that these arches are derived specializations of the second or third arches (Shirai, 1992a).

a. Family Hexanchidae. Cow sharks, seven-gill sharks, and six-gill sharks. *Heptanchias perlo* usually occurs at depths of about 30 to 720 m or more, but may be found at 1000 m. *Hexanchus griseus* occurs from the surface to a depth of about 2000 m.

b. Family Chlamydoselachidae. Frill shark. The frill shark, *Chlamydoselachus anguineus* is usually taken at depths of 120 to 1280 m. It has expandable jaws that allow capture and engulfment of very large prey.

6. ORDER SQUALIFORMES

Dogfish sharks, bramble sharks, saw sharks, sleeper sharks, and relatives. This order contains four families, 23 genera, and over 90 species. Many are known from deep waters of all oceans and seas.

a. Family Echinorhinidae. Bramble sharks. There are two species in one genus, *Echinorhinus*, of which one is recorded from depths of 900 m and may reach greater depths.

b. Family Dalatiidae. Sleeper sharks, lantern sharks, taillight shark, pygmy sharks, cookiecutter sharks, and relatives. The family has 18 genera and nearly 50 species, of which several reach depths of well below 1000 m (Last and Stevens, 1994). For example, *Centroscymus* reaches 3660 m in depth. Members of the genus *Etmopterus* possess light organs and reach depths of 920 m. All three species of cookiecutter sharks, *Isistius*, are known from depths below 500 m. *Isistius brasiliensis* reaches depths of 3500 m and has pitlike photophores on the ventral surface of the head and body; the other species probably also have photophores. *Zameus* occurs in depths of 550 to 2000 m. The pygmy shark, *Squaliolus aliae*, vertically migrates between 200 m (daytime) and 2000 m (nighttime). *Somniosus* has been filmed at 2200 m and a species of *Dalatias* is recorded from 1645 m deep.

c. Family Oxynotidae. Prickly dogfishes. Included in the Dalatiidae by Nelson (1994). One species, *Oxynotus bruniensis*, is known from depths of 350 to 650 m.

d. Family Centrophoridae. Centrophorid sharks. These are deep-water benthic sharks in two genera and perhaps 13 species, but only species of

Centrophorus are reported to reach depths to 1400 m. A record 6000 m is questionable.

    e. Family Squalidae. Dogfish sharks. This family has two genera, *Cirrhigaleus* and *Squalus,* and about 10 species. Only the spiny dogfish, *Squalus acanthias,* is known to reach depths of about 950 m, and may go deeper. Like most sharks, they have five gill arches, but lack an anal fin, and have two dorsal fins, each of which is usually preceded by a spine.

7. ORDER SQUATINIFORMES

Angel sharks. This order contains a single family, the Squatinidae, with a single genus, *Squatina.* These are ovoviviparous, shallow to often moderately deep-water benthic elasmobranchs sometimes mistaken for rays. They are nearly worldwide in distribution.

    a. Family Squatinidae. Angel sharks. *Squatina dumeril,* of the 12 to 15 species, is known to reach depths below 1300 m. These are raylike elasmobranchs with large pectoral and pelvic fins, no anal fin, two spineless dorsal fins, dorsally placed eyes, and five gill slits. The relationship of these sharklike fishes has been controversial, but they have been placed as a sister group of the Pristiophoriformes and Rajiformes and relatives (Shirai, 1992b).

8. ORDER PRISTIOPHORIFORMES

Saw sharks (one family).

    a. Family Pristiophoridae. Saw sharks. With two genera and over five species, one *Pristiophorus* sp. (Last and Stevens, 1994) is known from 150 to 630 m deep.

9. ORDER RAJIFORMES

Sawfishes, rays, guitarfishes, eagle rays, electric rays, stingrays, and relatives (four suborders and 12 families).

9(A). SUBORDER TORPRDINOIDEI

Electric rays (four families).

    a. Family Torpenidae. Electric rays or torpedo rays. One genus, *Torpedo,* with 13 species, reaches depths of 750 m.

    b. Family Narcinidae. Electric rays or numbfishes. With four genera, species of *Benthobatis* extend down to about 950 m, and some of *Narcine* reach depths of at least 640 m.

2. SYSTEMATICS OF DEEP-SEA FISHES

9(B). SUBORDER RAJOIDEI
Rays and skates (three families).

a. Family Rajidae. Skates. With 18 genera and some 200 species, many species of *Raja* live to depths of 1000 m and a few live as deep as 2000–3000 m. Some species of *Bathyraja* reach depths below 2300 m and some of *Notoraja* and *Anacanthobatis* occur to 1120 m.

9(C). SUBORDER MYLIOBATOIDEI
Myliobatoid rays (six families).

a. Family Plesiobatidae. Deep-water stingray. There is only one species, *Plesiobatis daviesi*, and it is known down to 460 m and probably extends to somewhat greater depths.

b. Family Hexatrygonidae. Hexatrygonids. *Hexatrematobatis longirostrum* has been captured down to 1000 m.

c. Family Dasyatidae. Stingrays. With six genera and about 50 species, *Dasyatis brevicaudata* is known from 182 to 476 m and may go deeper.

B. Class Actinopterygii—Ray-Finned Fishes

1. DIVISION TELEOSTEI — TELEOST FISHES

1(A). SUBDIVISION OSTEOGLOSSOMORPHA — OSTEOGLOSSOMORPHS

a. Order Osteoglossiformes. Bonytongues, butterflyfish, mooneyes, Old World knifefishes, elephantfishes, and gymnarchids (six families).

1(B). SUBDIVISION ELOPOCEPHALA

1(B1). INFRADIVISION ELOPOMORPHA — ELOPOMORPHS

a. Order Elopiformes. Tenpounders and tarpon (two families).

b. Order Albuliformes. Bonefish, pterothrissins, halosaurs, and spiny eels (three families). This order, with all member species having leptocephalus larvae, has two deep-water families.
Family Halosauridae. Halosaurs. Three genera with 21 species occur on continental slopes, living close to the bottom and ranging down to the

abyssal plains in the central and western Pacific as well as the Atlantic Ocean. At least one species of *Halosaurus* reaches depths of 3100 m; some species of *Aldrovandia* occur at depths of 1100 to possibly 5000 m, and *Halosauropsis* occurs between 1000 and 3200 m.

Family Notacanthidae. Spiny eels and toothless spiny eel. This apparently worldwide family includes three genera, *Lipogenys* (sometimes placed in its own family, the Lipogenyidae), *Polyacanthonotus*, and *Notacanthus*, totaling 10 species (Greenwood, 1977). Some but not all of the species in these genera are known to reach 3700 m in depth, the record being 4900 m. All are elongate benthopelagic fishes with ventral mouths, elongate bodies, and an elongate anal fin base connected to a reduced caudal fin (or the latter absent). Their biology is little known (Paulin and Moreland, 1979; Crabtree *et al.*, 1985).

c. *Order Anguilliformes.* Eels (15 families). The true eels, with 15 families, are a complex group much in need of phylogenetic study. This worldwide order comprises members occurring in a variety of marine habitats, from coral reefs to sandy bottoms; it also has about a dozen species that, as adults, are found in fresh water. Eel leptocephalus larvae are pelagic and distinct in certain characters from those of the Albuliformes. All extant eels lack pelvic fins and their supporting skeleton, and have a wide variety of derived internal features not found in other fishes (Nelson, 1994). This order has seven families that inhabit depths to 1000 m or more.

Family Anguillidae. Freshwater eels. These freshwater eels, all belonging to one genus, *Anguilla*, with 15 species, are usually catadromus and widely distributed except in the eastern Pacific and southern Atlantic oceans. The freshwater European eel, *Anguilla anguilla*, and the American eel, *Anguilla rostrata*, leave fresh water to spawn in the Sargasso Sea. They have been photographed at 2000 m depth near the Bahama Islands.

Family Synaphobranchidae. Cutthroat eels. The cutthroat and arrowtooth eels are noted for having larvae with elongate or "telescopic" eyes. These comprise perhaps 13 genera and about 25 species, many of them living at depths of 1000 to 3500 m. *Histiobranchus bathybius* has been captured at depths of at least 5400 m (Robins and Robins, 1989) and *Thermobiotes mytilogeiton* was taken near hydrothermal vents at 1750 m deep. *Diastobranchus, Histiobranchus, Ilyophis, Symenchelys,* and *Synaphobranchus* have species known to reach depths below 1000 m.

Family Colocongridae. Colocongrid eels. The colocongrids include one genus, *Cologconger,* and approximately five species. They are short-bodied eels occurring at depths of at least 900 m and probably deeper.

Family Derichthyidae. Narrowneck eels. The narrowneck eels occur at least to depths of 1800 m. Their appearance is remarkable in that there is

2. SYSTEMATICS OF DEEP-SEA FISHES

a relatively long distance between the posterior part of the head and gill openings and the pectoral girdle and fin. There are two genera: *Derichthys* has one species and *Nessorhamphus* has two species.

Family Nemichthyidae. Snipe eels. The snipe eels consist of three genera with a total of about nine species. They are fragile mesopelagic to bathypelagic fishes, with some said to occur at the ocean surface. *Nemichthys* occurs at least to depths of 1830 m; *Avocettina* reaches depths from 600 to at least 4500 m. Mature females and young males have elongate jaws and small teeth; the adult males lose these and develop large anterior nostrils apparently for detection of pheromones, presumably produced by the females.

Family Congridae. Conger eels. This large family has about 32 genera and approximately 150 species occurring in a variety of habitats. Most are shallow-water fishes, but some (for example, *Bassanago* from Australia) reach depths of 1100 m. *Bathymyrus*, with three species, reaches depths of at least 470 to 490 m. *Bathyuroconger* is known to reach a depth of 1318 m; *Promyllantor* has been found at 1800 m.

Family Nettastomatidae. Duckbill eels. The duckbill or witch eels, with six genera and about 30 species, have an elongate narrow head and mouth with long jaws. The pectoral fin is absent in all but one genus (Smith *et al.*, 1981). The adults occur near the bottom and the larvae are found in the open ocean. They inhabit depths of 100 to 2000 m.

Family Serrivomeridae. Sawtooth eels. The sawtooth eels are pelagic mostly midwater eels; the larvae and juveniles live near the surface and the adults are found at depths of 1200 m, although dubious reports indicate 4500 m. These eels also have very elongate jaws, but differ from most of the Nettastomatidae in having a pectoral fin. There are two genera, *Serrivomer* and *Stemonidium*, with 10 species.

d. *Order Saccopharyngiiformes.* Bobtail snipe eels and gulper eels or pelican eels. This order of deep-sea creatures occurring in the Atlantic, Indian, and Pacific oceans has many derived features, including extremely long jaws and a wide gape associated with many osteological modifications of the jaws and branchial apparatus. Thses fishes are eellike in body form, with long dorsal and anal fins. All have leptocephalus larvae.

Family Cyematidae. Bobtail snipe eels. These are relatively short-bodied, compressed bathypelagic eels. There are two known species: *Cyema atrum*, from all oceans, reaches estimated depths of 5100 m and *Neocyema erythrosoma*, from the central and eastern south Atlantic Ocean, reaches depths of at least 2200 m. These fishes, for many years placed among the anguiliform eels, are now placed in the saccopharyngiiform eels on the basis of several cranial features.

Family Saccopharyngidae. Swallower eels. Swallower eels are most frequently taken at depths of 1000 to 3000 m. One genus, *Saccopharynx*, and nine species are known.

Family Eurypharyngidae. Gulper eels. The single bathypelagic species, *Eurypharynx pelecanoides*, is thought to inhabit depths of 500 to about 3000 m, but the maximum depth that one has been taken, using a closing net, was at 1532 m. This species differs from those in *Saccopharynx* by having a huge mouth and many attendant differences in skull fractures.

Family Monognathidae. Singlejaw eels. These eels include one genus, *Monognathus*, and 15 species. They occur between 100 and 5400 m, with most specimens taken at depths below 2000 m. They lack an upper jaw and pelvic fins, but possess a poisonous rostral fang.

1(B2). INFRADIVISION CLUPEOCEPHALA — CLUPEOCEPHALANS

a. Group Otocephala—Otocephalans. See Johnson and Patterson (1996).

a(1). Subgroup Clupeomorpha—Clupeomorphs
i. Order Clupeiformes. Herrings, sardines, anchovies, and relatives (four families).

a(2). Subgroup Ostariophysi—Ostariophysans
i. Order Gonorynchiformes. Milkfish, beaked sandfishes, and relatives (four families).
ii. Order Cypriniformes. Minnows, suckers, hillstream fishes, loaches, and relatives (six families).
iii. Order Characiformes. Characins, tetras, piranhas, pencilfish, and relatives (12 families).
iv. Order Siluriformes. Catfishes, (34 families).
v. Order Gymnotiformes. Knifefishes, electric eel, and relatives (six families).

(b). Group Euteleostei—Euteleosteans

b(1). Subgroup Protacanthopterygii—Protacanthopterygians
i. Order Argentiniformes. Argentines, alepocephalids, and relatives. Deep-sea fishes occur in all of the included groups, in all major oceans and seas. The phylogeny of this order has recently been the focus of considerable research (Johnson and Patterson, 1996), and the arrangement of families and superfamilies here is based on that research.

Superfamily Argentinoidea. Argentines, barreleyes, deep-sea smelts, slender smelts, and relatives.

2. SYSTEMATICS OF DEEP-SEA FISHES

Family Argentinidae. Argentines. These fishes have eggs and larvae that are pelagic. The adults are commonly taken at the margins of the continental shelves. There are two genera, *Argentina* and *Glossanodon,* with a total of about 19 species. Some species of these large-eyed fishes apparently reach depths of nearly 1000 m, but most usually occur at shallower depths.

Family Opisthoproctidae. Barreleyes or spookfishes. These sometimes foreshortened somewhat laterally compressed creatures usually have dorsally oriented eyes with globes somewhat vertically elongate to tubular or barrel-shaped. Only some of the species are estimated to occur at depths of 1000 m. The family contains six genera (Cohen, 1964). The long and slender *Bathylychnops* is known from no deeper than 200 m whereas the elongate bathypelagic *Dolichopteryx* is taken as deep as 2700 m. It has a suborbital light organ. The short-bodied *Macropinna* occurs at depths of 100 to 914 m. *Opisthoproctus* has been caught between depths of 200 and 600 m. It has luminescent bacteria present in a rectal diverticulum and the bacteria spread over a flat reflecting organ along the ventral surface of the abdomen. *Rynchohyalus,* also with a suborbital light organ, is known from as deep as 550 m. *Winteria* has tubular eyes facing anteriorly, but its location in the water column is between 500 and 1250 m.

Family Microstomatidae. Slender smelts and deep-sea smelts. This family (Johnson and Patterson, 1997) is a nearly worldwide group of mesopelagic fishes with four genera, *Microstoma, Nansenia* (to 2750 m), *Xenophthalmichthys,* and *Bathylagus* (sometimes placed in a separate family), with a total of about 32 species. Reliable information about the depths these fishes reach is scarce, but most specimens have been taken at depths considerably less than 1000 m. Species of *Bathylagus* superficially resemble the argentines, but have some different skull features and lack a swim bladder. They have been caught at the surface and apparently to depths of 3600 m. Most have been taken between the surface and less than 1700 m. Larvae of some species of *Bathylagus* have their eyes on short stalks.

Superfamily Alepocephaloidea. Slickheads, tubeshoulders, and relatives.

Family Platytroctidae. Tubeshoulders and relatives. Most of the Platytroctidae (also called Searsiidae) are taken at depths of 200 to 2000 m. They have a modified tubelike scale behind the pectoral girdle just below the lateral line that connects to a sac containing a luminous fluid. The sac is located just medial to the pectoral girdle. The body is black and has many light organs. There are 13 genera: *Barbantus,* 525 to possibly 4500 m; *Holtbyrnia,* mesopelagic to bathypelagic, 0 to 3000 m; *Maulisia,* mesopelagic, 475 to about 1500 m; *Mirorictus,* to 1750 m; *Normichthys,* mesopelagic, 400 to usually below 1000 m; *Paraholtbyrnia,* 220 to 500 m; *Pellisolus,* bathypelagic, depths below 1000 to at least 1400 m; *Persparsia,* apparently mesopelagic; *Platytroctes,* 2500 to 5393 m; *Sagamichthys,* 37 to

1300 m; *Searsia*, 420 to 1000 m; *Searsioides*, 0 to 1500, mostly between 600 and 1000 m; and *Tragularius*, 1200 to 2000 m. There are 37 species (Matsui and Rosenblatt, 1987).

Family Bathylaconidae. Bathylaconids. There are three known species placed in two genera. The mesopelagic to bathypelagic *Bathylaco* extends to 4400 m and *Herwigia* to 2100 m (Nielsen, 1972). Photophores are absent.

Family Alepocephalidae. Slickheads. The slickheads, with a total of somewhat over 60 species, occur at depths of about 0 to 6000 m and consist of about 23 genera. Examples include *Alepocephalus*, from about 1000 to 3600 m; *Asquamiceps*, 0 to at least 2100 m; *Aulastomatomorpha*, 1717 to 2020 m; *Bajacalifornia*, 0 to 2000 m; *Bathyprion*, 1100 to 2100 m; *Bathytroctes*, benthopelagic, to 4900 m; *Bellocia*, from about 2865 to 5850 m; *Conocara*, benthopelagic; *Einara*, bathypelagic; *Ericara*, 2469 to 3990 m; and *Leptochilichthys*, with three species known from depths of 724 to 3000 m. *Leptoderma* is benthopelagic, to 2283 m; *Micrognathus*, bathypelagic, 1500 to 1600 m; *Narcetes*, bathypelagic, to at least 2000 m; *Photostylus*, bathypelagic depths exceeding 1000 to at least 1460 m; *Rinoctes*, bathybenthic, 2000 to 4156 m; *Rouleina*, benthopelagic, living near the bottom mainly between 1400 and 2100 m; *Talismania*, benthopelagic, to at least 1355 m; and *Xenodermichthys*, mesopelagic to benthopelagic and abyssal, to 6000 m (Tortonese and Hureau, 1979).

Order Salmoniformes. Trouts and smelts (four families).

ii(A). Suborder Salmonoidei. Salmons, trouts, and graylings (one family).

ii(B). Suborder Osmeroidei. Smelts, noodlefishes, aiu, southern smelts, salamanderfishes, galaxiids, and relatives (three families).

b(2). *Subgroup Neognathi—Neognathans*

b(2A). *Infragroup Haplomi*
i. Order Esociformes. Pikes and mudminnows (two families).

b(2B). *Infragroup Neoteleostei—Neoteleosts*

b(2Bi). *Section Stenopterygii—Stenopterygians*
i. Order Stomiiformes. Bristlemouths, light fishes, hatchetfishes, dragonfishes, and relatives. This oceanic clade includes four families of which the species are varyingly highly modified for a pelagic, especially mesopelagic, and/or sometimes bathypelagic existence. Several groups have members extending to depths well below 1000 m. Species of many of the genera are found in all oceans and many seas. The stomiiform fishes have been a focus of rather detailed phylogenetic (cladistic) revisions since 1974 (Harold and

Weitzman, 1996) and are hypothesized as monophyletic. They possess a complex set of photophores arranged in ways that characterize the various subgroups. Currently the mesopelagic, to occasionally bathypelagic *Diplophos*, to 2400 m deep, and *Manducus*, to 800 m deep, are "unplaced" in a family group, although they are the most primitive genera in the order and a sister group to all other stomiiforms. *Triplophos*, also currently "unplaced" to family, is mesopelagic and has been taken as deep as 800 m.

Family Photichthyidae. Lightfishes. The family Photichthyidae (Phosichthyidae) or lightfishes, a nonmonophyletic group (Harold and Weitzman, 1996), is in need of revision. Included are *Ichthyococcus*, mesopelagic to bathypelagic possibly to 1700 m or deeper; *Photichthys*, depth of capture uncertain; *Pollichthys*, mesopelagic to 900 m; *Polymetme*, benthopelagic to 580 m; *Vinciguerria*, mesopelagic to bathypelagic to perhaps 2000 m; *Woodsia*, mostly mesopelagic but to 1100 m; and *Yarrella*, mesopelagic to 870 m.

Family Gonostomatidae. Bristlemouths. There are four genera: *Bonapartia*, primarily mesopelagic, but also said to occur as deep as 2880 m; *Cyclothone*, mesopelagic and bathypelagic, but one species reaches to perhaps 5300 m deep [see Kashkin (1995) for depth records from the Pacific Ocean]; *Gonostoma*, a nonmonophyletic genus (Harold, 1997) that is primarily mesopelagic, but with some species bathypelagic to at least 2700 m; and *Margrethia*, ordinarily mesopelagic, but possibly taken as deep as 2744 m.

Family Sternoptychidae. Maurolicins and hatchetfishes. This anatomically diverse family includes 10 mostly mesopelagic genera (Harold and Weitzman, 1996). *Araiophos* has so far been taken with open nets from near the surface to a depth of 200 m. *Argyripnus* is possibly benthopelagic to 475 m. The primarily mesopelagic hatchetfish genus *Argyropelecus* has some species extending to 4060 m; *Danaphos* is mesopelagic, often between 300 and 699 m, but apparently was taken at depths of 2880 m; *Maurolicus* is mostly benthopelagic between 100 and 500 m and only occasionally taken in bathypelagic waters, with records extending down to approximately 1700 m; the hatchetfish genus *Polyipnus* is mesopelagic, but for the most part its species remain near continental shelves with some species extending to below a little over 1000 m (Harold, 1994); *Sonoda* is mesopelagic perhaps to 550 m; the hatchetfish *Sternoptyx* is mesopelagic to bathypelagic and abyssopelagic to depths of 3085 m; *Thorophos* and *Valenciennellus* are mesopelagic.

Family Stomiidae. Viperfishes, dragonfishes, snaggletooths, blackdragons, and relatives. The phylogeny and monophyly of this family and its genera were recently reviewed (Fink, 1985). All 26 genera and nearly 230 species are now placed in a single family. These chin-barbeled fishes with light organs on their barbels were previously scattered in several families, some being nonmonophyletic. These mostly mesopelagic to bathypelagic genera are as follows: *Aristostomias*, to depths of about 1800 m, but apparently most frequent between 30 and 300 m; *Astronesthes*, to at least 1500 m, but one species is known to migrate to the surface; *Bathophilus*, surface to at least 3500 m, but most common at 180 to 550 m, depending on the species; *Borostomias*, 320 to 2600 m, but most frequent at 450 to 1000 m; *Chauliodus*, most frequent between 75 and 1500 m, but one species occurs at least to 3500 m deep during the day and another has been taken at the surface at night; *Chirostomias*, from 75 to about 1300 m; *Echiostoma*, to at least 2000 m; *Eustomias*, near the surface to 2500 m depending on the species; *Flagellostomias*, between 75 and 1825 m; *Gramatostomias*, surface to 4500 m; *Heterophotus*, about 200 to 850 m; *Idiacanthus*, 1000 to 5000 m; *Leptostomias*, to 2700 m; *Malacosteus*, near the surface to at least 2000 m; *Melanostomias*, from 40 to 2000 m depending on the species; *Neonesthes*, 70 to about 1650 m; *Rhadinesthes*, 100 to about 600 m; *Odontostomias*, depth uncertain; *Pachystomias* to 4460 m; *Parabathophilus*, 400 to 600 m; *Photonectes*, surface to as deep as 1350 m; *Photostomias*, to depths of 3100, m; *Stomias*, 200 to 1500 m; but most common at 600 to 800 m; *Tactostoma*, from 30 to 1800 m; *Trigonolampa*, 0 to 950 m; and *Thysanactis*, 100 to 1000 m.

ii. Order Ateleopodiformes. Jellynose fishes. This order contains one widely distributed marine family of elongate fishes with a long anal fin and large nose and head.

Family Ateleopodidae. Jellynose fishes. This family contains four genera, *Ateleopus, Ijimaia, Parateleopus*, and *Guentherus*. All are primarily benthic. At least *Ateleopus* and *Guentherus* are found about as deep as 700 m and may go deeper (Smith, 1986).

b(2Bii). Section Eurypterygii—Euryptergyians

i. Order Aulopiformes. Lizardfishes and relatives. The relationships of the Aulopiformes have recently been studied (Baldwin and Johnson, 1996) and the classification presented here is based on their phylogenetic analysis. Their new data organize these fishes in a considerably different classification from that previously published (Nelson, 1994). Some species are known from all temperate and tropical oceans and seas.

i(A). Suborder Synodontoidei. Aulopids, lizardfishes, bombay duck, and relatives. This group consists of three marine families of tropical and subtropical shallow-water fishes, with some in deep water.

Family Aulopidae. Aulopids. There is only one genus, *Aulopus*, with about 10 benthic species. Some live from near shore down to about 1000 m.

i(B). Suborder Chlorophthalmoidei. Greeneyes, waryfishes, and spiderfishes (three families).

2. SYSTEMATICS OF DEEP-SEA FISHES

Family Chlorophthalmidae. Greeneyes. Of the two genera, *Chlorophthalmus* is known to depths of 1440 m and *Parasudis* is known to depths of 480 m. There are about 20 species. *Bathysauropsis*, with two species, taken between 2010 and 2600 m, was considered intermediate between this family and the Notosudidae/Ipnopidae (Baldwin and Johnson, 1997), but was not given family assignment subject to further investigation.

Family Notosudidae. Waryfishes. There are three mesopelagic to bathypelagic genera: *Ahliesaurus*, mesopelagic, deeper than 500 m to possibly the upper layers of the bathypelagic zone; *Luciosudis*, mesopelagic to 800 m; and *Scopelosaurus*, epipelagic to as deep as 1147 m. There are approximately 19 species in the family.

Family Ipnopidae. Spiderfishes and relatives. The Ipnopidae (Merrett and Nielsen, 1987) are widely distributed elongate bathypelagic to abyssopelagic benthic fishes. *Bathymicrops* has been caught between 3033–4225 and 5900 m; *Bathypterois*, the tripodfish, between 250 and 5150 m; *Bathytyphlops*, between 869 and 2265 m; *Discoverichthys*, from a trawl at 5440 m; and *Ipnops*, between 1392 and 4970 m. The eyes of the first four genera are minute whereas those of *Ipnops* are dorsally directed, peculiar platelike lenseless structures.

i(C). Suborder Alepisauroidei. Lancetfishes, sabertoothfishes, barracudinas, and relatives.

Family Alepisauridae. Lancetfishes. This group includes *Alepisaurus*, the pelagic lancetfishes, living from near the surface to apparently occasionally down to 1800 m, and *Omosudis*, which occurs between about 100 and 1800 m, but is most frequent near or below 1000 m.

Family Paralepididae. Barracudinas. The barracudinas have 50 to 60 epipelagic to mostly mesopelagic, but some bathypelagic, species. These are distributed in 13 genera: *Anotopterus*, epipelagic to mesopelagic (to 700 m); *Arctozenus*, epipelagic to 1459 m; *Dolichosudis*, taken at 1200 m; *Lestidiops*, between 50 and 2000 m; *Lestidium* (with light organs), between 50 and 2000 m; *Lestrolepis* (with light organs), epipelagic to bathypelagic; *Macroparalepis*, mesopelagic to bathypelagic; *Magnisudis*, mesopelagic to 1214 m; *Notolepis*, mesopelagic to bathypelagic; *Paralepis*, mesopelagic to 1073 m; *Stemonosudis*, epipelagic to mesopelagic; *Sudis*, mesopelagic to bathypelagic; and *Uncisudis*, mesopelagic, 170 to 660 m.

Family Anotopteridae. Daggertooth fish. The one species, *Anotopterus pharao*, occurs from 0 to 1200 m.

Family Evermannellidae. Sabertooth fishes. The three genera, *Coccorella*, *Evermannella*, and *Odontostomops*, with a total of 17 species, are mesopelagic, but some species occur down to 1000 m.

Family Scopelarchidae. Pearleyes. The pearleyes are mesopelagic, most living at 500 to 1000 m, and consist of four genera: *Benthalbella, Rosenblat-*

tichthys, Scopelarchoides, and *Scopelarchus,* with a total of 18 species. Scopelarchids have modified tubular eyes.

i(D). Suborder Giganturoidei. Giganturoids.

Family Bathysauridae. Bathysaurids. The benthic *Bathysaurus* has two species that look something like lizardfishes, but with a flatter head. They occur approximately between 1000 and 4400 m (Sulak *et al.,* 1985). A new genus, *Bathysauroides* (Baldwin and Johnson, 1996), with a single species *Bathysauroides gigas,* occurs from 480 m and probably extends deeper.

Family Giganturidae. Giganturids or telescopefishes. With two species in one genus, *Gigantura* has adults with anterior-projecting tubular eyes, indicating binocular vision. They are mesopelagic to bathypelagic, and have been taken between 500 and 2500 m.

b(2Biia). Subsection Ctenosqumata

i. Order Myctophiformes. Lanternfishes and blackchins (two families). Myctophiforms occur in all major oceans and seas. All except some neoscopelids and one myctophid species have photophores.

Family Neoscopelidae. Blackchin lanternfishes. Blackchin lanternfishes are mesopelagic to benthopelagic. Adults are most frequently taken at approximately 700 to 2000 m deep, and are placed in three genera: *Neoscopelus,* having light organs, and *Scopelengys* and *Solivomer,* without light organs. Their light organs, when present, are different in structure and arrangement from those in the Myctophidae.

Family Myctophidae. Lanternfishes. Myctophids include 32 genera and over 230 species of mesopelagic to sometimes bathypelagic fishes and are among the most common of oceanic fishes. Many undergo diurnal vertical migrations, with some species reaching the surface at night, but most come to within only about 100 m of the surface. During the day most species, so far as known, inhabit depths of about 300 to about 1200 m. There are exceptions; for example, a species of *Lampanyctus* has been taken at 3500 m. The genera of the Myctophidae have been arranged into subfamilies and tribes (Paxton, 1972). These genera are placed in two subfamilies. The first, the Myctophinae, includes 13 genera: *Benthosema, Centrobranchus, Diogenichthys, Electrona, Gonichthys, Hygophum, Krefftichthys, Loweina, Metaelectrona, Myctophum, Protomyctophum, Symbolophorus,* and *Tarletonbeania.* The second subfamily, Lampanyctinae, with 19 genera, consists of *Bolinichthys, Ceratoscopelus, Diaphus, Gymnoscopelus, Hintonia, Idiolychnus, Lampadena, Lampanyctodes, Lampanyctus, Lampichthys, Lepidophanes, Lobianchia, Notolychnus, Notoscopelus, Parvilux, Scopelopsis, Stenobrachus, Taaningichthys,* and *Triphoturus.*

b(2Biia1). Infrasection Acanthomorpha—Acanthomorphs. A monophyletic group not given formal rank in the latest documented fish classification (Nelson, 1994).

2. SYSTEMATICS OF DEEP-SEA FISHES 63

i. Order Lampridiformes. Oarfishes, opahs, crestfishes, tube-eyes, ribbonfishes, and relatives (seven families). Of the seven families of worldwide distribution, two are known to have species living deep enough to be included here.

Family Trachipteridae. Ribbonfishes. Of the three genera, *Desmodema, Trachipterus,* and *Zu,* only one species of *Trachipterus* occurs from near the surface to at least 1000 m deep.

Family Stylephoridae. Tube-eyes or threadtails. The single known species, *Stylephorus chordatus,* is oceanic, primarily mesopelagic, although stated abyssal (Nelson, 1994). It undergoes vertical migrations, reaches depths of somewhat over 800 m, and has an elongate compressed body, anteriorly directed tubular eyes, and a small mouth for feeding on crustaceans.

ii. Order Polymixiiformes. Beardfishes (one family).

iii. Order Percopsiformes. Trout-perches, pirate-perch, and cavefishes (three families).

iv. Order Ophidiiformes. Cusk-eels, pearlfishes, and relatives. This order is currently thought related to gadiforms. The species, like some cod relatives, have long dorsal and anal fin bases and most have reduced slender tails. There are about 350 species distributed among 90 genera and these fishes occur in all oceans from shallow waters to abyssal depths.

Family Carapidae. Pearlfish and other carapids. There are seven genera and about 32 species in this family of mostly shallow to some deep-water fishes that live in the body cavities of invertebrates such as holothurians and bivalves. A few are deep-water species; for example, species of *Snyderidia* occur as deep as 1500 m; *Pyramodon,* to 730 m; and *Echiodon,* to 2000 m.

Family Ophidiidae. Cusk-eels and brotulas. The cusk-eels consist of approximately 165 species distributed in about 50 genera. All are oviparous so far as known. Many species are benthopelagic at depths between 2000 and 6600 m and one, *Abyssobrotula galatheae,* was taken at 8370 m, the greatest depth known for any fish. Space does not permit listing all the genera with their estimated depths, but these data are available (Cohen and Nielsen, 1978).

Family Bythitidae. Ventfishes, viviparous brotulas, and cave brotulas. The viviparous brotulas, with perhaps 86 known species and about 25 genera, are mostly shallow-water fishes, with a few living in freshwater caves. However, some are benthopelagic, to at least 2600 m, examples being *Diplacanthopoma, Cataetyx,* and the ventfish, *Bythites hollisi* (Cohen *et al.,* 1990).

Family Aphyonidae. Aphyonids. The viviparous aphyonids consist of about 20 small species distributed in six genera, *Aphyonus, Barathronus,*

Meteoria, Nybelinella, Parasciadonus, and *Sciadonus.* Most are benthic or benthopelagic to abyssal, living between 1000 and 6000 m (Nielsen, 1984).

Family Parabrotulidae. False brotulas. The relationships of the so-called false brotulas have been controversial, some relating them to zoarciforms, others suggesting they are derived aphyonid ophidiiforms. Further research is needed to settle this issue. These mesopelagic to bathypelagic fishes are known from depths of about 600 to 1500 m, with two species in *Parabrotula* and one in *Leucobrotula* (Miya and Nielsen, 1991).

v. *Order Gadiformes.* Cods, hakes, morid cods, codlets, grenadiers or rattails, pelagic cods, southern hakes, eel cods, and relatives (12 families). Most of the families are distributed in all oceans and major seas and have recently been reviewed (Cohen *et al.,* 1990).

Family Euclichthyidae. Eucla cod. A single benthopelagic species, *Euclichthys polynemus,* is known from 250 to 800 m.

Family Macrouridae. Rattails and grenadiers. Most species of the rattails or grenadiers are primarily benthopelagic, living between 200 and 2000 m, but one is known from 6000 m. There are four subfamilies, 38 genera, and over 300 species. The deep-sea genera are as follows. For the Bathygadinae, species in the genera *Bathygadus* and *Gadomus* are slope dwellers living at depths of about 200 to 2700 m. The Macrouroidinae, with two genera, *Macrouroides* and *Squalogadus,* are benthopelagic to bathypelagic and abyssopelagic to 5300 m. The Trachyrincinae, with two genera, *Idiolophorhynchus* and *Trachyrincus,* have benthopelagic species occurring between 400 and 2500 m. The Macrourinae consist of the bulk of the family, with 32 genera and over 255 species. Most are on continental slopes, benthopelagic, occurring at various depths between 50 and 1000 m, depending on the species; many, including the genera *Coelorinchus, Coryphaenoides, Cynomacrurus, Macrourus, Nezumia,* and *Trachonurus,* also occur below 1000 to 2500 m and beyond, depending on the genus and the species. Some are abyssal. For example, *Coryphaenoides* has been recorded from 6380 to 6450 m. Many species bear a ventral light organ.

Family Merlucciidae. Merlucciid hakes. The Merlucciidae, or hakes, are mostly shallow to deep-water continental shelf and slope fishes, but a few—for example, *Merluccius australis* and *Merluccius productus*—reach depths of 1000 m.

Family Moridae. Morid cods. The morid cods are shallow coastal to pelagic and benthopelagic fishes. Some for have ventral light organs. *Antimora* occurs as deep as 3000 m and species in *Austrophycis, Halargyreus, Laemonema, Lepidion, Mora, Momonatira, Physiculus,* and *Tripterophycis* are known to reach depths of 1000 to 2700 m.

Family Melanonidae. Pelagic cods. These little known mesopelagic to bathypelagic cods extend down to at least 1100 m. Two species occur in a single genus, *Melanonus.*

2. SYSTEMATICS OF DEEP-SEA FISHES

Family Muraenolepididae. Eel cods. The eel cods are cold-water southern hemisphere fishes mostly living near the bottom at moderate continental shelf depths, but *Muraenolepis microps* is known from 10 to 1600 m deep.

Family Phycidae. Phycid hakes and rocklings. There are five genera and 27 species, of which one in *Gaidropsarus* is known to live between 360 and 2000 m. *Phycis chesteri* is found down to at least 1370 m and *Urophycis tenuis* is known from 980 m, although it is usually found around 180 m.

Family Gadidae. Codfishes, haddocks, and cuskfishes. With 15 genera and about 30 species, this family has some species of *Brosme* and *Molva*, subfamily Lotinae, and *Gadiculus*, and *Micromesistius,* subfamily Gadinae, found to 1000 m deep.

vi. Order Batrachoidiformes. Toadfishes and midshipmen (one family).

vii. Order Lophiiformes. Goosefishes, frogfishes, batfishes, anglerfishes, and relatives (16 families). The lophiiforms consist of approximately 300 species distributed among five suborders, 18 families, and 64 genera. They live in a wide variety of habitats, from those near the shoreline to bathypelagic regions. These fishes have a highly derived anterior dorsal fin spine modified as a flexible "fishing rod" (the illicium) with a fleshy lure called the esca.

vii(A). Suborder Lophioidei. Goosefishes or lophioids.

Family Lophiidae. Goosefishes. Goosefishes have the least modified dorsal fin. There are four genera and 45 species; some species of *Lophius* reach a depth of 1000 m or more.

vii(B). Suborder Antennarioidei. Frogfishes (two families). This tropical shore suborder contains no deep-sea fishes.

vii(C). Suborder Ogcocephalioidei. Sea toads, batfishes, and deep-sea ceratioids (13 families).

Superfamily Chaunacioidea—Family Chaunacidae. Sea toads. The benthic 14 species in two genera, *Bathychaunax* (1000 to 2200 m) and *Chaunax* (80 to a little over 1000 m), as a family, live in most oceans, except in polar regions (Caruso, 1989).

Superfamily Ogcocephalioidea—Family Ogcocephalidae. Batfishes. The benthic batfishes have a shortened illicium contained completely within the esca; they live in tropical and subtropical regions between approximately 20 and 1000 m. There are nine genera and about 60 species. The species of *Halieutopsis* have a bathymetric range of 391 to 2487 m, with one species from 3800 to 4000 m. *Malthopsis* and *Dibranchus* have species known to live to about 2300 m.

vii(D). Suborder Ceratioidei. Anglerfishes. There are 11 families, 34 genera, and about 150 species of deep-sea anglerfishes. Some species have the tip of the illicium as a bulbous light organ that employs luminescent bacteria. The species are mesopelagic to bathypelagic, markedly sexually dimorphic, and in some families the relatively tiny males have no illicium and attach themselves to a female and become parasitic, drawing all nourishment from the female, and able to fertilize eggs when needed (Bertelsen, 1951). This is apparently obligatory in the Ceratiidae, Linophrynidae, and Neoceratiidae, but facultative in the Caulophrynidae and one oneirodid genus (Pietsch, 1976). Males are not parasitic in the other families.

Family Caulophrynidae. Fanfin anglers. These fishes are primarily bathypelagic, with specimens collected at depths between 500 and 3000 m. They consist of two genera, *Robia* and *Caulophryne,* with two, perhaps three, species.

Family Neoceratiidae. Needlebeard angler. This family has one species, *Neoceratias spinifer,* with derived elongate moveable jaw teeth. This species has been taken at depths of 1700 to perhaps 2500 m.

Family Melanocetidae. Blackdevils. The species of the single genus, *Melanocetus,* are taken from depths of 600 to 4790 m.

Family Himantolophidae. Footballfishes. At least some of the 18 species of *Himantolophus* are known to range from 250 to possibly 4000 m.

Family Diceratiidae. Double anglers. *Diceratias* and *Phrynichthys* are the only ceratioids to have a small second modified dorsal fin ray in juveniles. They occur at depths of 300 to 2000 m.

Family Oneirodidae. Dreamers. The dreamers, with about 60 species distributed in 16 genera, have a little over half of the species in *Oneirodes.* As a family they are distributed in all oceans and have been taken between 300 and 3000 m, but are most common between 800 and 1500 m.

Family Thaumatichthyidae. Wolftrap angler. The wolftrap anglers, with six species in one genus, *Thaumatichthys,* and three in another, *Lasiognathus,* are known at least from 780 to 3680 m.

Family Centrophrynidae. Halloween angler. The Halloween angler (only one species) occurs from near the surface (larvae) to depths of 2500 m (adults).

Family Ceratiidae. Seadevils. Three species are known in *Ceratias* and one in *Cryptopsaras.* Although they occur at depths between 150 and 3400 m, they may occur at depths from near the surface to possibly as deep as 4400 m.

Family Gigantactinidae. Whipnose anglers. Whipnose anglers are most commonly collected between about 1000 and 2500 m. There are two genera, *Gigantactis,* with 17 species, and *Rhynchactis,* with a single species.

Family Linophrynidae. Netdevils. Netdevils, like many other ceratioids, have mesopelagic larvae and adults that are most often collected at depths between 1000 and 4000 m. There are four monotypic genera, *Acentrophryne, Borophryne, Haplophryne,* and *Photocorynus.* A fifth genus, *Lynophryne,* has 21 species.

2. SYSTEMATICS OF DEEP-SEA FISHES 67

viii. Order Mugiliformes. Mullets (one family).

ix. Order Atheriniformes. Rainbow fishes, blue eyes, silversides, sailfin silversides, topsmelts, grunions, phallostethids, and relatives (eight families).

x. Order Beloniformes. Ricefishes, flyingfishes, needlefishes, halfbeaks, and sauries (five families).

xi. Order Cyprinodontiformes. Rivulines, poeciliids, goodeids, pupfishes, killifishes, four-eyed fishes, and relatives (eight families).

xii. Order Beryciformes. Beardfishes, squirrelfishes, flashlight fishes, pineapple or pinecone fishes, alfonsinos, and relatives (seven families). This order contains several shallow-water families and a few that occur in deep waters.

Family Berycidae. Alfonsinos. There are nine species distributed in two genera, *Beryx* and *Centroberyx.* The former has some species found to at least 1000 m deep, but most of the species occur between 200 and 600 m.

Family Anoplogasteridae. Fangtooths. Two deep-sea species in *Anoplogaster* live from near the surface as juveniles to about 2000 to 3000 m as adults.

Family Diretmidae. Spinyfins. There are three genera, *Diretmichthys, Diretmoides,* and *Diretmus,* with a total of four species occurring down to over 2000 m.

Family Trachichthyidae. Slimeheads or roughies. There are 35 species in seven genera. *Hoplostethus* includes about half of the species that are benthopelagic, with some species known to occur down to 1500 m. The species of *Soroichthys* and *Aulotrachichthys* have light organs.

xiii. Order Stephanoberycoidei. Bigscale fishes, whalefishes, and relatives (nine families).

xiii(A). Suborder Stephanoberycoidei. Bigscale fishes, gibberfishes, and pricklefishes (three families).

Family Melamphaeidae. Bigscale fishes. This family has five genera: *Melamphaes,* with 19 species, *Poromitra* and *Scopeloberyx,* with about five; *Scopelogadus,* with three; and *Sio,* with about 33. The species are mesopelagic, bathypelagic to sometimes abyssopelagic. The adults are known from 800 m to as deep as 5000 m in some species.

Family Gibberichthyidae. Gibberfishes. These midwater fishes with two species in *Gibberichthys* live between 50 m (juveniles) and 400–1000 m (adults).

Family Stephanoberycidae. Pricklefishes. The pricklefishes are bathypelagic, but live near the bottom at 1000 to 2700 m and below. There are three monotypic genera, *Melacosarcus, Stephanoberyx,* and *Acanthochaenus.* The last occurs at depths between 2176 and 5308 m.

Family Hispidoberycidae. Bristlyskin. This family has one species, *Hispidoberyx ambagiosus,* known from five specimens caught between 580 and 1020 m deep.

xiii(B). Suborder Cetomimoidei. Whalefishes (five families).

Family Rondeletiidae. Orangemouth whalefishes. This family consists of two species in *Rondeletia* taken from depths of 500 to 1500 m.

Family Barbourisiidae. Redvelvet whalefishes. Known from one species, *Barbourisia rufa,* that has been collected between 550 and 1500 m.

Family Cetomimidae. Flabby whalefishes. The deep-sea pelagic flabby whalefishes have about 35 species distributed among nine genera, *Cetichthys, Cetomimus, Cetostomus, Dannacetichthys, Ditropichthys, Gyrinomimus, Notocetichthys, Procetichthys,* and *Rhamphocichthys;* nearly all were captured below 1000 m (between 3200 and 4000 m) (Paxton, 1989).

Family Mirapinnidae. Hairyfish and tapetails. The hairyfish, *Mirapinna esau,* the single known species of the Mirapinninae, occurs in the Atlantic from surface to 200 m (small specimens) and between 700 and 1400 m (large specimens). The Eutaeniophorinae, or tapetails, have three species in two genera, *Eutaeniophorus,* from surface to 200 m, and *Parataeniophorus,* from 700 to 1400 m. It is assumed that the unknown mature specimens of this subfamily live in deep seas (Bertelsen, 1986).

Family Megalomycteridae. Mosaicscale fishes. The mosaicscale or big-nosefishes are in four genera; *Cetomimoides, Megalomycter,* and *Vitiaziella* are monotypic, and *Ataxolepis* has two species. Only males are known, and occur at depths at least to 1829 m.

xiv. Order Zeiformes. Dories, oreos, boarfishes, and relatives (six families). These fishes occur in all major oceans and seas. Of the six families, only the Oreosomatidae contains deep-water fishes.

Family Macrurocyttidae. Macrurocyttids. *Macrurocyttus acanthopodus* is known from about 1000 m deep.

Family Oreosomatidae. Oreos. Nine species are distributed among four genera: *Allocyttus,* occurring between 360 and 1900 m; *Neocyttus,* to about 1000 m; *Pseudocyttus,* between about 460 and 1160 m; and *Oreosoma,* with some species extending below 1000 m and a few to 1800 m.

xv. Order Gasterosteiformes. Sand eels, tubesnouts, sticklebacks, ghost pipefishes, pipefishes, seahorses, seamoths, trumpetfishes, cornetfishes, snipefishes, shrimpfishes, and relatives (11 families).

Family Macroramphosidae. Snipefishes. The snipefishes are the only family in this order to have a species, *Centriscops humerosus,* known to occur as deep as 1000 m.

xvi. Order Synbranchiformes. Swamp eels, spiny eels, and relatives (three families).

2. SYSTEMATICS OF DEEP-SEA FISHES 69

xvii. Order Perciformes. Perciforms. This largest order of vertebrates, with over 150 families, is not the only order of fishes with spiny fin rays. Although the synapomorphies characteristic of this order still remain to be explored fully, the significance of their modified dorsal fin spines and associated myology has confirmed that this order is likely a monophyletic group (Mooi and Gill, 1995).

xvii(A). Suborder Scorpaenoidei. Flying gurnards, rockfishes, scorpionfishes, stonefishes, velvetfishes, prowfishes, searobins, flatheads, greenlings, sculpins, poachers, lumpfishes, snailfishes, and relatives (25 families). This large group is equivalent to the order Scorpaeniformes (Nelson, 1994). Until recently the Scorpaenoidei were usually considered derived independently of the Perciformes. But the suborder has recently been shown to have a complex synapomorphy that unites the Perciformes (Mooi and Gill, 1995).

Superfamily Scorpaenoidea. Scorpaenoids. This superfamily is equivalent to the suborder Scorpaenoidei of (Nelson, 1994) and consists of seven families, of which only one has species known to occur below 1000 m.

Family Scorpaenidae. Scorpionfishes, rockfishes, stonefishes, and relatives. With 11 scorpaenid subfamlies, the benthic *Sebastolobus*, Sebastolobinae, is one of the few genera known to contain species reaching depths of 1000 m and more, possibly to 2200 m. One species in each of *Ectreposebastes*, *Pontinus*, and *Trachyscorpia* of the Scorpaeninae is known from depths of about 1000 m.

Superfamily Platycephaloidea. Flatheads or platycephaloids (three families).

Family Bembridae. Deep-water flatheads. Of the four genera, the depths recorded for species of *Bembradium* are 150 to 950 m, and for *Parabembras*, 80 to 600 m.

Family Holichthyidae. Ghost flatheads. This group has about 10 species in *Hoplichthys*. All are benthic, living from about 10 to 1500 m.

Superfamily Anoplopomatoidea—Family Anoplomatidae. Sablefishes. The sablefish, *Anoplopoma fimbria*, and the skilfish, *Erilepis zonifer*, from the northern parts of the Pacific Ocean, are the only members of this family. The sablefish is most abundant between 300 and 900 m, but is known from a depth of 1830 m. The skilfish is not known to reach such depths.

Superfamily Cottoidea. Sculpins and relatives. This group has about 137 genera and perhaps about 630 species. Only a few reach very deep waters.

Family Cottidae. Sculpins. This is a large family of about 70 genera and 300 species of mostly marine coastal shallow to deep-water fishes; almost all are northern hemisphere in distribution. Many species are tidepool fishes but others, for example, in *Icelus* and one species in *Artediellus*, commonly occur at depths of 300 to 500 m or more. A species of *Zesticelus* may reach depths of 2000 m. *Antipodocottus* from the southern hemisphere has at least one species that may reach 765 m or deeper. Species of *Cottus* are common in fresh water. The freshwater pelagic plankton feeder, *Cottocomephorus comephoroides*, sometimes placed in a separate family, the Cottocomephoridae, occurs from shoreline to depths of 1000 m in Lake Baikal, Russia.

Family Comephoridae. Lake Baikal oilfishes. This family has two viviparous species. *Comephorus dybowskii* reaches a depth of over 1000 m.

Family Abyssocottidae. Lake Baikal sculpins. This family has six genera, *Abyssocottus, Asprocottus, Cottinella, Limnocottus, Neocottus,* and *Procottus,* with a total of 31 species. Most live below 170 m. Some species of *Abyssocottus* reach a depth of over 1000 m or more, and the benthic *Cottinella boulengeri* occurs from 700 to over 1000 m.

Family Agonidae. Poachers. The poachers, with perhaps 50 benthic species distributed among 20 genera, occur from inshore habitats down to at least 1280 m, mostly in the North Pacific Ocean, but a few reach the North Atlantic and even fewer reach southern South America. Most species appear to be adapted to varying depths between 10 and 300 m, but some live considerably deeper, for example, *Bathyagonus nigripinnis* is recorded from 1250 m deep.

Family Psychrolutidae. Fatheads and blobfishes. There are seven genera and about 30 species. The benthic blobfishes in *Psychrolutes* occur at depths of 100 to 1600 m, but *Psychrolutes phrictus* is recorded at 2800 m. *Cottunculus thompsoni* of the North American Atlantic coast reaches depths of nearly 1500 m, and an African Atlantic species, *Cottunculus spinosus,* is recorded at 2180 m. *Malacocottus* is known to occur from 100 to 1980 m.

Family Bathylutichthyidae. Bathylutichyids. One species, *Bathylutichthys taranetzi,* was taken from the South Atlantic Ocean, South Georgia Island, reportedly at a depth of 1650 m.

Superfamily Cyclopteroidea. Lumpfishes and snailfishes, marine, worldwide in cold waters.

Family Cyclopteridae. Lumpfishes. With seven genera and 28 species, some, for example, *Cyclopterus,* may be caught down to about 1000 m, but usually occur above 200 m.

Family Liparidae. Snailfishes. The Liparidae, or snailfishes, occur in shallow shoreline to mesopelagic and bathypelagic zones. Most are benthic. There are about 20 genera and nearly 200 species so far described. Some species of *Careproctus* occur at depths down to 3600 m, *Notoliparis* at least to 5474 m, *Rhodichthys* to 2415 m, and *Paraliparis* species are taken at depths from near the surface and others to 7500 m. However, most species in these genera are captured at depths of about 1000 m to 2000 m.

2. SYSTEMATICS OF DEEP-SEA FISHES

xvii(B). Suborder Percoidei. Perches, seabasses, sunfishes, snappers, butterflyfishes, croakers, goatfishes, tilefishes, angelfishes, and many other perchlike fishes (approximately 70 families, of which only a few live in the deep-sea environment).

Superfamily Percoidea. Percoids.

Family Acropomatidae. Temperate ocean basses. This provisional group of perhaps 40 species and 11 genera of ocean basses, along with another provisional family, the Percichthyidae or south temperate basses, are in need of phylogenetic study. The placement of *Brephostoma* in this family is questionable, but species of this genus occur down to about 2800 m. The genus *Howella,* perhaps better placed in its own family Howellidae, has perhaps 10 species, with *Howella brodiei* extending down to 2000 m.

Family Epigonidae. Deep-water cardinalfishes. The approximately 15 species in five genera of deep-water cardinalfishes are usually found between depths of about 130 to 425 m, but some in *Epigonus* may have been captured at 2000 to 3000 m. *Rosenblattia robusta* is recorded from 700 to 2000 m.

Family Bramidae. Pomfret and fanfishes. Epipelagic to mesopelagic, one species of *Brama* has been stated to reach as deep as 1000 m, but depths lower than about 400 m seem unlikely for any member of this genus.

Family Malacanthidae. Tilefishes. This family of five genera and about 40 species from all oceans has one subfamily, Latilinae, with a species in *Branchiostegus* extending down to a little over 600 m in depth.

Family Caristiidae. Manefishes. This oceanic family of one genus, *Caristius,* and four species has been taken at depths of 500 m, possibly to 1100 m.

Family Emmelichthyidae. Rovers. The tropical oceanic 24 species of *Plagiogeneion, Erythrocles,* and *Emmelichthys* have been recorded as deep as 500 m.

Family Bathyclupeidae. Bathyclupeids. The bathyclupeids in one genus, *Bathyclupea,* with perhaps four species, are recorded from depths of 400 to 3000 m.

xvii(C). Suborder Elassomatoidei. Pygmy sunfishes (one family).

xvii(D). Suborder Labroidei. Cichlids, surfperches, damselfishes, wrasses, and parrotfishes, (six families).

xvii(E). Suborder Zoarcoidei. Ronquils, eelpouts, gunnels, wolffishes, and relatives (nine families).

Family Zoarcidae. Eelpouts. Mostly benthic eelpouts consist of about 46 genera and 220 species. Most occur in the North Pacific and North Atlantic oceans. They live from near the shoreline to abyssal depths; for example, *Dieidolycus* occurs from 2273 to 3040 m, and *Lycenchelys* has some species at 300 to 700 m, but others from 2000 to 5300 m. Other examples include the pelagic *Lycodapus,* which is known from 323 to

1200 m; *Melanostigma,* which is also pelagic, to possibly 2561 m; *Oidiphorus,* between 1300 and 3000 m; *Lycogramma,* from 322 to 1952 m; *Lycodapus,* possibly pelagic to 1150 m; *Pachycara,* between 200 and 1800 m; and *Taranetzella,* to 3000 m in depth.

Family Stichaeidae. Pricklebacks. With about 40 genera and over 65 species, some, such as *Lumpanella,* may occur as deep as about 700 m.

Family Anarhichadidae. Wolffish. There are two genera, *Anarrhichthys* and *Anarhichas,* with four species, but only *Anarhichas* occurs at depths of 40 to 1500 m.

xvii(F). Suborder Notothenioidei. Cod icefishes, spiny plunderfishes, barbeled plunderfish, and antarctic dragonfishes (six families).

Family Nototheniidae. Cod icefishes. This group of 17 genera and about 50 species comprises mostly antarctic benthic shelf fishes from near shore to 700 to 800 m, but some pelagic and benthic species occur at 1600 m (for example, species of *Dissostichus*).

Family Artedidraconidae. Barbeled plunderfishes. Of four genera and 24 species, *Dolloidraco longedorsalis* is caught to 1145 m deep, and a few species of *Pogonophryne* occur below 1000 m, including one to 2542 m.

Family Bathydraconidae. Antarctic dragonfishes. This family has about 15 species distributed in 10 genera. Some species of *Bathydraco* range from 340 to 2950 m in depth, but most in that genus and in the other genera live in shallower waters.

Family Channichthyidae. Crocodile icefishes. There are 11 genera and about 17 species, but only *Chionobathyscus dewitti* is truly deep living, between 500 and 2000 m. The members of this family, so far as known, lack red blood cells as well as myoglobin in their muscle tissue.

xvii(G). Suborder Trachinoidei. Chiasmodons, convict blennies, sandfishes, sand perches, duckbills, and lances, weaverfishes, stargazers, and relatives (13 families).

Family Chasmodontidae. Swallowers. These are worldwide bathypelagic fishes with large extendable mouths capable of eating prey as large as or larger than themselves. There are about 15 species in four genera, *Chiasmodon* is known from 550 to 2745 m, *Dysalotus* down to over 3000 m, and *Kali* between 500 and 2500 m. The other genera have similar ranges. *Pseudoscopelus,* with similar depth ranges, possesses photophores.

xvii(H). Suborder Blennioidei. Triplefin blennies, labrisomids, sand stargazers, clinids, combtooth blennies, and relatives (six families).

xvii(I). Suborder Icosteoidei. Ragfishes (one family).

xvii(J). Suborder Gobiesocoidei. Clingfishes (one family).

xvii(K). Suborder Callionymoidei. Dragonets (one family).

xvii(L). Suborder Gobioidei. Loach gobies, sleepers, gobies, and relatives (eight families).

2. SYSTEMATICS OF DEEP-SEA FISHES

xvii(M). Suborder Kurtoidei. Nurseryfishes (one family).

xvii(N). Suborder Acanthuroidei. Spadefishes, scats, rabbitfishes, louvar, Moorish idol, surgeonfishes, and relatives (six families).

xvii(O). Suborder Scombrolabracoidei—Family Scombrolabracidae. Deep-water mackerel. This single-family suborder contains the worldwide deep-water mackerel, *Scombrolabrax heterolepis,* which is known to live approximately between 150 and 900 m.

xvii(P). Suborder Scombroidei. Barracudas, snake mackerels, cutlassfishes, mackerels, tunas, swordfishes, sailfishes, marlins, and relatives (five families).

Family Gempylidae. Snake mackerels. Some of the 23 species of this worldwide family with 16 genera are known to reach depths exceeding 1000 m. The genera *Diplospinus* and *Nesiarchus* have been caught at the surface and down to a depth of about 1200 m; the larvae of some other genera, for example, *Paradiplospinnus,* have been taken as deep as 1000 to 2800 m. Species of several other genera extend to depths of over 500 m.

Family Xiphidae. Billfishes. With a total of four genera and 12 species, the swordfish, *Xiphias gladius,* is known to dive to 550 m and perhaps deeper, to 650 m.

Family Trichiuridae. Cutlassfishes. This oceanic family with 16 genera and about 34 mostly little-known species are from deep water. Some species of *Aphanopus* extend down to 2000 m; some of *Benthodesmus* occur down to about 1000 m.

xvii(Q). Suborder Stromateoidei. Medusafishes, driftfishes, squaretails, butterfishes, and relatives (six oceanic families of worldwide distribution).

Family Centrolophidae. Medusafishes. There are seven genera and about 27 species. Of these, species of *Schedophilus* are known to inhabit waters as deep as 1000 m and the juveniles of *Icichthys australis* have been caught between the surface and 2000 m.

Family Nomeidae. Driftfishes. This family includes about 15 species in three genera, *Cubiceps, Nomeus,* and *Psenes.* Of these, species of *Psenes* are known from 200 to 1000 m.

xvii(R). Suborder Anabantoidei. Gouramies, fightingfishes, pikeheads, and paradisefishes (five families)

xvii(S). Suborder Channoidei. Snakeheads (one family).

xviii. Order Pleuronectiformes. Flatfishes. This large worldwide undoubtedly monophyletic order has somewhat more than 570 benthic species in about 123 genera and 11 families. Most species are marine, but several also enter fresh water and a few apparently are found only in fresh water.

xviii(A). Suborder Psettodoidei. Psettodids (one family).

xviii(B). Suborder Pleuronectoidei. Citharids, lefteye flounders, southern flounders, righteye flounders, soles, tonguefishes, and relatives (10 families, benthic, worldwide).

Family Pleuronectidae. Righteye flounders. The righteye flounders, probably not a monophyletic group (Nelson, 1994), consist of about 40 genera and over 90 species of benthic shallow to moderately deep-water fishes. However, a few genera have deep-sea species; for example, *Hippoglossus* and *Reinhardtius* both occur to about 2000 m.

Family Bothidae. Lefteye flounders. One species of *Chascanopsetta* is known from 120 to 977 m.

Family Scophthalmidae. Scophthalmids. One species of *Mancopsetta* is known from 190 to 840 m.

Family Soleidae. Soles. With 20 genera and about 90 species, this family has a few species living at considerable depths; for example, *Bathysolea profundicola* is found from about 200 to 1300 m.

Family Cynoglossidae. Tongue soles. Among the 60 species of tongue soles in *Symphurus,* some are known from 300 to 1900 m; in *Cynoglossus,* with about 50 species, some extend down to 1000 m.

xix. Order Tetraodontiformes. Tetraodontiforms.

xix(A). Suborder Triacanthoidei. Spikefishes (two families).

Family Triacanthodidae. Spikefishes. With 11 genera and about 20 species in the family, only the following species are known to occur below 500 m: *Atrophacanthus japonicus,* to about 1500 m; *Hollardia hollardi* and *Macrohamphsoides platycheilus,* to 740 m; *Johnsonina eriomma,* to about 730 m; and *Parahollardia schmidti,* to about 550 m.

xix(B). Suborder Tetraodontoidei. Triggerfishes, filefishes, boxfishes, puffers, porcupinefishes, molas, and relatives (seven families).

ACKNOWLEDGMENTS

I thank Carole C. Baldwin, John R. Burns, Bruce B. Collette, Daniel M. Cohen, Antony S. Harold, G. David Johnson, Leslie Knapp, Thomas A. Munroe, Joseph S. Nelson, Jørgen Nielsen, Randall K. Packer, Lisa F. Palmer, Lynne Parenti, David G. Smith, Kenneth A. Tighe, Victor G. Springer, and Marilyn J. Weitzman for providing helpful comments on all or parts of the manuscript. Karsten Hartel and provided certain unpublished depth records.

REFERENCES

Baldwin, C., and Johnson, D. (1996). Aulopiform interrelationships. *In* "Interrelationships of Fishes" (M. L. J. Stiassny, L. R. Parenti, and G. D. Johnson, eds.), pp. 355–404. Academic Press, San Diego.

Berg, L. S. (1955). Classification of fishes, both recent and fossil. *Trudy. Zool. Inst. Akad. Nauk SSSR,* **20,** 286 (2nd Ed., in Russian).

Bertelsen, E. (1951). The ceratioid fishes. *DANA Rep.* **39,** 1–276.
Bertelsen, E. (1986). Family No. 125: Mirapinnidae. *In* "Smith's Sea Fishes" Macmillan South Africa, (M. M. Smith and P. C. Heemstra, eds.), pp. 406–407.
Caruso, J. H. (1989). Systematics and distribution of the Atlantic chaunacid anglerfishes (Pisces: Lophiiformes). *Copeia* **1,** 153–165.
Clark, E., and Kristof, E. (1990). Deep-sea elasmobranchs observed from submersibles off Bermuda, Grand Cayman, and Freeport, Bahamas. *In* Elasmobranchs as living resources: Advances in the biology, ecology, systematics and the status of the Fisheries. *NOAA Tech. Rep.* 90.
Cohen, D. M. (1964). Family Opisthoproctidae. *In* "Fishes of the Western North Atlantic, Part 4. Soft-rayed Bony Fishes, Order Isospondyli (Part)" (H. B. Bigelow, ed.), Sears Foundation for Marine Research, Memoir 1, pp. 48–68. Yale University, New Haven, Connecticut.
Cohen, D. M., ed. (1989). Papers on the systematics of gadiform fishes. *Nat. Hist. Mus. Los Angeles Co. Sci. Ser.* **32,** ix and 1–262.
Cohen, D. M., and Nielsen, J. G. (1978). Guide to the identification of genera of the fish order Ophidiiformes with a tentative classification of the order. *NOAA Tech. Rep. NMFS Circ.* **417,** 1–77.
Cohen, D. M., Inada, T., Iwamoto, T., and Scialabba, N. (1990). FAO species catalog, volume 10, gadiform fishes of the world. *FAO Fisheries Synopsis No. 125* **10,** i–x and 1–442.
Compagno, L. J. V. (1984). FAO species catalogue, volume 4. Sharks. An annotated and illustrated catalogue of shark species known to date. Part 1. Hexanchiformes to Lamniformes. *FAO Fish. Synop. No. 125* **4**(1), 1–249.
Compagno, L. J. V. (1988). "Sharks of the order Carcharhiniformes." pp. 1–486. Princeton Univ. Press, Princeton, New Jersey.
Crabtree, R. E., Sulak, K. J., and Musick, J. A. (1985). Biology and distribution of species of *Polyacanyhonotus* (Pisces: Notacanthiformes) in the western North Atlantic. *Bull. Mar. Sci.* **36**(2), 235–248.
Fink, W. L. (1985). Phylogenetic interrelationships of the stomiid fishes (Teleostei: Stomiiformes). *Misc. Publ. Mus. Zool. Univ. Mich.* **171,** 1–127.
Gon, O., and Heemstra, P. C. (1990). "Fishes of the Southern Ocean," pp. i–xviii and 1–462. J. L. B. Inst. of Ichthyol. Grahamstown, South Africa.
Grey, M. (1956). The distribution of fishes found below a depth of 2,000 meters. *Fieldiana: Zool.* **36**(2), 73–337.
Greenwood, P. H. (1977). Notes on the anatomy and classification of elopomorph fishes. *Bull. Br. Mus. Nat. Hist. (Zool.)* **32,** 65–102.
Greenwood, P. H., Rosen, D. E., Weitzman, S. H., and Myers, G. S. (1966). Phyletic studies of teleostean fishes, with a provisional classification of living forms. *Bull. Am. Mus. Nat. Hist.* **131,** 399–456.
Greenwood, P. H., Miles, R. S., and Patterson, C. (1973). Interrelationships of fishes. *Zool. J. Linnean Soc.* **53**(Suppl 1), i–xvi and 1–536.
Hall, B. (1982). Bone in the cartilaginous fishes. *Nature (London)* **298,** 324.
Harold, A. S. (1994). A taxonomic revision of the sternoptychid genus *Polyipnus* (Teleostei: Stomiiformes) with an analysis of phylogenetic relationships. *Bull. Mar. Sci.* **54**(2), 428–534.
Harold, A. S. (1997). Phylogenetic relationships of the Gonostomatidae (Teleostei: Stomiiformes. *Bull. Mar. Sci.*
Harold, A. S., and Weitzman, S. H. (1996). Interrelationships of stomiiform fishes. *In* "Interrelationships of Fishes" (M. L. J. Stiassny, L. R. Parenti, and G. D. Johnson, eds.), pp. 333–353. Academic Press, San Diego.

Jamieson, B. G. M. (1991). "Fish Evolution and Systematics: Evidence from Spermatozoa." Cambridge Univ. Press, Cambridge.

Johnson, G. D., and Anderson, W. D., eds. (1993). Proceedings of the symposium on phylogeny of Percomorpha, June 15–17, 1990, held in Charleston, South Carolina at the 70th annual meetings of the American Society of Ichthyologists and Herpetologists. *Bull. Mar. Sci.* **52,** 1–626.

Johnson, G. D., and Patterson, C. (1996). Relationships of lower euteleostean fishes. *In* "Interrelationships of Fishes" (M. L. J. Stiassny, L. R. Parenti, and G. D. Johnson, eds.). Academic Press, San Diego.

Kashkin, N. I. (1995). Vertical distribution of *Cyclothone* (Gonostomatidae) in the Pacific Ocean (brief review). *J. Ichthyol.* **35**(8), 53–60 [*Vopr. Ikhtiol.* **35**(4), 440–444].

Last, P. R., and Stevens, J. D. (1994). "Sharks and Rays of Australia." CSIRO, Melbourne, Australia.

Lauder, G. V., and Liem, K. F. (1983). The evolution and interrelationships of the actinopterygian fishes. *Bull. Mus. Comp. Zool.* **150,** 95–197.

Maisey, J. G. (1986). Heads and tails: A chordate phylogeny. *Cladistics* **2,** 201–256.

Marshall, N. B. (1954). "Aspects of Deep Sea Biology," pp. 1–380. Hutchinson's, London.

Marshall, N. B. (1979). "Deep-sea Biology, Developments and Perspectives," pp. i–x and 1–566. Blandford, London.

Matsui, T., and Rosenblatt, R. (1987). Review of the deep-sea fish family Platytroctidae (Pisces: Salmoniformes). *Bull. Scripps Inst. Oceanogr., Univ. Calif. San Diego* **19,** 159.

Merrett, N. R., and Nielsen, J. G. (1987). A new genus and species of the family Ipnopidae (Pisces, Teleostei) from the eastern North Atlantic, with notes on its ecology. *J. Fish. Biol.* **31,** 451–464.

Miya, M., and Nielsen, J. (1991). A new species of the deep-sea fish genus *Parabrotula* (Parabrotulidae) from Sagami Bay, with notes on its ecology. *Jpn. J. Ichthyol.* **38**(1), 1–5.

Mooi, R. D., and Gill, A. C. (1995). Association of epaxial musculature with dorsal-fin pterygiophores in acanthomorph fishes, and its phylogenetic significance. *Bull. Nat. Hist. Mus. London (Zool.)* **61,** 121–137.

Moser, H. G., Richards, W. J., Cohen, D. M., Fahay, M. P., Kendall, A. W., and Richardson, S. L., eds. (1984). Ontogeny and systematics of fishes. *Am. Soc. Ichthyol. Herpetol. Spec. Publ.* **1,** 1–760.

Naylor, G. P. (1992). The phylogenetic relationships among requiem and hammerhead sharks: Inferring phylogeny when thousands of equally most parsimonious trees result. *Cladistics* **8,** 295–318.

Nelson, J. S. (1994). "Fishes of the World," 3rd Ed., pp. i–xvii and 1–600. Wiley, New York.

Nielsen, J. (1972). Ergebnisse der Forschungsreisen FFS "Walther Herwig" nach Südamerika XX, Additional notes on Atlantic Bathylaconidae (Pisces, Isospondyli) with a new genus. *Arch. Fisch. Wiss.* **23**(1), 29–36.

Nielsen, J. (1984). Two new abyssal *Barathronus* spp. from the North Atlantic (Pisces: Aphyonidae). *Copeia* **3,** 579–584.

Patterson, C., and Johnson, G. D. (1995). The intermuscular bones and ligaments of teleostean fishes. *Smithson. Contrib. Zool.* **559,** 1–85.

Paulin, C. D., and Moreland, J. M. (1979). Halosauridae of the south-west Pacific (Pisces: Teleostei: Notacanthiformes). *N. Z. J. Zool.* **6,** 267–271.

Paxton, J. R. (1972). Osteology and relationships of the Lanternfishes (Family Myctophidae). *Bull. Nat. Hist. Mus. Los Angeles* **13,** i–v and 1–81.

Paxton, J. R. (1989). Synopsis of the whalefishes (family Cetomimidae) with descriptions of four new genera. *Rec. Aust. Mus.* **41,** 135–206.

Paxton, J. R., and Eschmeyer, W. N., ed. (1995). "Encyclopedia of Fishes," pp. 1–240. Academic Press, San Diego.
Pietsch, T. W. (1976). Dimorphism, parasitism and sex: Reproductive strategies among deep-sea ceratioid anglerfishes. *Copeia* **4,** 781–793.
Regan, C. T. (1929). Fishes. *Encyclopaedia Britannica* **9,** (14th ed.) 305–329.
Robins, C. R. (Chairman), Bailey, R. M., Bond, C. E., Brooker, J. R., Lachner, E. A., Lea, R. N., and Scott, W. B. (1991a). Common and scientific names of fishes from the United States and Canada. *Am. Fish. Soc. Spec. Publ.* **20**(5th ed.).
Robins, C. R. (Chairman), Bailey, R. M., Bond, C. E., Brooker, J. R., Lachner, E. A., Lea, R. N., and Scott, W. B. (1991b). World Fishes important to North Americans exclusive species from the continental waters of the United States and Canada. *Am. Fish. Soc. Spec. Publ.* **21**(1st Ed.).
Robins, C. H., and Robins, C. R. (1989). Family Synaphobranchidae. *In* "Fishes of the western North Atlantic, Part 9. 1. Orders Anguilliformes and Saccopharyngiformes" (E. B. Böhlke, ed.), pp. 207–253. Sears Foundation for Marine Research. Memoir, Yale University, New Haven, Connecticut.
Shirai, S. (1992a). Identity of extra branchial arches of Hexanchiformes (Pisces, Elasmobranchii). *Bull. Fac. Fish. Hokkaido Univ.* **43,** 24–32.
Shirai, S. (1992b). Phylogenetic relationships of the angel sharks, with comments on elasmobranch phylogeny (Chondrichtyes, Squatinidae). *Copeia,* 505–518.
Smith, M. M. (1986). Family Number 124: Ateleopodidae. *In* "Smith's Sea Fishes" (M. M. Smith and P. C. Heemstra, eds.), pp. 404–406. Macmillan South Africa, Johannesburg.
Smith, D. G., Böhlke, J. E., and Castle, P. H. (1981). A revision of the nettastomatid genera *Nettastoma* and *Nettenchelys* (Pisces: Anguilliformes), with descriptions of six new species. *Proc. Bio. Soc. Washington* **94,** 533–560.
Springer, S. (1979). A revision of the catsharks, family Scyliorhinidae. *NOAA Tech. Rep. NMFS Circular* **422,** i–v and 1–152.
Springer, V. G., and Gold, J. P. (1989). "Sharks in Question," pp. 1–187. Smithsonian, Washington, D.C.
Springer, V. G., and Gold, J. P. (1992). "Sharks in Question," p. 1–276. (in Japanese). Heibonsha, Tokyo.
Stiassny, M. L. J., Parenti, L. R., and Johnson, G. D., eds. (1996). "The Interrelationships of Fishes Revisited." Academic Press, San Diego.
Sulak, K. J., Wenner, C. A., Sedberry, G. R., and Guelpen, L. V. (1985). The life history and systematics of deep-sea lizard fishes, genus *Bathysaurus* (Synodontidae). *Can. J. Zool.* **63,** 623–642.
Tortonese, E., and Hureau, J. C. (1979). Check-list of the fishes of the north-eastern Atlantic and Mediterranean, supplement to Clofnam. *Cybium 3rd Ser.* **5,** 333–394.

3

DISTRIBUTION AND POPULATION ECOLOGY

RICHARD L. HAEDRICH

I. How Many Deep-Sea Species Are There?
II. Pelagic Habitats
III. Demersal Fauna: Shelf, Slope, and Rise
IV. Distribution Patterns
   A. Trends in Diversity
   B. Broad-Scale Horizontal Patterns
   C. Vertical Zonation
V. Feeding Relationships
VI. Age Determination
VII. Reproductive Strategies
   A. Reproduction and Development
   References

I. HOW MANY DEEP-SEA SPECIES ARE THERE?

The global fish fauna comprises something over 25,000 species. Of these, perhaps 10–15% are found in the deep sea (Fig. 1). Such limited diversity is perhaps surprising considering that the watery living space available in the deep sea is more than 100 times greater than the collective volume of the rest of the world's waters. Consequently, deep-sea species are likely to be very widespread, and their populations are very large. Consideration of their distribution patterns, of the relationship of these patterns to the physical environment, and of the ecology of deep-sea fish communities is the subject of this chapter.

There are two main deep-sea habitats, the pelagic and benthic realms, and the deep-sea fish faunas that live in these two habitats are quite different. Fishes that live in the water column of open waters are termed *pelagic;* those that live on the seafloor are termed *demersal* or *benthic,* whereas those that live just above the seafloor are termed *benthopelagic.* The differences

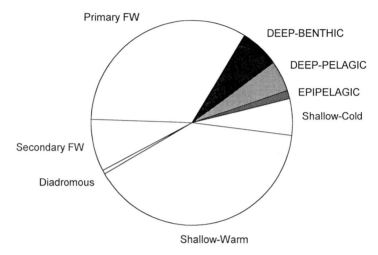

Fig. 1. The relative numbers of fish species in the world, according to their habitat. FW, Freshwater species. Based on data from Cohen (1970).

between pelagic and benthic fishes are strongly represented in the taxonomy of the various species and species groups, for the regions are faunally distinct from each other at the family and even higher levels.

The differences between the pelagic and benthic habitats are reflected in the overall ecology (Campbell, 1983), as well as in the morphological adaptations and taxonomic affinities of the fishes that live there. N. B. Marshall, a foremost student of general deep-sea biology and fishes, spent a lifetime examining the diversity and adaptations of both pelagic and demersal species. Marshall (1971) described the wide variety of systematic groups of fishes represented in the deep sea, ranging from chimaeras, sharks and rays, through eels, salmonoids, rattails and other codlike fishes, to the diverse lilliputian mesopelagic lanternfish, predatory stomiiform fishes, and the deep-sea anglers. He found that in some groups only single species or genera occur in the deep sea, whereas in others whole families and orders are found solely there.

Despite the fact that there is a clear taxonomic distinction between pelagic and demersal fish faunas, deep-sea fishes do share one attribute: all the dominant taxa tend to be representative of groups that appeared rather early in the evolution of modern fishes. Thus, deep-sea fishes can all be considered highly and quite specifically evolved and adapted to the particular environment and ecological conditions of the deep sea. In fact, finely tuned adaptations, i.e., those most evident in the pelagic fishes, with their specialized eyes, highly complex bioluminescent organs, elaborate gas

3. DISTRIBUTION AND POPULATION ECOLOGY

glands and swim bladder construction, and often remarkable jaws and teeth, have allowed deep-sea fishes to succeed very well. The fauna has persisted over very long time spans and has resisted competitive invasion from more recently evolved forms, most particularly the spiny-rayed (perciform) fishes, which dominate the fish fauna everywhere else in the world (Marshall, 1979).

Knowledge of deep-sea fish diversity has grown steadily over time. The early discoveries came mostly from beach strandings in regions where strong currents brought oceanic fishes to the surface. Even by 1775, 26 species of deep demersal fishes had been described from the Atlantic, and the rate of description continued at a steady rate for the next 100 years. The latter part of the nineteenth century saw a burgeoning of deep-sea exploration, and the rate of description increased dramatically. At the end of this era, the rate returned to its former level and is maintained to the present.

As to actual numbers of species, Cohen (1970) surveyed a large number of ichthyologists, and from their responses concluded that there were something like 1010 deep demersal fish species and 1280 pelagic species in the world's oceans. Merrett later (1994) considered only the North Atlantic Basin and found 505 demersal and 589 pelagic species in that relatively small part of the ocean. Clearly, the fauna may be somewhat more speciose than it was even fairly recently thought to be, and the earlier guesses must certainly be considered underestimates. Even at this most basic of levels, the knowledge of deep-sea fishes is far from complete. The description of new species of deep-sea fishes has gone on for almost 200 years, and new species continue to be found. How well the fauna is actually known will be discussed a bit more fully below.

Early students of marine biology thought that the deep ocean was a haven for many evolutionarily older and archaic groups that were competitively inferior to more recently evolved, modern taxa. Woodward (1898) summed up this argument for fishes. However, this concept was rejected when Andriyashev (1953) looked more closely at the question. Instead, he recognized two groups that had colonized the depths at different times. To reach this conclusion he considered the evolutionary sequence of adaptation of fishes to life in the deep sea and drew on his knowledge of their morphology, distribution, and biology. According to his analyalysis, "ancient deep-water forms" moved into the deep sea early on, and underwent their primary evolution and radiation there. Most ancient deep-water forms, he argued, are found only in the deep sea, and exhibit clear structural adaptations to deep-sea life, with highly specialized light organs, modified eyes, remarkable teeth, fins, and flotation devices. The "secondary deep-water" fishes, on the other hand, were considered to have undergone their primary evolution and radiation on the shallow continental shelves, where most are still found today, in various common and generally more derived families. Species that belong to secondary deep-water groups were thought to have moved into the deep sea much later, and for that reason do not display such marked morphological adaptation to the deep-sea environment as do the ancient deep-water species. Broadly speaking, pelagic deep-sea fishes belong mostly to "ancient" groups whereas the affinities of demersal deep-sea fishes can be considered more "secondary."

II. PELAGIC HABITATS

The ocean is a layered system, and the living spaces are subdivided in a vertical sense. Thus, the pelagic region can be subdivided into epi-, meso-, and bathypelagic zones. The epipelagic region is the uppermost of the pelagic realm, extending perhaps to about 200 m depth. Seasonal effects are felt most keenly there and, because 200 m is about the depth to which light sufficient to support photosynthesis can reach, this is where virtually all the ocean's primary production occurs. Below the epipelagial, from about 200 to 1000 m depth, is the mesopelagial. This region comprises most of the main thermocline, where the variable conditions experienced in the surface layers become damped to the far more stable and invariant conditions of the true deep sea. Light from the surface penetrates, but there is too little for plant growth. Below the mesopelagial lies the bathypelagic realm, the greatest watery living space on earth. The bathypelagial is dark, cool, and still.

Many fishes of the epipelagic regions belong to well-known and familar groups, for example, the tunas (Thunnidae), swordfish (*Xiphias*) and marlins, flying fish (Exocoetidae), and jacks (Carangidae). Even less familar epipelagic oceanic fishes, such as the Stromateoidei (the suborder to which the Portuguese Man-o'-War fish *Nomeus* belongs), are in general quite ordinary and unexceptional in their appearance. All of these are perciform relatives or derivatives and, by Andriyashev's (1953) argument, are "secondary" fish in the deep ocean. Merrett's (1994) very useful and wide-ranging study of reproductive strategies found that 43% of North Atlantic upper pelagic (<400 m) fish species belong to perciform groups. The overall number of species there is not large, however, and Merrett (1994) reports 80 upper pelagic species in 28 families for the North Atlantic; of these, 89% were found only within this depth zone.

Mesopelagic and bathypelagic fishes are taxonomically quite different from those associated with the epipelagic regions. They are more speciose as well. Merrett (1994) reports 66 families with 509 species, and of these 79% are found only at those depths. Characteristic deep pelagic families include the speciose lanternfish (Myctophidae), silver hatchetfishes and gonostomatids (Sternoptychidae, Gonostomatidae), viperfish (Chauliodontidae), and an entire suborder of predatory black stomiatoids (Stomiatoidei) in mesopelagic depths. In the deeper, bathypelagic regions are found the unique deep-sea anglers (Ceratioidei), whalefish (Barbourisidae and relatives), and gulper eels (Saccopharyngidae). None of these groups is important in the demersal fauna.

Most species of the deeper (>400 m) pelagic regions in the North Atlantic were found by Merrett (1994) to be stomiiforms (with 29% of the species), myctophids and their relatives (17%), and anglers (14%). Only 6% of the species in this region are perciform (Merrett, 1994).

III. DEMERSAL FAUNA: SHELF, SLOPE, AND RISE

The dominant families of the deep-demersal fauna are for the most part distant relatives of the shallow-water codfish—the rattails (Macrouridae), the deep-water cods (Moridae), the brotulids (Ophidioidei), eelpouts (Zoarcidae) and their relatives, slickheads (Alepocephalidae), and a diverse group of sharks, rays, and chimaeras (Chondrichthyes). The species diversity found in the North Atlantic is comparable to that of pelagic regions, with about 505 species in 72 families (Merrett, 1994). The demersal regions, however, have species from a larger number of orders than does the pelagic region, 22 as opposed to 13. Important demersal groups unknown from deep pelagic regions are the spiny eels (Notacanthiformes), John Dories (Zeiformes), flounders (Pleuronectiformes), and, with just three exceptions, chondrichthyans (~21%). Most species are found in gadiform groups (with 19% of the species), ophidiiforms (12%), and sculpins and their relatives (scorpaeniforms; 8%); perciform species, largely zoarcids, comprise 9% of the North Atlantic deep-demersal fish fauna (Merrett, 1994).

Figure 2 summarizes the discovery of new species of deep-demersal fishes over time. Based on this information, Haedrich and Merrett (1992) concluded that the deep-demersal fish fauna is still far from fully known.

IV. DISTRIBUTION PATTERNS

The distribution patterns recognized in the ocean are based on relatively few samples, particularly when one considers the vast area and volume they are meant to represent. Even so, oceanographers for many years have been quite comfortable with the idea that one or a few cruises provided

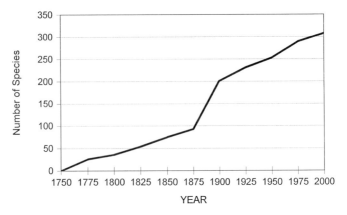

Fig. 2. Cumulative number of deep-demersal fish species described vs. year (25-year groups). Data based on the Atlantic fauna (Based on data from Haedrich and Merrett, 1988).

an adequate basis for generalization, especially in regard to the deep ocean and central gyres. This essentially static view of the deep sea underlies much of the biological literature of the deep-sea ecology and biogeography. In fact, the recurrent suite of questions concerned with deep-sea biodiversity is one example where this is so (e.g., McGowan and Walker, 1985; Haedrich, 1985; Rex et al., 1993; Gray, 1994), and the same can be said of work dealing with distribution patterns (Backus et al., 1977; McGowan, 1986). For example, the summary atlas by Haedrich and Merrett (1988), describing the state of knowledge of deep-demersal fish distribution in the North Atlantic Basin, falls into this category. The science of the deep sea has mostly operated under the assumption of stable environments and unchanging situations over very broad areas. Even though the cycle of primary production in the surface layer of the ocean is everywhere somewhat seasonal, the deep sea was for many years considered by biological oceanographers as an unvarying, benign, and aseasonal environment (Tyler, 1988).

We now realize that the deep ocean is not static (see Chapter 1, this volume). Accordingly, we must modify our views of the fauna living there. Evidence that this is so has been accumulating for some time. A very clear demonstration is to be found in the pictures taken with the Bathysnap camera made in the Porcupine Seabight (Lampitt and Burnham, 1983; Rice et al., 1986). These pictures showed that the epipelagial spring plankton bloom, certainly recognized as a strongly seasonal and dynamic phenomenon, was rapidly translated through the water column to the deep seabed, and that the fauna there responded at a comparable pace. The rapid accumulation and dispersion of floc from surface production (Billett et al.,

1983) showed that a clear seasonal signal exists in the deep sea, and other investigations have subsequently shown this to be true even at abyssal depths under the central ocean gyres (Rice *et al.,* 1991). Even on time scales as short as a few weeks or months, conditions are not constant in the deep sea (Gage and Tyler, 1991).

Unfortunately, details on the nature of such changes are largely lacking because time-series data of any kind are rare for the deep ocean. Deep-sea exploration has been most often motivated by pure scientific curiosity. However, deep-ocean investigations are costly, requiring specialized ships and dedicated time commitments. Except in the case of mineral resources, which require relatively simple one-time "look–see" sampling, there has been little economic incentive to undertake long-term monitoring of the kind necessary to produce time series adequate for ecological study. So observations on deep-sea fishes are not large in number. Even the most comprehensive oceanwide biogeographic treatments are based on fewer than a 1000 samples, and these samples usually come from a single gear type.

Sampling techniques and gear have evolved little, and then mostly with the aim to better quantify or determine patterns in the plankton or infaunal benthos rather than in fishes. Indeed, the need for consistent sampling protocols requires that gear development cannot go hand-in-hand with collection of long-term data series. For example, as Merrett *et al.* (1991a) and Gordon and Bergstad (1992) have shown, even slight modifications to trawl nets can result in very different catches. There has also been an understandable bias in directing studies toward the more economically important groups or areas, for example, potentially commercial fishes in the demersal fauna of the upper continental slope or the mesopelagic sound-scattering layers.

Quite in contrast to the situation in deep-sea studies, large numbers of samples and an emphasis on time series have been a keystone of fisheries research in shallow seas since the late 1800s. Fisheries science is fundamentally concerned with the dynamics of natural populations. Data from regular monitoring surveys are used to study those dynamics, to predict the health of stocks, and to set catch quotas. Commonly, survey data take the form of species numbers (abundance) and weight (biomass) taken in standard net samples from a preestablished grid or stratified set of standard stations. Nonetheless, time-series-derived data are mostly confined to the continental shelf regions or, on the high seas, to a few commercially important pelagic species such as tunas and billfishes. Work in the fisheries tends also to focus only on the species of interest, taking an autecological perspective as opposed to the synecological view that most ocean ecologists strive for. But, because of the perspective on change over time that fisheries data offer, they are being increasingly examined by ocean ecologists interested in community dynamics and the persistence of pattern (e.g., Gomes *et al.*, 1995).

Myctophids are the only group from the lower pelagic assemblage that are commercially exploited at present (Gjøsæter and Kawaguchi, 1980). Commercial species are much more commonly found in demersal assemblages, but even there comprise only a small proportion of the total species richness. Regularly exploited deep-sea demersal fish species occur in the orders Squaliformes (*Centroscymnus coelolepis*), Rajiformes (*Raja hyperborea*), Gadiformes (*Coryphaenoides rupestris, Macrourus berglax, Molva dypterygia, Macruronus novaezelandiae, Mora moro*), Beryciformes (*Hoplostethus atlanticus*), Scorpaeniformes (*Sebastes* and *Sebastolobus* spp.), Pleuronectiformes (*Glyptocephalus cynoglossus, Reinhardtius hippoglossoides*), and Perciformes (*Aphanopus carbo*). Fisheries directed to these species have, for the most part, developed only as those on the continental shelves worldwide are collapsing, and thus they have been operating for only a fairly short time. Nonetheless, there is every indication that unregulated deep-sea trawling has the potential to wipe out stocks very quickly and will very likely do so (Hopper, 1995).

A. Trends in Diversity

In the deep ocean, the greatest diversity occurs at middepths. This is true in both pelagic and demersal fish faunas. Merrett (1994) lists 80 species in the pelagic regions at depths less than 400 m, and 505 species in deeper water. For demersal regions, he found 74 species in the range of 200–400 m, 347 species from 400 to 2000 m, 64 species between 2000 and 4000 m, and 20 species below 4000 m. But, as suggested previously, even these recent numbers must be considered estimates as long as the taxonomic knowledge of deep-sea fishes remains incomplete.

Still, the patterns of biodiversity and community structure seen in deep-sea fishes do not appear to be very different from those observed in biological communities in other environments. In areas where primary production is high, for example, in high latitudes or in upwelling areas, dominance of the fauna by one or a few species is common. In areas where production is low, which comprise in fact a very large part of the ocean, species are much more likely to be present in roughly comparable abundances, and evenness is the rule. This situation is the same in both pelagic and demersal deep-sea fish faunas.

A part of the admitted imperfect knowledge of the biodiversity of deep-sea fish stems from the fact that each of the few samplers employed offers only a limited window on reality (Angel, 1977; Merrett *et al.*, 1991a). Moreover, newly discovered deep-sea fish species surprisingly are not cryptic novelties from remote regions.

Included in the new species are the megamouth shark (*Megachasma pelagios*), described first from relatively shallow midwater depths off Hawaii (Taylor *et al.*, 1983), and a large pelagic ray from off South Africa (Heemstra and Smith, 1980). Both species belong to entirely new families. The 4.5-m long, 750-kg megamouth shark appears adapted to feeding on small pelagic prey by sucking in large volumes of water with its bellowslike jaws (Compagno, 1990).

Clearly, the discovery of new large fish species in the deep sea is quite likely. These animals are certainly more capable of avoiding conventional sampling gear than are smaller ones, and aspects of their behavior and ecology can also make them less susceptible to notice and sampling. New sampling gear will pave the way to new discoveries. For example, the recent use of baited cameras and other free vehicles in the deep sea has revealed the presence there of very large fishes, and also has allowed ingenious experiments relating to foraging behavior and abundance (Desbruyères *et al.*, 1985; Priede *et al.*, 1990).

B. Broad-Scale Horizontal Patterns

Ideas about oceanic biogeography and the nature of community patterns in the ocean, comprehensively spelled out by McGowan (1974), first began to develop from the national oceanographic expeditions of the late nineteenth and early twentieth centuries. Ekman (1953) summarized the state of knowledge at the time, and his book has become the classic starting point for consideration of distribution patterns, mostly horizontal, in the sea. The basic view is that animals tend to be widespread, that patterns are relatively simple, and that the physics of the ocean largely determines species' limits (faunal boundaries) and also their paths of dispersal. Pelagic biogeography has been treated more fully than has the biogeography of the sea bottom.

1. PELAGIC

Although Ekman (1953) was able to compose only a short chapter on the pelagic biogeography, the modern era of ocean exploration that began about that time provided sufficient new material to enable the publication by van der Spoel and Pierrot-Bults (1979) of an important book "Zoogeography and Diversity of Plankton." This important landmark deals with mesopelagic fishes as well as invertebrates. Follow-up conferences in 1985 (Pierrot-Bults *et al.*, 1986) and 1995 (Pierrot-Bults and van der Spoel, 1997) have kept the subject alive.

Many deep-sea pelagic fish species are widespread. The distribution of individual species groups and of community assemblages, which forms the basis for biogeographic schemes, suggests but a few large pelagic faunal regions. These regions are characteristic of different parts of the ocean, and generally follow the topography and the overall temperature structure and circulation patterns of the ocean (see Chapter 1, this volume). There are, for example, faunal groupings that characterize semienclosed basins such as the Norwegian Sea, the Mediterranean Sea, and the Gulf of Mexico and Caribbean Sea. There are also pelagic assemblages that characterize tropical, subtropical, temperate, and cold-water parts of the open sea, for example, the Sargasso Sea, the Rockall Trough, the Labrador Sea, and the Southern Ocean. Most classifications (e.g., Backus, 1986) suggest that the number of pelagic faunal regions in the world ocean is remarkably few, perhaps about 20 (Fig. 3). This number is significantly lower than the number of separate biomes recognized by community ecologists on the relatively much smaller land area.

2. DEMERSAL

Knowledge of the deep-demersal fish fauna has improved considerably over the past 10 years. Surveys with comparable gear now have been conducted over a considerable depth range and in a number of widely spread locations around the North Atlantic Basin. Moreover, a uniform taxonomy has been applied to the collections. Regional studies by teams of investigators have been carried out and are on-going in a number of places in the world ocean, for example, the Norwegian Deep (Bergstad, 1990), the Rockall Trough (Mauchline, 1990), the Porcupine Seabight (Rice et al., 1991), the Bay of Biscay (Mahaut et al., 1990), the eastern South Atlantic (Golovan, 1978; Macpherson, 1989), off Tasmania (May and Blaber, 1989), southeastern Australia (Koslow et al., 1994), the Great Australian Bight (Newton and Klaer, 1991), and the abyssal plains of the eastern Atlantic (Merrett, 1987). These studies have yielded valuable information on deep-demersal fish ecology (e.g., Stein and Pearcy, 1982; Crabtree et al., 1985).

Some generalities seem to hold for deep-sea fish communities. For example, there is a diversity maximum in many taxa, not just the fish, in demersal regions somewhere on the lower part of the continental slope, in

---

Fig. 3. Suggested pelagic faunal regions of the world ocean. 1, 9: Polar; 2, 8: subpolar; 3, 7: temperate; 4, 6: subtropical; 5: tropical. From Backus (1986), *UNESCO Technical Papers in Marine Science* **49.**

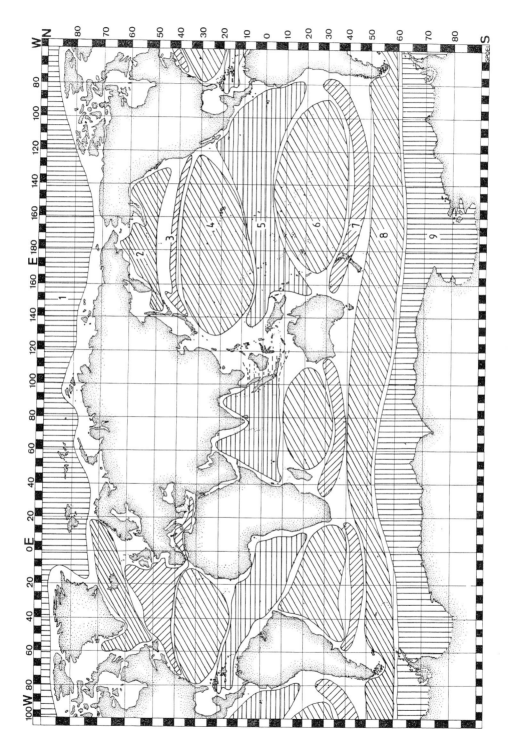

depths of the order of 1500 m (Rex *et al.*, 1993). The question as to exactly what this diversity stems from is a matter that has intrigued deep-sea biologists for many years, and although theories abound, no clear explanation has been advanced (Gage and Tyler, 1991). One explanation might be that diversity is related to the level of primary production. Indeed, the maps that contour oceanic production give a first approximation of local diversity. In addition, species/area relationships, for both pelagic and deep-demersal regions, hold best when production is integrated into the area term (Haedrich, 1985).

Studies suggest that there is relatively little faunal similarity at the species level from one part of the rim of an ocean basin to another. Each area studied has proved to be somewhat different in terms of its faunal composition. This view is contrary to the idea that deep-demersal fishes are very widely distributed over broad areas, and therefore that deep-sea demersal fish community structure is simple and predictable. For example, Haedrich and Merrett (1990) analyzed catch data from 692 trawl hauls between depths of 204 and 5345 m (96,779 specimens and 325 species), with samples coming from the Atlantic continental margin from the Bahamas north to Canada, Iceland, and the British Isles and around to northwest Africa. They concluded that the demersal fishes present in any one particular area were rarely, if ever, strongly associated with any other species in the sense of a community that could be identified elsewhere. Haedrich and Merrett (1988, 1990) could find very little evidence to support continuity in horizontal zones, a view supported by Campbell (1983, 1990) based on the occurrence of deep-sea fish parasites. In fact, many demersal fish species appear to have comparatively small geographical ranges.

As such, Haedrich and Merrett (1990) suggested that the community concept for deep-sea benthic fishes seemed untenable, and suggested its abandonment. Their action was perhaps hasty, as Koslow (1993) has pointed out, for certain insights are to be gained from analyzing fish species associations in the deep ocean. Moreover, the scale of investigation that is selected is of great importance when examining demersal fish distributions.

On a basin-wide basis (thousands of kilometers)—the scale on which debates about diversity and other matters dear to the hearts of deep-ocean ecologists rage (e.g., Rex *et al.*, 1993, Gray, 1994)—strict adherence to the community concept, although ignoring spatial scale issues, has befogged the issue (cf. Haedrich, 1985). On a species by species basis, assemblages compared between widely scattered locales at the basin scale are clearly not the same. Quite often the presence of a few widespread and dominant forms has diverted attention from the fact that many of the less abundant species seem to be quite restricted in their distribution (Haedrich and Merrett, 1990).

On a smaller mesoscale of banks and eddies (hundreds of kilometers), the criterion of comparability seems to be met. Species lists tend to be very similar, and the distribution patterns of individual species overlap broadly. In the fisheries data for continental shelves, the only datasets for which faunal composition over time is available, stability and persistence over at least a few generations are characteristic of assemblages identified on those scales (Overholtz and Tyler, 1985; Gabriel, 1992; Gomes, 1993). These data provide valuable material for a community to be analyzed and understood within an holistic ecological framework (Sherman, 1994). Alverson (1993) expects that fisheries management within such a framework must be the way of the future.

3. Relationship to Ocean Production

Primary production is at the base of all oceanic food webs. However, production in the oceans varies a great deal, and the efficiency and rate with which carbon moves through food webs, measured as secondary and tertiary production, become more difficult to determine the closer one gets to the fishes. In fact, production can vary by orders of magnitude, and certain kinds of food webs involving fishes are associated with particular production regimes (Ryther, 1969; Longhurst, 1981). The food chains of central ocean gyres, for example, those in which tuna are the top predators, are relatively long, involving five or more links. In contrast, food chains of high-production regimes, for example, on the northern fishing banks or in areas of seasonal upwelling, can be very short.

Although mesopelagic fish practice some feeding selectivity, there is broad overlap in the diets of many species (see Chapter 4, this volume). Crustaceans, most especially copepods and euphausiids, seem to constitute the dominant prey species almost everywhere (Hopkins and Torres, 1989). When variations do occur, they are interpretable in terms of local abundances of the prey species involved. Midwater fish at times appear to switch to whatever is most available without much regard for remaining at the appropriate level of the classical food chain. For example, there are observations of mesopelagic lanternfish, which normally feed on crustaceans, feeding on phytoplankton (Robison, 1984).

Selectivity by size may be more important than selectivity by species. In general, larger predators consume larger prey, not necessarily ceasing to take smaller size classes (Young and Blaber, 1986). As such, the size/feeding relationship offers one avenue for modeling fish production (Sheldon *et al.*, 1977; Gorelova, 1983). A number of demersal faunal studies provide data with sufficient detail in well-defined geographic areas to formulate and test such an approach.

Size is of additional interest because most biological processes, including production, are scaled to the size of the organism (Peters, 1983), a generalization that has grown out of physiological studies. Calder (1985) develops the argument as to how such empirical correlations can be used in a holistic approach to the determination of growth and production in natural populations. This is possible because annual production/average biomass (P/B) ratios appear to follow the allometric rule and are remarkably constant within major taxa.

Banse and Mosher (1980) determined the P/B ratio in fishes scaled to the -0.26th power according to the relationship:

$$P/B = 0.38 \times M^{-0.26}$$

where M is expressed as size in grams wet weight. Their study used information mostly from small freshwater fishes, with M (kilocalorie equivalents, and therefore using a coefficient of 0.44 in the equation) based on the adult size.

This relationship may have broader applications to groups well outside the taxa used to derive the equation. For example, Haedrich (1986) used this allometric relationship to produce Atlantic mesopelagic fish production maps that mirror the familiar map of primary production in the sea (e.g., Koblentz-Mishke *et al.*, 1970). Applied to deep-demersal assemblages, the same allometric relationship gives annual fish production levels (Fig. 4) that are quite close to those predicted by Mann (1984) using a very different, and much more conventional, fisheries-type approach.

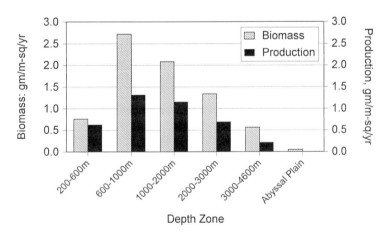

Fig. 4. Demersal fish biomass and annual production, calculated from an allometric equation, for faunal depth zones in the northeast Atlantic Ocean. Based on data from Haedrich and Merrett (1992).

3. DISTRIBUTION AND POPULATION ECOLOGY   93

Another indirect approach to estimating production is to use the biochemical composition of mesopelagic fish to infer rates of processes in their populations (e.g., Childress et al., 1980; Reinhardt and Van Vleet, 1986; Stickney and Torres, 1989). Attention in this area originally focused on relatively few taxa (Baird and Hopkins, 1981) and on depth as the controlling factor for metabolic rates in the deep sea (Siebenaller et al., 1982; Graham et al., 1985). It appears, however, that simple food availability may be the most significant factor in determining rates, regardless of depth (Bailey and Robison, 1986). With increased investigation, there seemed to be few if any universal physiological trends with depth and there was considerable variation in the responses of different taxonomic groups (Childress and Thuesen, 1992). What did emerge from comparative studies of Antarctic midwater fishes is that the observed decrease in metabolic rates with depth is an adaptive trait, and is not simply a consequence of declining temperatures (Torres and Somero, 1988). The hemoglobins of deep-sea demersal fish from different depths are also appropriately adapted (Noble et al., 1986).

C. Vertical Zonation

The abundance and overall biomass of both pelagic and demersal deep-sea fishes decline with depth. The deeper one goes, the fewer fish there are and the less is their biomass. Thus, fish follow the general rule of a decline in amount with depth that is well-documented for other biota. Even so, the biomass of demersal fish generally seems to be considerably greater than that of the oceanic midwater fishes (Marshall and Merrett, 1977; Mann, 1984; Merrett, 1986; Gauldie et al., 1989). There are important exceptions to this generality. Regions exist where pelagic fish biomass over the slope can at times greatly exceed the benthic fish biomass (May and Blaber, 1989), but this situation would seem to be unusual. Furthermore, there is an admitted bias in the general conclusion because most demersal fish studies have been conducted at the edge of continental margins, rather than well out on the abyssal plains. Still, the apparent pattern remains similar to that found for oceanic zooplankton by Wishner (1980), Angel (1989, 1990), and others.

A recurring theme in deep-sea benthic studies is that the fauna is vertically zoned to form communities at different depths that are identifiable over rather broad geographical areas (see Carney et al., 1983). It is true that there is a change in assemblage composition with depth that is evident in the demersal deep-sea fish fauna. Vertical zonation, to some degree or another, has been identified in many parts of the ocean (Haedrich and Merrett, 1988); it can be seen quite clearly, for example, in Fig. 5. As shown

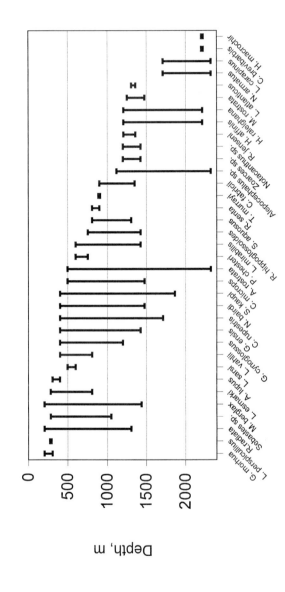

Fig. 5. Depth ranges (vertical lines) for demersal fish species on the Newfoundland continental slope. Data from Snelgrove and Haedrich (1985).

by Merrett (1994) and others, there is a rather clear distinction between the fish fauna found on the continental shelves, on the continental slope from about 200 to 2000 m depth, and on the continental rise and then out across the abyssal plains. Shelf faunas appear somewhat restricted within the so-called large marine ecosystems (Sherman, 1994), slope faunas extend in long ribbonlike bands around the rim of the deep ocean, and rise and abyssal faunas are widespread across deep basins.

This well-established view of clearly defined communities zoned by depth appears to be too simplistic and somewhat out of date. Instead, it appears more likely that each species, in any one area, occurs over its own particular depth range. Species replacement with depth, although regular, follows no strict pattern and is not repeatable in detail from place to place. The rule summed up in Heincke's Law (bigger fish live in deeper water, the "bigger–deeper" trend) has been shown in general to be a sampling artifact resulting from net avoidance by larger fish at shallow depths (Pearcy *et al.,* 1982; Merrett *et al.,* 1991b). Within species, however, this law may hold in some, but certainly not in all, instances (Macpherson and Duarte, 1991; Uiblein *et al.,* 1996).

The vertical ranges displayed by certain continental slope fish species can show considerable variation. For example, the deep-sea eel *Synaphobranchus kaupi* in the Rockall Trough has a wide depth range (500 to 2000 m) and shows a well-marked "bigger–deeper" distribution (Gordon and Bergstad, 1992). *Coryphaenoides rupestris* is another species that has almost as broad a depth range as *Synaphobranchus,* but the size distribution with depth, at least in the Rockall Trough, is much more complex (Gordon, 1979). Most continental slope species have more restricted depth ranges than do these two important slope fishes. Figure 5 shows the overall depth ranges for deep-demersal fish on the continental slope off Newfoundland (Snelgrove and Haedrich, 1985). Each species is rather unique in its depth range, and the number that resemble one another even to some extent is rather small. The same picture is seen in most areas of the world where such data are available.

The vertical distribution of deep-water demersal species that enter the water column is poorly understood because large midwater trawls have seldom been used near the bottom in deep water (Merrett *et al.,* 1986). The reasons for the apparently relatively low abundance of juvenile stages compared with the adults of many species, such as deep-water sharks and the black scabbardfish (*Aphanopus carbo*), in bottom trawl catches is probably because, in their early life history, they are unavailable to demersal trawls fishing only to a height of a few meters off the bottom. Diurnal vertical migrations have also been described for some commercially exploited deep-water species, such as *Coryphaenoides rupestris* (Savvatimskii, 1987). The orange roughy (*Hoplostethus atlanticus*) in New Zealand waters can be found at depths of at least 50 m off the bottom, mainly associated with spawning aggregations (Clark and Tracey, 1992).

1. Relationship to Ocean Structure

The strong dependence of any biological pattern in the ocean on the regional physical circumstances is a central theme in ocean ecology. The explanation for patterns and variability in primary production, fishery recruitment, biodiversity, and species distributions are all sought through this link. The area of the Rockall Trough has been especially well-studied from this perspective. Much of the following account, which basically describes the situation in that oceanic region, is based on Gordon *et al.* (1995).

The percentages of the total area of the world ocean occupied by the continental slope (approximately within water depths of 200 to 2000 m) and the continental shelf (depths <200 m) are relatively similar at 7.5 and 8.8%. The slope is generally narrower in horizontal extent compared to the shelf, and marks the true limit of the continental land masses and the beginning of the deep sea. Gradients on the slope are much steeper than they are on the shelf, and the relief is far more accentuated. Continental slope regions are often referred to, quite accurately, as the oceanic rim.

As a general rule temperature decreases with depth in the ocean, but the rate of change can vary from area to area. There is generally a well-defined summer thermocline to depths of several hundred meters; this breaks down due to wind mixing during the winter and early spring. The deep mixing, usually occurring in winter, replenishes the nutrients in the well-illuminated surface waters, leading to a spring phytoplankton bloom. In most higher latitudes, the burst of primary production, the so-called spring bloom, can be very dramatic (Prasad and Haedrich, 1993, 1994), and the growth achieved at that time then provides the food energy that drives the deep-ocean system for the rest of the year. Almost all the food energy reaching the seabed on the slope, for example, is derived from this surface production.

The deep sea, at least at depths below about 500 m, is an environment of relatively broad physical constancy, and yet many of the animals, both fish and invertebrates, appear to have seasonal cycles of reproduction or growth (Tyler, 1988). The seasonal signal from the sinking of the spring bloom must play an important role.

The Rockall Trough in the eastern North Atlantic is one of the best studied ocean regions, and provides good specific examples that illustrate well the differences between oceanic areas. There, the water temperature below 500 declines gradually, but the annual variation in most places is negligible (<0.5°C). The Trough is separated from the Norwegian Basin by a number of underwater ridges, such as the Wyville Thomson Ridge between Shetland and the Faroe Islands. At depths down to the top of the ridge the temperature regime is similar on either side of the ridge. Below the sill the temperature decreases gradually to the west, but to the east the temperature decreases rapidly to below 0°C. These differing temperature regimes have a clear effect on the composition, abundance, and biomass of the fish fauna found there.

In the Rockall Trough species diversity and biomass peak between 1000 and 1500 meters, whereas in the Norwegian Basin diversity, abundance and biomass decrease rapidly below about 500 m. Off Norway the fish biomass at 1000 appears to be only 1% of that on the upper slope (400 to 500 m), and the number of species decreases from 10 to 3 (Bakken *et al.*, 1975). There is also a change with depth from a boreal faunal composition similar to that in the Rockall Trough to a boreo-arctic/arctic fauna (Bergstad and Isaksen, 1987); another similar faunal transition occurs across the ridge at the Davis Strait west of Iceland between the Arctic Ocean and the Irminger Sea (Haedrich and Krefft, 1978).

Annual changes in salinity on the slope are generally small and are not thought to affect the fishes living there. Long-term changes in the surface salinity of the Rockall Trough, for example, are well-documented and can be related to changes in the distributions of the water masses that constitute the Atlantic Ocean generally (Gordon *et al.*, 1995). One such change was the Great Salinity Anomaly, a large area of unusually low surface salinity that persisted in the central Rockall Trough and in the North Atlantic, generally during much of the 1970s. Although this anomaly was described for surface waters, its effect could also be detected in deeper water (Ellett, 1993), and it has even been seized as a possible cause of the decline of fish stocks around the Atlantic rim, although no mechanism has been suggested.

In general, oxygen levels are close to saturation in continental slope waters. Oxygen concentrations change with depth in the Rockall Trough. The lowest values occur at about 1000 m, indicating a mixing with northward-flowing Mediterranean water that originates from the Straits of Gibraltar. Because of its characteristic high salinity and low oxygen, it is detectable over a wide area of the Atlantic. Distinctive water types such as this one can help to describe the distribution of certain oceanic animals. The black scabbardfish (*Aphanopus carbo*), for example, seems to be associated with the Mediterranean water in the deep Atlantic.

The patterns of temperature, salinity, and oxygen distribution in the ocean result from movements by the oceanic currents and circulation. The gradients so established, combined with the depth-related diminution of light and increase of pressure, are the main environmental characteristics of the upper parts of the ocean—epipelagic and mesopelagic regions and the continental slopes. Because of the linkages established by the circulation of water, conditions at any one place can only be understood by reference to situations that may occur at other, quite distant, places. An example is the arctic conditions that occur off Newfoundland (at the latitude of Paris) because of the strong southward-flowing Labrador Current. Another is the input of food energy to the upper slope from a nearby continental shelf or bank, where the spring bloom referred to previously may be most well-developed.

2. DIEL MIGRATION

The physical properties of water related to temperature make the ocean a layered system, so vertical distribution of animals is an important consideration. In open ocean waters, the usual breakdown identifies epipelagic (photic, mixed layer), mesopelagic (disphotic, main thermocline), and bathypelagic (aphotic, relatively unvarying) realms. The mesopelagic region, extending from about 200 to 1000 m depth, is a region where the habit of many resident pelagic fishes is to undergo extensive vertical migrations on a daily basis. These migrations can extend well into the epipelagic region, where the fish do most of their feeding, and even to the surface at night.

Because many pelagic species do move up and down in the water column on a daily basis, the study of vertical distribution in the open ocean has tended to focus on such movements. In doing so, attention has been on individual species and the different patterns that each may display. Because of this focus, scientific preoccupation with strict vertical zonation in the fauna as a whole has been unlikely and of less interest.

3. ONTOGENETIC MIGRATION

A knowledge of the vertical distribution of all growth stages is essential for understanding the life histories of deep-sea fishes. This is because the life cycle of deep-water fish may include an ontogenetic migration whereby the larvae and juveniles are found in epipelagic regions. Here there is greater biological production and therefore food is more available. Early development in the epipelagic is followed by a descent in later life into mesopelagic, bathypelagic, or even deep-demersal regions. Vertical migration, both diel and ontogenetic, tends to break down an easy characterization of deep-sea fishes as belonging uniquely to any of the particular subdivisions of the ocean in a vertical sense.

Angel (1986) more fully shows how changes in the vertical depth distribution meet the different requirements at different life stages in a species. In addition, the Loeb (1986) study of the depth ranges occupied by larval and adult mesopelagic fish (myctophids and gonostomatids) provides details on how complex the picture can be. Where primary production is relatively

3. DISTRIBUTION AND POPULATION ECOLOGY   99

high and concentrated at shallow depths, such as in the eastern tropical Pacific, those fish that vertically migrate concentrate at night in the upper layers but have larvae that tend to remain at depth. The larvae of nonmigratory species, on the other hand, predominate in the productive layers and the adults are found at depth.

The corollary to this is where production is less and more diffuse, such as in the North Pacific central gyre. Here vertical migration is diminished and both larvae and adults appear more widely distributed in the vertical sense. The physiological changes that must accompany the development and associated changes in habitat and vertical distribution of deep-sea fishes have not been studied.

V. FEEDING RELATIONSHIPS

Even though the density of oceanic animals tends to be low, the simple fact that the oceans cover much of the globe suggests that most biomass carbon occurs in the food webs of the sea. In these food webs, fishes are usually the dominant top carnivores, and certainly in the deep sea. The biomass of all mesopelagic fish in the ocean is estimated to be at least 9.5×10^8 tonnes (Gjøsæter and Kawaguchi, 1980). The global biomass of the deep-demersal macrourid species *Coryphaenoides armatus* and *Coryphaenoides yaquinae* is estimated to be 1.5×10^7 tonnes, a figure about the same as the total world catch of demersal fish (Gage and Tyler, 1991).

Food webs process carbon. Organic matter provides the energy that drives the components of the web and the raw material for reproducing them. For fishes, there is an established methodology—most commonly combining field measurements of size and age and the application of generalized models—for determining the expected yield of fisheries. The ultimate measure is in terms of production. Quantification of this dynamic aspect of food webs in the oceans is a far more difficult task than determining what fisheries biologists refer to as the size of the stock, i.e., the average biomass over the year in a particular area.

Food webs are most commonly determined by the time-honored but tedious method of stomach contents analysis. Other, newer methods have been tried. Feller *et al.* (1985) and Fry *et al.* (1984) suggest immunological and stable isotope approaches, which can be used to determine pathways and relative trophic positions of constituent species within deep-sea food webs. Dickson (1986) used stable isotope ratios to study and compare pelagic food webs in two deep fjord systems, and Fry (1988) did the same for a major fishing bank on the continental shelf. Williams *et al.* (1987) used radiocarbon activity to infer rates of food energy input. Advances in understanding deep-sea parasites and their life cycles through various hosts (Campbell *et al.*, 1980; Campbell, 1983, 1990; Houston and Haedrich, 1986) have also yielded information on feeding relationships. Even with the best of data, however, the indeterminacy that results from the way in which a food web is structured and modeled probably cannot be overcome (Gomes and Haedrich, 1992).

Important quantitative data on feeding relationships in midwater fish assemblages have been presented by Dalpadado and Gjøsæter (1988), Hopkins and Baird (1985a,b), Hopkins and Torres (1989), Kinzer and Schulz (1985, 1988), and Young and Blaber (1986). These studies, and others to which they refer, have been conducted from the equator to the ice edge in high latitudes, and from the relatively high-production areas over continental slopes and in upwelling regions to the oligotrophic centers of ocean gyres. The studies range oceanwide and the generalization they offer can thus be accepted with a high degree of confidence.

The pelagic fish studies show that Vinogradov's "ladder of migrations" has stood the test of time. The diel vertical migrations of mainly mesopelagic fish species into the epipelagial to feed at night each comprise a ladder; the mechanism continues to be recognized as an important mechanism for moving organic material rapidly from the surface layers into the deep sea (Willis and Pearcy, 1982; Roe and Badcock, 1984). It is not just particulate matter that is moved. Because of respiration at depth by the animals that comprise Vinogradov's ladders, the pool of dissolved organic matter there is enhanced as well. Measurements from a very few stations, but over broad areas, indicate that this respiratory flux rate could range from 12 to 53% of the measured small-particle flux (Longhurst *et al.*, 1990). Active vertically migrating fishes—mainly myctophids—certainly play an important role here.

Within local areas, feeding studies have also shown the dependence of demersal fish on food from the water column (Bulman and Balber, 1986). Euphausiids can be very important in the diets of a large number of demersal fishes on northern continental shelves (Astthorsson and Pálsson, 1987). This is also true on the upper continental slope (Blaber and Bulman, 1987; Bergstad, 1991b), where there is frequently an ontogenetic feeding shift. Smaller representatives of a species may feed mostly on the bottom, but with growth change to more benthopelagic/pelagic prey (Eliassen and Jobling, 1985). The idea that demersal fish feed on pelagic prey is supported by data from the Porcupine Seabight (Haedrich and Merrett, 1992), collected over a depth range of several thousand meters with a variety of gear. There, 35% of the demersal fish species fed on pelagic prey and 52% fed on a mixed diet; only 13% depended on the benthos for food. Of the 11 dominant species, seven relied on pelagic sources and five had a mixed

3. DISTRIBUTION AND POPULATION ECOLOGY

diet; none relied on benthic animals alone. Data summarized by 200-m depth increments show that the picture is maintained throughout the full range of depths (Fig. 6).

From the standpoint of species interactions and food web complexity, the abundance data are of greater interest. There is a rather regular decline in number of individuals with depth. From the standpoint of overall energy flow, the biomass data should be considered. There is an increase to a peak at depth, around 2200 m, and then a decline to a rather uniform level that is maintained over a broad range of depths.

There can be considerable overlap in deep-demersal fish diets at continental shelf and upper slope depths (Mattson, 1981; Mauchline and Gordon, 1985, 1986; Gordon and Mauchline, 1990), depending on the area. The overlap is strongest within feeding guilds (Campbell *et al.*, 1980; Blaber

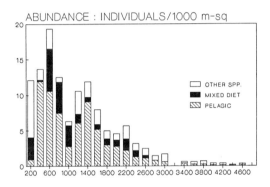

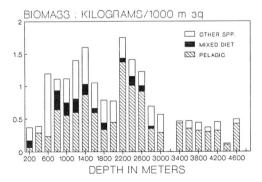

Fig. 6. Mean abundance (top) and biomass (bottom) of dominant species of deep-demersal fish by 200-m depth increments, with the proportion of pelagic and mixed (pelagic + benthic) feeders indicated. "Other" species are nondominant species not categorized according to diet. Based on data from Haedrich and Merrett (1992), using data from the Porcupine Sea Bight.

and Bulman, 1987; Bergstad, 1991b). This suggests, for the areas where these studies were carried out, that there is an excess of resources, and that selectivity, if practiced, is more likely to be on the basis of size rather than taxon (Mattson, 1981). In areas with lesser pelagic food resources, less dietary overlap is observed (Macpherson, 1981) and trophic groups within the demersal fish assemblage are well-defined and reasonably discrete (Macpherson and Roel, 1987).

Direct measurement of feeding rates and bioenergetics has only just begun, largely through new sorts of deep-sea free vehicle-mounted instruments (e.g., Armstrong *et al.*, 1992; Bagley *et al.*, 1990; Priede *et al.*, 1990). Daily rations of upper continental slope demersal fishes have been measured for a few species; values range from 0.5 to 2% of the wet body weight day^{-1} (Macpherson, 1985). Calorimetric analysis of the diet indicates that although euphausiids predominate in numbers, fish supply 90% of the energy to the upper slope demersal fish *Macruronus novaezelandiae* (Bulman and Blaber, 1986). Pelagic prey, especially vertically migrating fish, are good energy sources (Childress *et al.*, 1980; Bailey and Robison, 1986) and are important in demersal fish diets (Mauchline and Gordon, 1986; Blaber and Bulman, 1987; Bergstad, 1991a).

VI. AGE DETERMINATION

Although the data are rather sparse, it also seems that growth rates in deep-demersal fishes are slow, and deep-sea species reach ages that are relatively, in comparison to fish found on the shelf, quite old.

Fish age is customarily determined by counting checks (discontinuities) in the banding patterns seen in bony structures, especially otoliths and scales. Surprisingly little attention has been paid to the physiology of check formation, but the assumption is that the checks reflect variations in growth rate. Otoliths are considered somewhat more reliable than scales for age estimation, and usually the larger sagitta is used. Viewed by transmitted light the otolith shows concentric opaque and hyaline zone patterns, which usually conform with the concentric surface sculpturing seen under reflected light. Under higher magnification, however, even smaller units can be seen—the so-called microincrements that correspond to daily growth rings.

Seasonal temperature changes in the shallow ocean slow growth in the cold winter, and the growth change shows up as a check. In contrast, temperatures at depth below the seasonal thermocline in the deep sea are permanently low and, together with increased pressure, may retard general metabolic rate processes despite special enzymatic adaptations to deep-sea conditions (Siebenaller *et al.*, 1982; Somero *et al.*, 1983). Food availability varies also and, although supplies may be relatively enhanced over the continental slope compared to the open ocean, the supply still drops off logarithmically with depth. Conditions in the cold deep sea seem to rule against achieving the growth rates possible in warmer, shallower oceans.

Hyaline and opaque circuli occur in the otoliths of most deep-sea fish. However, it is not clear to what rhythm the banding patterns observed should be related, because these fish live in areas where seasonal signals are assumed to be weak or nonexistent. Is it diurnal and annual, or is it set by lunar tides or even by some uneven pattern of food availability? Confirmation that the rhythm is annual has been achieved in only a few species. Massuti *et al.* (1995) used the evolution over time of the opaque rings in the otolith margin of several macrourid species to link the highest percentage of opaque rings at the margin of the otolith with the fastest growth rate, assumed to occur in summer. In all species one opaque ring was formed each year, suggesting that the rings in these deep-sea fish were formed annually.

Age estimation from the alternating pattern of opaque and hyaline rings, each pair assumed to reflect an annual growth cycle, often shows poor agreement with counts of microincrements. The former tend to give relatively older ages, up to 75–100 years in some deep-sea fish, than does the microincrement approach, which has commonly found ages to be in the range of 15–30 years. Validation is an important part of any age determination scheme and standards in the validation of deep-sea fish ages are regularly and often hotly debated (cf. Gauldie, 1994; Bergstad, 1995).

Wilson (1988) examined the microincrement structure in the otoliths of two abyssal macrourids from the North Pacific abyss. These conformed with the general description of daily growth rings and indicated relatively rapid growth, as suggested for *Bathysaurus ferox* by Sulak *et al.* (1985) and *Conocara macropterum* by Crabtree and Sulak (1986). Morales-Nin (1990) has also shown relatively high growth rates in several slope-dwelling fishes from the Mediterranean. She suggests that the low caloric density of deep-sea fish (Childress and Nygaard, 1973), combined with their low metabolic rate (Smith, 1978), could in fact result in relatively high growth efficiencies.

VII. REPRODUCTIVE STRATEGIES

The contrast in phylogenetic composition of the midwater and demersal ichthyofauna is broadly reflected in their various reproductive styles (Merrett, 1994). The reduced resources available in midwaters have necessitated the evolution of many reproductive adaptations, such as hermaphroditism, extremes in sexual dimorphism, sex ratio adjustment, and so on. These and other strategies were well-described by Mead *et al.* (1964) and further elaborated on by Marshall (1971, 1979) and most recently examined by Merrett (1994). An impressive and diverse set of reproductive strategies is found among deep-sea fishes. The fish represent the full spectrum, from R-strategists (those that produce large numbers of young and that grow relatively quickly) to K-strategists (those that produce only a few young that grow slowly but live to relatively great ages, perhaps receiving parental care). The sensory problems of mate location in the deep sea are alluded to in Chapter 8 (this volume).

The roving benthopelagic ichthyofauna of the deep-sea floor display few striking reproductive adaptations. Such specializations tend to be found more commonly among the sedentary, benthic families. Noteworthy are the deep-sea tripodfishes (Ipnopidae) and the synodontids, which display synchronous hermaphroditism, an adaptation that increases the chances of reproductive success despite the reduced encounter rate compared with wider ranging benthopelagic fishes. *Sebastes* and certain ophidioids, on the other hand, are livebearers, but with high fecundity. The sharks display adaptations in reproductive style similar to those of their counterparts in midwater. The most diverse and abundant chondrichthyan family, the Squalidae, are low-fecundity livebearers, whereas some, such as the false catshark, *Pseudotriakis microdon,* are oviphagous.

Where adequate data exist, deep-sea fishes display seasonal reproductive cycles as well (Gordon, 1979). Seasonal spawning must be tied to a strategy where the young fish, once they hatch, enter the food web at a time coordinated with the seasonal peak in primary production so characteristic of regions such as the Porcupine Seabight. As previously mentioned, the pattern of small eggs and pelagic early development typifies both gadiforms and notacanths. The fact that species in these groups show clear seasonal spawning cycles appropriate to the spring phytoplankton bloom supports predictions that would be made based on a consideration of their egg sizes and presumed life history strategies.

Turnover times for many deep-sea fish populations will be slow because deep-sea fishes are relatively old. This, of course, makes perfect sense when considered in terms of the generally low food availability in the deep ocean.

A. Reproduction and Development

Many of the details of the reproductive styles found in deep-demersal fishes remain unknown. It has been generally accepted that fish fecundity increases with body size. Duarte and Alcaraz (1989) assessed the advantages of producing many small or few large eggs from among 51 species of mixed marine (mostly neritic) and freshwater fishes. They found no evidence of a phylogenetic trend that suggested a tendency to evolve toward the production of larger eggs. In reworking the data used by Duarte and Alcaraz (1989), Elgar (1990) concluded that the partitioning of reproductive output between the size and number of offspring varies independently of body size. The general rule seems to be that pelagic spawning marine fish produce many small eggs and demersal spawners produce fewer large eggs. Duarte and Alcaraz (1989) argued that colonization of the oceanic environment involved the production of pelagic eggs, with the consequence that small eggs should be important in the deep sea. The data given by Crabtree and Sulak (1986) for deep-demersal teleosts is consistent with this view, for they observed that the predominant pattern among this group is one of high fecundity with small egg size. They could report only a few families characterized by low fecundity and large eggs.

In long-lived shelf fishes, such as the Gadidae, females become mature while they are still in their early fast-growth stage, and continue to spawn, perhaps annually, into old age. This does not appear to be the situation in at least some deep-demersal fishes. In *Bathysaurus ferox*, Sulak *et al.* (1985) found that the length frequency was bimodal and numerically dominated by large adults. Intermediate-sized juveniles were rare. The conclusion was that growth in *Bathysaurus* must be relatively rapid to full adult size, with selective predation occurring on larger juveniles and young adults. Crabtree and Sulak (1986) found that the size distribution of the alepocephalid *Conocara macropterum* was also bimodal. The large size achieved by both species was hypothesized as being advantageous in feeding success and in predator avoidance. Crabtree and Sulak (1986) suggest that the relative advantages gained by increased size might ultimately be offset by the energetic demands of reproduction on achieving maturity. Thus a well-defined maximum size for deep-sea fish species could result from a diversion of energy away from somatic growth and into gamete production.

Observations by Gordon *et al.* (1995) on slope-dwelling fishes in the Rockall Trough and the Porcupine Seabight tend to confirm that in many deep-sea species females become mature only after they reach adult size and when somatic growth has slowed or ceased. The implication is strong that a choice is imposed on deep-demersal fish in general; energy is available for either growth or reproduction, but not both.

As part of an attempt to understand life histories in a phylogenetic framework, Merrett (1989) developed an early life history model for macrourids. He found that persistent larval characters were absent and that adult features (i.e., dorsal, anal and pelvic fins, vertebral ossification, a functional swim bladder, and light organs) developed early on. Thus the early life history style of macrourids seems to typify a pattern of relatively direct development that, following the Balon (1980, 1984) scheme of saltatory ontogeny, classifies the youngest stages of these deep-sea fish as alevins rather than larvae.

In the Balon model a sequence of rapid changes in form and function alternates with prolonged intervals of slower development, during which complex structures are prepared for the next rapid change. These are absent in fishes that develop definitive adult organs throughout the larval developmental period. At most, such fishes may have an intermediate state with mixed feeding and some persistent temporary organs, vestiges from the larval form. This situation seems to typify the macrourids.

Although the duration of prejuvenile life cannot be determined as yet, embryonic steps in macrourids are accomplished with relatively little growth, because there are no large yolk reserves to support it. Yet permanent organs are developed directly, at the onset of exogenous feeding, facilitating intensive foraging and relatively rapid growth during the alevin period. The number of macrourid alevins found gorged with copepods is ample evidence of such foraging success.

Thus, and in contrast to the larval developmental pattern seen in Myctrophidae (the lanternfishes that dominate mesopelagic waters), evolutionary adaptation for dispersal in the early life history phase of Macrouridae has been minimized and is ecologically consistent with the adult demersal lifestyle. Most of the 350 or so species of macrourid are slope dwellers, living in ribbonlike distributions of varying bathymetric range around the oceanic rim. Were such species to produce larval stages that developed pelagically within the seasonal thermocline, currents would regularly sweep them away from any suitable adult habitat, an evolutionarily untenable situation.

REFERENCES

Alverson, D. L. (1993). The management challenge, In "The Newfoundland Groundfish Fisheries: Defining the Reality" (K. Storey, ed.), pp. 78–93. Conference Proceedings, Institute of Social and Economic Research, Memorial University, St. John's, Newfoundland.

Andriyashev, A. P. (1953). Ancient deep-water and secondary deep-water fishes and their importance in a zoogeographical analysis. *In* "Notes on Special Problems in Ichthyology," pp. 58–64. *Izd. Akad. Nauk SSSR, Ikhtiol. Kom. Moscow.* (English translation by A. R. Gosline. Issued by Bureau of Commercial Fisheries, U.S. National Museum, Washington, D.C. pp. 1–9.)

Angel, M. V. (1977). Windows into a sea of confusion: Sampling limitations to the measurement of ecological parameters in oceanic mid-water environments. *In* "Oceanic Sound Scattering Prediction" (N. R. Andersen and B. J. Zahuranec, eds.), pp. 217–248. Plenum, New York.

Angel, M. V. (1986). Vertical distribution: Study and implications. Pelagic biogeography. *UNESCO Tech. Papers Mar. Sci.* **49**, 3–8.

Angel, M. V. (1989). Vertical profiles of pelagic communities in the vicinity of the Azores Front and their implications to deep ocean ecology. *Prog. Oceanogr.* **22**, 1–46.

Angel, M. V. (1990). Life in the benthic boundary layer: Connections to the mid-water and sea floor. *Philos. Trans. R. Soc. London* **331A**, 15–28.

Armstrong, J. D., Bagley, P. M., and Priede, I. G. (1992). Photographic and acoustic tracking observations of the behaviour of the grenadier *Coryphaenoides* (*Nematonurus*) *armatus,* the eel *Synaphobranchus bathybius,* and other abyssal demersal fish in the North Atlantic Ocean. *Mar. Biol.* **112**, 535–544.

Astthorsson, O. S., and Pálsson, O. K. (1987). Predation on euphausiids by cod, *Gadus morhua,* in winter in Icelandic subarctic waters. *Mar. Biol.* **96**, 327–334.

Backus, R. H. (1986). Biogeographic boundaries in the open ocean. Pelagic biogeography. *UNESCO Tech. Papers Mar. Sci.* **49**, 9–13.

Backus, R. H., Craddock, J. E., Haedrich, R. L., and Robison, B. H. (1977). "Atlantic Mesopelagic Zoogeography, Fishes of the Western North Atlantic," Mem. 1, Part 7, pp. 266–287. Sears Foundation for Marine Research, New Haven, Connecticut.

Bagley, P. M., Priede, I. G., and Armstrong, J. D. (1990). An autonomous deep ocean vehicle for acoustic tracking of bottom living fishes. *In* "Monitoring the Sea," IEE Colloquium, Digest No. 182, Ref. E15/E11, pp. 211–213. Institute of Electrical Engineers, London.

Bailey, T. G., and Robison, B. H. (1986). Food availability as a selective factor on the chemical compositions of midwater fishes in the eastern North Pacific. *Mar. Biol.* **91**, 131–141.

Baird, R. C., and Hopkins, T. L. (1981). Trophodynamics of the fish *Valenciennellus tripunctulatus.* III. Energetics, resources, and feeding strategy. *Mar. Ecol. Prog. Ser.* **5**, 21–28.

Bakken, E., Lahn-Johannesson, J., and Gjøsaeter, J. (1975). Bunnfisk paa den norske kontinentalskraning. *Fiskets Gang* **61**, 557–565.

Balon, E. K. (1980). Early ontogeny of the lake charr, *Salvelinus* (*Cristovomer*) *namaycush. In* "Charrs: Salmonid Fishes of the Genus *Salvelinus*" (E. K. Balon, ed.), Junk, The Hague.

Balon, E. K. (1984). Reflections on some decisive events in the early life of fishes. *Trans. Am. Fish. Soc.* **113**, 172–185.

Banse, K., and Mosher, S. (1980). Adult body mass and annual production/biomass relationships of field populations. *Ecol. Monogr.* **50**, 355–379.

Bergstad, O. A. (1990). Ecology of the fishes of the Norwegian Deep: Distribution and species assemblages. *Neth. J Sea Res.* **25**, 237–266.

Bergstad, O. A. (1991a). Distribution and trophic ecology of some gadoid fish of the Norwegian Deep. 1. Accounts of individual species. *Sarsia* **75**, 269–313.

Bergstad, O. A. (1991b). Distribution and trophic ecology of some gadoid fish of the Norwegian Deep. 2. Food-web linkages and comparison of diets and distributions. *Sarsia* **75**, 315–325.

Bergstad, O. A. (1995). Age determination of deep-water fishes; experiences, status and challenges for the future. *In* "Deep Water Fisheries of the North Atlantic Oceanic Slope" (A. G. Hopper, ed.), pp. 267–283. Kluwer Academic Publishers, Dordrecht, The Netherlands.

Bergstad, O. A., and Isaksen, B. (1987). Deep-water resources of the Northeast Atlantic: Distribution, abundance and exploitation. *Fisken og Havet,* **3**, 1–56.

Billett, D. S. M., Lampitt, R. S., Rice, A. L., and Mantoura, R. F. L. (1983). Seasonal sedimentation of phytoplankton to the deep-sea benthos. *Nature* (*London*) **302**, 520–522.

Blaber, S. J. M., and Bulman, C. M. (1987). Diets of fishes of the upper continental slope of eastern Tasmania: Content, calorific values, dietary overlap and trophic relationships. *Mar. Biol.* **95**, 345–356.

Bulman, C. M., and Blaber, S. J. M. (1986). Feeding ecology of *Macruronus novaezelandiae* (Hector) (Teleostei: Merlucciidae) in south-eastern Australia. *Aust. J. Mar. Freshwater Res.* **37,** 621–639.

Calder III, W. A. (1985). Size and metabolism in natural systems. Ecosystem theory for biological oceanography. *Can. Bull. Fish. Aquat. Sci.* **213,** 65–75.

Campbell, R. A. (1983). Parasitism in the Deep Sea. *In* "Deep-sea Biology, The Sea" (G. T. Rowe, ed.), Vol. 8, Chapt. 12, pp. 473–552. Wiley (Interscience), New York.

Campbell, R. A. (1990). Deep water parasites. *Ann. Parasitol. Hum. Comp.* **65**(Suppl. 1), 65–68.

Campbell, R. A., Haedrich, R. L., and Munroe, T. A. (1980). Parasitism and ecological relationships among deep-sea fishes. *Mar. Biol.* **57,** 301–313.

Carney, R. S., Haedrich, R. L., and Rowe, G. T. (1983). Zonation of fauna in the deep sea. *In* "Deep-Sea Biology" (G. T. Rowe, ed.), Chapt. 7, pp. 371–398. Wiley (Interscience), New York.

Childress, J. J., and Nygaard, M. H. (1973). The chemical composition of midwater fishes as a function of depth of occurrence off southern California. *Deep-Sea Res.* **20,** 1093–1109.

Childress, J. J., and Thuesen, E. V. (1992). Metabolic potential of deep-sea animals: Regional and global scales. *In* "Deep-Sea Food Chains and the Global Carbon Cycle" (G. T. Rowe and V. Pariente, eds.), pp. 217–236. Kluwer Academic Publishers, Dordrecht, The Netherlands.

Childress, J. J., Taylor, S. M., Caillet, G. M., and Price, M. H. (1980). Patterns of growth, energy utilization and reproduction in some meso- and bathypelagic fishes off southern California. *Mar. Biol.* **61,** 27–40.

Clark, M. R., and Tracey, D. M. (1992). Trawl survey of orange roughy in southern New Zealand waters, June–July 1991. *N. Z. Fish. Tech. Rep.* **32,** 2.

Cohen, D. M. (1970). How many recent fishes are there? *Proc. Calif. Acad. Sci.* **38,** 341–346.

Compagno, L. J. V. (1990). Alternative life-history styles of cartilaginous fishes in time and space. *Environ. Biol. Fishes* **28,** 33–75.

Crabtree, R. E., and Sulak, K. J. (1986). A contribution to the life history and distribution of Atlantic species of the deep-sea fish genus *Conocara* (Alepocephalidae). *Deep-Sea Res.* **33,** 1183–1201.

Crabtree, R. E., Sulak, K. J., and Musick, J. A. (1985). Biology and distribution of species of *Polyacanthonotus* (Pisces: Notacanthiformes) in the western North Atlantic. *Bull. Mar. Sci.* **36,** 235–248.

Dalpadado, P., and Gjøsæter, J. (1988). Feeding ecology of the lanternfish *Benthosema pterotum* from the Indian Ocean. *Mar. Biol.* **99,** 555–567.

Desbruyères, D., Deming, J., Dinet, A., and Khripounoff, A. (1985). Réactions de l'ecosystème benthique profond aux perturbations: nouveaux résultats expérimentaux. *In* "Peuplements Profonds du Golfe de Gascogne" (L. Laubier and C. Manniot, eds.), pp. 121–142. IFREMER, Brest.

Dickson, M.-L. (1986). A comparative study of the pelagic food webs in two Newfoundland fjords using stable carbon and nitrogen isotope tracers. M. Sc. Thesis, Department of Biology, Memorial University of Newfoundland, St. John's.

Duarte, C. M., and Alcaraz, M. (1989). To produce many small or few large eggs: A size-dependent reproductive tactic of fish. *Oecologia* **80,** 401–404.

Ekman, S. (1953). "Zoogeography of the Sea." Sidgwick & Jackson, London.

Elgar, M. A. (1990). Evolutionary compromise between a few large and many small eggs: Comparative evidence in teleost fish. *Oikos* **59,** 283–287.

Eliassen, J.-E., and Jobling, M. (1985). Food of the roughhead grenadier, *Macrourus berglax*, Lacépède in North Norwegian waters. *J. Fish Biol.* **26,** 367–376.

Ellett, D. J. (1993). Transit times to the NE Atlantic of Labrador Sea water signals. International Council for the Exploration of the Seas; Council Meeting, C:25.

Feller, R. J., Zagursky, G., and Day, E. A. (1985). Deep-sea food web analysis using cross-reacting antisera. *Deep-Sea Res.* **32,** 485–497.

Fry, B. (1988) Food web structure on Georges Bank from stable C, N, and S isotopic compositions. *Limnol. Oceanogr.* **33,** 1182–1189.

Fry, G., Anderson, R. K., Entzeroth, L., Bird, J. L., and Parker, P. L. (1984). ^{13}C enrichment and oceanic food web structure in the northwestern Gulf of Mexico. *Contrib. Mar. Sci.* **27,** 49–63.

Gabriel, W. L. (1992). Persistence of demersal fish assemblages between Cape Hatteras and Nova Scotia, northwest Atlantic. *J. Northwest Atlantic Fish. Sci.* **14,** 29–46.

Gage, J. D., and Tyler, P. A. (1991). "Deep-Sea Biology: A Natural History of Organisms at the Deep-Sea Floor." Cambridge Univ. Press, Cambridge.

Gauldie, R. W. (1994). The morphological basis of fish age estimation methods based on the otolith of *Nemadactylus macropterus. Can. J. Fish. Aquat. Sci.* **51,** 2341–2362.

Gauldie, R. W., West, I. F., and Davies, N. M. (1989). K-selection characteristics of orange roughy (*Hoplostethus atlanticus*) stocks in New Zealand waters. *J. Appl. Ichthyol.* **5,** 127–140.

Gjøsæter, J., and Kawaguchi, K. (1980). A review of the world resources of mesopelagic fish. *FAO Fish. Tech. Papers* **193,** 1–151.

Golovan, G. A. (1978). Composition and distribution of the ichthyofauna of the continental slope off North-Western Africa. *Trudy Inst. Okeanol.* **111,** 195–258.

Gomes, M. C. (1993). "Predictions Under Uncertainty: Fish Assemblages and Food Webs on the Grand Bank of Newfoundland." ISER Books, St. John's, Newfoundland.

Gomes, M. C., and Haedrich, R. L. (1992). Predicting community dynamics from food web structure. *In* "Deep-Sea Food Chains and the Global Carbon Cycle" (G. T. Rowe and V. Pariente, eds.), Kluwer Academic Publishers, Dordrecht, The Netherlands.

Gomes, M. C., Haedrich, R. L., and Villagarcía, M. G. (1995). Spatial and temporal changes in the groundfish assemblages on the Northeast Newfoundland/Labrador Shelf, Northwest Atlantic, 1978–1991. *Fish. Oceanogr.* **4,** 85–101.

Gordon, J. D. M. (1979). Lifestyle and phenology in deep sea anacanthine teleosts. *Symp. Zool. Soc. London* **44,** 327–359.

Gordon, J. D. M., and Bergstad, O. A. (1992). Species composition of demersal fish in the Rockall Trough, north-eastern Atlantic, as determined by different trawls. *J. Mar. Biol. Assoc. U. K.* **72,** 213–230.

Gordon, J. D. M., and Mauchline, J. (1990). Depth-related trends in diet of a deep-sea bottom-living fish assemblage of the Rockall Trough. *In* "Trophic Relations in the Marine Environment" (M. Barnes and R. N. Gibson, eds.), Proceedings of the 24th European Marine Biological Symposium, pp. 439–452. Aberdeen Univ. Press, Aberdeen, Scotland.

Gordon, J. D. M., Merrett, N. R., and Haedrich, R. L. (1995). Environmental and biological aspects of slope-dwelling fishes. *In* "Deep Water Fisheries of the North Atlantic Oceanic Slope" (A. G. Hopper, ed.), pp. 1–30. Kluwer Academic Publishers, Dordrecht, The Netherlands.

Gorelova, T. A. (1983). A quantitative assessment of consumption of zooplankton by epipelagic lanternfishes (family Myctophidae) in the equatorial Pacific Ocean. *J. Ichthyol.* **23,** 106–113.

Graham, M. S., Haedrich, R. L., and Fletcher, G. L. (1985). Hematology of three deep-sea fishes: A reflection of low metabolic rates. *Comp. Biochem. Physiol.* **80A,** 79–84.

Gray, J. S. (1994). Is deep-sea diversity really so high—Species diversity of the Norwegian continental shelf. *Mar. Ecol. Prog. Ser.* **112,** 205–209.

Haedrich, R. L. (1985). The species number/area relationship in the deep sea. *Mar. Ecol. Prog. Ser.* **24,** 303–306.

Haedrich, R. L. (1986). Size spectra in mesopelagic fish assemblages. Pelagic biogeography. *UNESCO Tech. Papers Mar. Sci.* **49,** 107–111.

Haedrich, R. L. (1997). Biogeography of an exploited fish assemblage. In "Pelagic Biogeography ICoPB II. Proceedings of the Second International Conference" (A. C. Pierrot-Bults and S. van der Spoel, eds.). IOC/UNESCO, Paris.

Haedrich, R. L., and Krefft, G. (1978). Distribution of bottom fishes in the Denmark Strait and Irminger Sea. *Deep-Sea Res.* **25,** 705–720.

Haedrich, R. L., and Merrett, N. R. (1988). Summary atlas of deep-living demersal fishes in the North Atlantic Basin. *J. Nat. Hist.* **22,** 1325–1362.

Haedrich, R. L., and Merrett, N. R. (1990). Little evidence for faunal zonation or communities in the deep demersal fish fauna. *Prog. Oceanogr.* **24,** 239–250.

Haedrich, R. L., and Merrett, N. R. (1992). Production/biomass ratios, size frequencies, and biomass spectra in deep-sea demersal fishes. In "Deep-Sea Food Chains and the Global Carbon Cycle" (G. T. Rowe and V. Pariente, eds.), pp. 157–182. Kluwer Academic Publishers, Dordrecht, The Netherlands.

Heemstra, P. C., and Smith, M. M. (1980). Hexatrygonidae, a new family of stingrays (Myliobatiformes, Batoidei) from South Africa, with comments on the classification of batoid fishes. *Ichthyological Bulletin of the J. L. B. Smith Institute of Ichthyology* **43.**

Hopkins, T. L., and Baird, R. C. (1985a). Aspects of the trophic ecology of the mesopelagic fish *Lampanyctus alatus* (family Myctophidae) in the eastern Gulf of Mexico. *Biol. Oceanogr.* **3,** 285–313.

Hopkins, T. L., and Baird, R. C. (1985b). Feeding ecology of four hatchetfishes (Sternoptychidae) in the eastern Gulf of Mexico. *Bull. Mar. Sci.* **36,** 260–277.

Hopkins, T. L., and Torres, J. J. (1989). Midwater food web in the vicinity of a marginal ice zone in the western Weddell Sea. *Deep-Sea Res.* **36,** 543–560.

Hopper, A. G., ed. (1995). "Deep Water Fisheries of the North Atlantic Oceanic Slope." Kluwer Academic Publishers, Dordrecht, The Netherlands.

Houston, K. A., and Haedrich, R. L. (1986). Food habits and intestinal parasites of deep demersal fishes from the upper continental slope east of Newfoundland, northwest Atlantic Ocean. *Mar. Biol.* **92,** 563–574.

Kinzer, J., and Schulz, K. (1985). Vertical distribution and feeding patterns of midwater fish in the central equatorial Atlantic I. Myctophidae. *Mar. Biol.* **85,** 313–322.

Kinzer, J., and Schulz, K. (1988). Vertical distribution and feeding patterns of midwater fish in the central equatorial Atlantic II. Sternoptychidae. *Mar. Biol.* **99,** 261–269.

Koblentz-Mishke, O. J., Volkovinsky, V. V., and Kabanova, J. G. (1970). "Scientific Exploration of the South Pacific" (W. S. Wooster, ed.), pp. 183–193. National Academy of Science, Washington, D.C.

Koslow, J. A. (1993). Community structure in North Atlantic deep-sea fishes. *Prog. Oceanogr.* **31,** 321–328.

Koslow, J. A., Bulman, C. M., and Lyle, J. M. (1994). The mid-slope demersal fish community off southeastern Australia. *Deep-Sea Res. 1,* **41**(1), 113–141.

Lampitt, R. S., and Burnham, M. P. (1983). A free-fall time-lapse camera and current meter system 'Bathysnap' with notes on the foraging behaviour of a bathyal decapod shrimp, *Deep-Sea Res.* **30A,** 1009–1017.

Loeb, V. J. (1986). Importance of vertical distribution studies in biogeographic understanding: Eastern Tropical Pacific vs. North Pacific Central Gyre ichthyoplankton assemblages. *UNESCO Tech. Papers Mar. Sci.* **49,** 177–181.

Longhurst, A. R., ed. (1981). "Analysis of Marine Ecosystems." Academic Press, New York.

Longhurst, A. R., Bedo, A. W., Harrison, W. G., Head, E. J. H., and Sameoto, D. D. (1990). Vertical flux of respiratory carbon by oceanic diel migrant biota. *Deep-Sea Res.* **37,** 685–694.

McGowan, J. A. (1974). The nature of oceanic ecosystems. *In* "The Biology of the Oceanic Pacific" (C. B. Miller, ed.), pp. 7–28. Oregon State Univ. Press, Corvallis.

McGowan, J. A. (1986). The biogeography of pelagic ecosystems. Pelagic Biogeography. *UNESCO Tech. Papers Mar. Sci.* **49,** 191–200.

McGowan, J. A., and Walker, P. W. (1985). Dominance and diversity maintenance in an oceanic ecosystem. *Ecol. Monogr.* **55**(1), 103–118.

Macpherson, E. (1981). Resource partitioning in a Mediterranean demersal fish community. *Mar. Ecol. Prog. Ser.* **4,** 183–193.

Macpherson, E. (1985). Daily ration and feeding periodicity of some fishes off the coast of Namibia. *Mar. Ecol. Prog. Ser.* **26,** 253–260.

Macpherson, E. (1989). Influence of geographical distribution, body size and diet on population density of benthic fishes off Namibia (South West Africa). *Mar. Ecol. Prog. Ser.* **50,** 295–299.

Macpherson, E., and Duarte, C. M. (1991). Bathymetric trends in demersal fish size; is there a general relationship? *Mar. Ecol. Prog. Ser.* **71,** 103–112.

Macpherson, E., and Roel, B. A. (1987). Trophic relationships in the demersal fish community off Namibia. The Benguela and Comparable Ecosystems. *S. Afr. J Mar. Sci.* **5,** 585–596.

Mahaut, M.-L., Geistdoerfer, P., and Sibuet, M. (1990). Trophic strategies in carnivorous fishes: Their significance in energy transfer in the deep-sea benthic ecosystem (Meriadzek Terrace, Bay of Biscay). *Prog. Oceanogr.* **24,** 223–237.

Mann, K. H. (1984). Fish production in open ocean ecosystems. *In* "Flows of Energy and Materials in Marine Ecosystems" (M. J. R. Fasham, ed.), pp. 435–458. Plenum, New York.

Marshall, N. B. (1971). "Explorations in the Life of Fishes." Harvard Univ. Press, Cambridge.

Marshall, N. B. (1979). "Developments in Deep-sea Biology." Blandford, Poole, U.K.

Marshall, N. B., and Merrett, N. R. (1977). The existence of a benthopelagic fauna in the deep-sea. A Voyage of Discovery: George Deacon 70th anniversary volume. *Deep-Sea Res.* **24**(Suppl.), 483–497.

Massuti, E., Morales-Nin, B., and Stefanescu, C. (1995). Distribution and biology of 5 grenadier fish (Pisces, Macrouridae) from the upper and middle slope of the northwestern Mediterranean. *Deep-Sea Res. 1,* **42**(3), 307–318.

Mattson, S. (1981). The food of *Galeus melastomus, Gadiculus argenteus thori, Trisopterus esmarkii, Rhinonemus cimbrius,* and *Glyptocephalus cynoglossus* (Pisces) caught during the day with shrimp trawl in a West-Norwegian fjord. *Sarsia* **66,** 109–127.

Mauchline, J. (1990). Aspects of production in a marginal oceanic region, the Rockall Trough, northeastern Atlantic Ocean. *Aquat. Sci.* **2**(2), 167–183.

Mauchline J, and Gordon, J. D. M. (1985). Trophic diversity in deep-sea fish. *J. Fish Biol.* **26,** 527–535.

Mauchline, J., and Gordon, J. D. M. (1986). Foraging strategies of deep-sea fish. *Mar. Ecol. Prog. Ser.* **27,** 227–238.

May, J. L., and Blaber, S. J. M. (1989). Benthic and pelagic fish biomass of the upper continental slope off eastern Tasmania. *Mar. Biol.* **101,** 11–25.

Mead, G. W., Bertelsen, E., and Cohen, D. M. (1964). Reproduction among deep-sea fishes. *Deep-Sea Res.* **11,** 569–596.

Merrett, N. R. (1986). Biogeography and the oceanic rim: A poorly known zone of ichthyofaunal interaction. Pelagic biogeography. *UNESCO Tech. Papers Mar. Sci.* **49,** 201–209.

Merrett, N. R. (1987). A zone of faunal change in assemblages of abyssal demersal fish in the eastern North Atlantic: A response to seasonality in production? *Biol. Oceanogr.* **5,** 137–151.

Merrett, N. R. (1989). The elusive macrourid alevin and its seeming lack of potential in contributing to intrafamilial systematics. Papers on the systematics of gadiform fishes. *Los Angeles County Nat. Hist. Mus. Sci. Ser.* **32,** 175–185.

Merrett, N. R. (1994). Reproduction in the North Atlantic oceanic ichthyofauna and the relationship between fecundity and species' sizes. *Environ. Biol. Fishes* **41,** 207–245.

Merrett, N. R., Ehrich, S., Badcock, J., and Hulley, P. A. (1986). Preliminary observations on the near-bottom ichthyofauna of the Rockall Trough: A contemporaneous investigation using commercial-sized midwater and demersal trawls to 1000 m depth. *Proc. Roy. Soc. Edinburgh* **88B,** 312–314.

Merrett, N. R., Gordon, J. D. M., Stehmann, M., and Haedrich, R. L. (1991a). Deep demersal fish assemblage structure in the Porcupine Seabight (eastern North Atlantic): Slope sampling by three different trawls compared. *J. Mar. Biol. Assoc. U. K.* **71,** 329–358.

Merrett, N. R., Haedrich, R. L., Gordon, J. D. M., and Stehmann, M. (1991b). Deep demersal fish assemblage structure in the Porcupine Seabight (eastern North Atlantic): Results of single warp trawling at lower slope to abyssal soundings. *J. Mar. Biol. Assoc. U. K.* **71,** 359–373.

Morales-Nin, B. (1990). A first attempt at determining growth patterns of some Mediterranean deep-sea fishes. *Sci. Mar.* **54**(3), 241–248.

Newton, G., and Klaer, N. (1991). "Deep-Sea Demersal Fisheries Resources of the Great Australian Bight: A Multivessel Trawl Survey." Bureau of Rural Resources Bulletin No. 10, Australian Government Publishing Service, Canberra.

Noble, R. W., Kwaitkowski, L. D., De Young, A., Davis, B. J., Haedrich, R. L., Tam, L.-T., and Riggs, A. F. (1986). Functional properties of hemoglobins from deep-sea fish: Correlations with depth distribution and presence of a swimbladder. *Biochim. Biophys. Acta* **870,** 552–563.

Overholtz, W. J., and Tyler, A. V. (1985). Long-term responses of the demersal fish assemblages of Georges Bank. *Fish. Bull. U.S.* **83,** 507–520.

Pearcy, W. G., Stein, D. L., and Carney, R. S. (1982). The deep-sea benthic fish fauna of the northeastern Pacific Ocean on Cascadia and Tufts Abyssal Plains and adjoining continental slopes. *Biol. Oceanogr.* **1,** 375–428.

Peters, R. H. (1983). "Empirical Limitations of Body Size." Cambridge Univ. Press, New York.

Pierrot-Bults, A. C., van der Spoel, S., Zahuranec, B. J., and Johnson, R. K., eds. (1986). Pelagic biogeography. *UNESCO Tech. Papers Mar. Sci.* **49.**

Prasad, K. S., and Haedrich, R. L. (1993). Primary production estimates on the Grand Banks of Newfoundland, north-west Atlantic Ocean, derived from remotely-sensed chlorophyll. *Int. J. Remote Sensing* **17,** 3299–3304.

Prasad, K. S., and Haedrich, R. L. (1994). Satellite-derived primary production estimates from the Grand Banks: Comparison to other oceanic regimes. *Continental Shelf Res.* **14**(15), 1677–1687.

Priede, I. G., Smith, K. L., Jr., and Armstrong, J. D. (1990). Foraging behavior of abyssal grenadier fish: Inferences from acoustic tagging and tracking in the North Pacific Ocean. *Deep-Sea Res.* **37,** 81–101.

Reinhardt, S. B., and Van Vleet, E. S. (1986). Lipid composition of twenty-two species of Antarctic midwater zooplankton and fish. *Mar. Biol.* **91,** 149–159.

Rex, M. A., Stuart, C. T., Hessler, R. R., Allen, J. A., Sanders, H. L., and Wilson, G. D. F. (1993). Global-scale latitudinal patterns of species diversity in the deep-sea benthos. *Nature (London)* **365,** 636–639.

Rice, A. L., Billett, D. S. M., Fry, J., John, A. W. G., Lampitt, R. S., Mantoura, R. F. C., and Morris, R. J. (1986). Seasonal deposition of phytodetritus to the deep-sea floor. *Proc. R. Soc. Edinburgh B* **88,** 265–279.

Rice, A. L., Billett, D. S. M., Thurston, M. H., and Lampitt, R. S. (1991). The Institute of Oceanographic Sciences biology programme in the Porcupine Seabight: Background and general introduction. *J. Mar. Biol. Assoc. U. K.* **71,** 281–310.

Robison, B. H. (1984). Herbivory by the myctophid fish *Ceratoscopelus warmingii. Mar. Biol.* **84,** 119–123.

Roe, H. S. J., and Badcock, J. (1984). The diel migrations and distributions within a mesopelagic community in the North East Atlantic. 5. Vertical migrations and feeding of fish. *Prog. Oceanogr.* **13,** 389–429.

Ryther, J. H. (1969). Photosynthesis and fish production in the sea. *Science* **166,** 72–76.

Savvatimskii, P. I. (1987). Changes in species composition of trawl catches by depth on the continental slope from Baffin Island to Northeastern Newfoundland, 1970–85. *NAFO Sci. Council Studies* **11,** 43–52.

Sheldon, R. W., Sutcliffe, W. H., and Paranjape, M. A. (1977). Structure of pelagic food chains and relationship between plankton and fish production. *J. Fish. Res. Bd. Can.* **34,** 2344–2353.

Sherman, K. L. (1994). Sustainability, biomass yields, and health of coastal ecosystems: An ecological perspective. *Mar. Ecol. Prog. Ser.* **112,** 277–301.

Siebenaller, J., Somero, G. N., and Haedrich, R. L. (1982). Biochemical characteristics of macrourid fishes differing in their depths of distribution. *Biol. Bull.* **163,** 240–249.

Smith, K. L., Jr. (1978). Metabolism of the abyssopelagic rattail *Coryphaenoides armatus* measured *in situ. Nature (London)* **274,** 362–364.

Snelgrove, P. V. R., and Haedrich, R. L. (1985). Structure of the deep demersal fish fauna off Newfoundland. *Mar. Ecol. Prog. Ser.* **27,** 99–107.

Somero, G. N., Siebenaller, J. F., and Hochachka, P. W. (1983). Biochemical and physiological adaptations of deep-sea animals. *In* "Deep-Sea Biology, The Sea" (G. T. Rowe, ed.), Vol. 8, Chapt. 8, pp. 261–330. Wiley (Interscience), New York.

Stein, D. A., and Pearcy, W. G. (1982). Aspects of reproduction, early life history, and biology of macrourid fishes off Oregon, U.S.A. *Deep-Sea Res.* **29,** 1313–1329.

Stickney, D. G., and Torres, J. J. (1989). Proximate composition and energy content of mesopelagic fishes from the eastern Gulf of Mexico. *Mar. Biol.* **103,** 13–24.

Sulak, K. J., Wenner, C. A., Sedberry, G. R., and van Guelpen, L. (1985). The life history and systematics of deep-sea lizard fishes, genus *Bathysaurus* (Synodontidae). *Can. J. Zool.* **63,** 623–642.

Taylor, L. R., Jr., Compagno, L. J. V., and Struhsaker, P. J. (1983). Megamouth—A new species, genus, and family of lamnoid shark (*Megachasma pelagios*, Family Megachasmidae) from the Hawaiian Islands. *Proc. Calif. Acad. Sci.* **43,** 87–110.

Torres, J. J., and Somero, G. N. (1988). Metabolism, enzyme activities and cold adaptation in Antarctic mesopelagic fishes. *Mar. Biol.* **98,** 169–180.

Tyler, P. A. (1988). Seasonality in the deep sea. *Oceanogr. Mar. Biol. Annu. Rev.* **18,** 227–258.

Uiblein, F., Bordes, F., and Castillo, R. (1996). Diversity, abundance, and depth distribution in demersal deep-water fishes off Lanzarote and Fuerteventura, Canary Islands. *J. Fish Biol.* **49** (Supplement A), 75–90.

van der Spoel, S., and Pierrot-Bults, A. C., eds. (1979). "Zoogeography and Diversity of Plankton." Bunge, Utrecht.

Williams, P. M., Druffel, E. R. M., and Smith, K. L., Jr. (1987). Dietary carbon sources for deep-sea organisms as inferred from their organic radiocarbon activities. *Deep-Sea Res.* **34,** 253–266.

Willis, J. M., and Pearcy, W. G. (1982). Vertical distribution and migration of fishes of the lower mesopelagic zone off Oregon. *Mar. Biol.* **70,** 87–98.

Wilson, R. R., Jr. (1980). Analysis of growth zones and microstructure in otoliths of two macrourids from the North Pacific abyss. *Environ. Biol. Fishes* **21**(4), 251–261.

Wishner, K. F. (1980). The biomass of the deep-sea benthopelagic plankton. *Deep-Sea Res.* **27A,** 203–216.

Woodward, A. S. (1898). The antiquity of the deep-sea fish-fauna. *Nat. Sci.* **12,** 257–260.

Young, J. W., and Blaber, S. J. M. (1986). Feeding ecology of three species of midwater fishes associated with the continental slope of eastern Tasmania, Australia. *Mar. Biol.* **93,** 147–156.

4

FEEDING AT DEPTH

JOHN V. GARTNER, Jr., ROY E. CRABTREE, AND KENNETH J. SULAK

I. Introduction
  A. General Introduction to Feeding—A Brief Review of Current Knowledge
  B. Definition of Terms
II. Feeding Habits of Deep-Sea Fishes
  A. Direct Evidence
  B. Indirect Evidence
III. Patterns in the Diets of Deep-Sea Fishes
  A. Categories of Trophic Specialization
  B. Morphological and Behavioral Specializations among Trophic Guilds of Deep-Sea Fishes
  C. Congruent Patterns in Morphological Specialization among Benthic and Demersal Fish Species: Common Themes on the Shelf and in the Deep Sea
  D. Inferences from the Morphology of Deep-Sea Fishes: Trophic Strategies and Prey Selection
  E. Diel and Seasonal Feeding Patterns
IV. Sources of Food in the Deep Sea
  A. Marine Snow and Foodfalls
  B. Benthopelagic Interface
V. Deep-Sea Energetics Related to Feeding
  A. Chemical Composition Data
  B. Energetics
VI. Future Directions in Deep-Sea Fish Research
  References

I. INTRODUCTION

A. General Introduction to Feeding—A Brief Review of Current Knowledge

It is well documented that relatively little energy is available to deep–ocean macrofaunas, especially in waters underlying the great central ocean gyres. Biomass available as energy at depths exceeding 1000 m drops to less than 5% of that available in surface waters (<200 m, Marshall, 1980). An important question in deep-sea fish ecology is how enough energy is located and acquired in such apparently depauperate environments to meet the metabolic needs of individuals, as well as maintain species population size. Another meaningful question is how energy is transferred from the productive epipelagic zones of the ocean to the bathypelagic and deep benthic zones.

The physiology of feeding in deep-ocean fish remains problematic because of the nature of the environment and the difficulties inherent in attempting both *in situ* and laboratory observations. Digestion and gut evacuation rates in deep-ocean fishes have been addressed by few published papers, and those generally relied heavily on assumptions based on studies of shallow-water fishes. The literature available on feeding in deep-sea fishes is primarily focused on the nature of the diet (feeding habits), feeding chronology, and analyses of selectivity patterns, with relatively fewer contributions on chemical composition, feeding behaviors, and structural and physiological adaptations to feeding at depth.

Daily rations have been estimated for various mesopelagic fishes, but the methods of calculation are widely variable. Similarly, some studies have estimated potential trophic impacts of fishes on their prey populations, but these also vary in methods used and assumptions made. Few studies have attempted to apply the bioenergetics question (Q_c) to deep-sea fishes and are reliant on numerous assumptions and extrapolations to quantify most of the parameters (see Hopkins and Baird, 1977; Baird and Hopkins, 1981b).

Many of the published studies on feeding in deep-sea fishes have discussed pelagic species (see Section I,B). The difficulties and costs of sampling bathypelagic and deep benthic/demersal (see Section I,B) habitats have resulted in considerably fewer published reports on the dominant species of these zones.

In this chapter, we review the current state of knowledge regarding feeding in deep-ocean fishes. Deep-sea pelagic fish species are also discussed because there is evidence to suggest that pelagic fishes are responsible for a significant transfer of energy to the deep benthos (Marshall and Merrett, 1977; Robison and Bailey, 1981). Furthermore, some pelagic species may spend considerable amounts of time near the bottom.

B. Definition of Terms

Because the adaptations related to feeding that are evident in feeding structures, behavior, and physiology differ in the pelagial and benthic envi-

4. FEEDING AT DEPTH 117

ronments, we define here the habitat-related terms that we will employ throughout this chapter.

Some terms, particularly *benthopelagic* and *demersal*, have been undergoing transformations in their current use to include both pelagic and benthic forms. The definitions that we present below reflect the terms as used in the existing literature we have reviewed for this chapter.

1. BENTHIC

As used throughout this chapter, the term *benthic* applies to fish species that are in physical contact with the bottom and are not very mobile. Examples of such fishes are members of the Bathysauridae, Bathypteroidae, and Zoarcidae.

2. DEMERSAL/BENTHOPELAGIC

Marshall (1980) used the term *benthopelagic* as a synonym for the more widely used term *demersal*. We prefer to retain the use of the word demersal for fishes that spend most of their lives near (<5 m) the bottom and that move actively over the bottom. Demersal fishes are morphologically quite dissimilar to the pelagic forms that spend only part of their life cycle near the bottom, for which we retain the word benthopelagic. Examples of demersal fish families are the Macrouridae, Synaphobranchidae, Halosauridae, and Ophidiidae.

3. PELAGIC

For pelagic species, we use the terms *mesopelagic* (species residing primarily between 200 and 1000 m) and *bathypelagic* (species residing primarily below 1000 m) *sensu* Marshall (1971). The common term *midwater* is often used as a collective synonym for both groups. Some of the most important mesopelagic fish families are the Myctophidae, Stomiidae, Gonostomatidae, and Sternoptychidae, whereas representative bathypelagic families include many of the ceratioid anglerfish families and the "gulper eel" families Eurypharyngidae and Saccopharyngidae (see Chapter 2, this volume, for taxonomy).

There are two other terms applicable to midwater fishes. One is *benthopelagic*, which includes pelagic species that spend part of their life cycle near the bottom (<10 m). Many of the mesopelagic fish families mentioned previously have benthopelagic members (Marshall and Merrett 1977).

Another term, *pseudoceanic* (e.g., Hulley and Lutjeharms, 1989), is applied to mesopelagic fish species consistently found associated with submerged land features such as islands or continental shelf edges. A number of lanternfish species (Hulley and Lutjeharms, 1989; Reid *et al.*, 1991) and sternoptychids of the genus *Polyipnus* are pseudoceanic.

II. FEEDING HABITS OF DEEP-SEA FISHES

A. Direct Evidence

1. BENTHIC AND DEMERSAL SPECIES

   a. Stomach Contents and Diet. Direct evidence concerning the feeding habits of deep-sea fishes comes principally from the analysis of gut contents. This approach is limited by several factors. Often, only small sample sizes have been available for study; consequently, the diets of abyssal and less common species have been characterized based on the examination of relatively few guts. Compounding the problem of small sample sizes, everted swim bladders are common among some macrourids, morids, and other deep-sea fishes brought up from great depths, and few of the specimens examined contain prey (Sedberry and Musick, 1978; Mauchline and Gordon, 1984a). Additionally, many large predators feed only infrequently, and are often found with empty stomachs. In most cases, little or no information has been available on prey availability in the deep sea, so few conclusions are possible regarding feeding selectivity. In addition, experimental studies have not been possible in the deep sea, so we know little regarding the effects of competition and predation on the foraging habits of deep-sea fishes.

   The absence of data on prey availability and competitive interactions makes it difficult or impossible to evaluate the extent to which deep-sea fishes may be regarded as "generalized" or "specialized" predators. Since publication of the classic papers by Sanders (1968) and Dayton and Hessler (1972) there has been considerable discussion of the role of deep-sea fishes in the maintenance of the diversity of deep-sea communities and the extent to which deep-sea fishes are selective or nonselective predators. However, beyond a characterization of a species' diet as narrow or diverse, it is difficult to assess the degree of selection exercised because we have no data on the type of acceptable prey available to a predator at any given time. Thus, a flexible opportunistic species may appear to have a specialized diet as it feeds opportunistically on an abundant prey species. Under different circumstances, however, the same species' diet may be quite different.

   Existing data on the feeding habits of deep-sea fishes are subject to several biases. Opportunistic feeding on prey items by fishes after they have been captured in a net ("net feeding") is often suspected, particularly at abyssal depths where nets are often towed for up to 3 hr before recovery. The feeding habits of many demersal fishes are based on their capture or attraction to baited traps and long lines (Table I). Taxonomic difficulties identifying various groups of invertebrates are often encountered in the

Table I
Demersal Deep-Sea Species Routinely Captured on Baited Long Lines and in Baited Traps, or Photographed at Baited Cameras and Free Vehicles

Family/species	Method[a]	Reference
Macrouridae		
Coryphaenoides armatus	LL	Forster (1968, 1973)
	BC	Jannasch and Wirsen (1977)
	BC	Jannasch (1978)
Coryphaenoides yaquinae	FV	Smith *et al.* (1979)
	FV	Priede and Smith (1986)
	FV	Priede *et al.* (1990)
Coryphaenoides spp.[b]	FV	Wilson and Smith (1984)
Moridae		
Antimora rostrata	LL	Forster (1968, 1973)
Mora moro	LL	Forster (1964, 1968, 1973)
Ophidiidae		
Spectrunculus grandis	LL	Forster (1968)
	BC	Jannasch and Wirsen (1977)
	BC	Jannasch (1978)
Synaphobranchidae		
Synaphobranchus kaupii	LL	Forster (1964, 1973)
Simenchelys parasitica	TR	Solomon-Raju and Rosenblatt (1971)
Chimaeridae		
Hydrolagus affinis	LL	Forster (1964, 1968, 1973)
	LL	Clarke and Merrett (1972)
Squalidae		
Centrophorus squamosus	LL	Forster (1964, 1968, 1971)
Centroscymnus coelolepis	LL	Forster (1964, 1968, 1973)
	LL	Clarke and Merrett (1972)
Deania calceus	LL	Forster (1964, 1968, 1973)
Etmopterus princeps	LL	Forster (1968, 1971, 1973)
Rajidae		
Bathyraja richardsoni	LL	Forster (1968)
	BC	Jannasch (1978)
Myxinidae		
Myxine glutinosa	BC	Isaacs and Schwartzlose (1975)

[a] LL, Baited long line; BC, baited camera; FV, free vehicle; TR, baited trap.
[b] Probably includes both *C. armatus* and *C. yaquinae*.

deep sea. Furthermore, the degree and direction of the expertise of those identifying gut contents can be a form of taxonomic bias that exaggerates the importance of some groups in the diet while underestimating that of others (Mauchline and Gordon, 1985).

A more general problem in all feeding habits studies, in both the deep sea and shallow waters, is that of the differential rates at which various types of prey are digested (Gerking, 1994). In many cases, soft-bodied and gelatinous organisms may be quickly digested and rendered unrecognizable, whereas hard parts such as squid beaks and teleost eye lenses can be quite durable. This has probably resulted in an underestimate of the importance of gelatinous prey in many species' diets.

b. Foraging Modes. Deep-sea fishes feed extensively on demersal prey whose distributions are closely associated with the bottom; however, many demersal fishes feed principally on vertically migrating mesopelagic organisms such as myctophids and cephalopods. The presence of pelagic prey in the diets of demersal fishes has been interpreted both as evidence of the occurrence of mesopelagic prey near the bottom (Sedberry and Musick, 1978) and of off-bottom migrations by some demersal species into the mesopelagic realm to feed (Haedrich, 1974; Haedrich and Henderson, 1974). An alternative explanation suggested by Merrett and Domanski (1985) is that dead mesopelagic prey are scavenged after sinking to the bottom.

The presence of pelagic prey in the diet of species typically regarded as demersal has caused considerable discussion regarding vertical migrations of prey and predator. Vertical movements by both prey and predator have been implicated as important mechanisms of transporting organic matter from near surface waters to slope and abyssal depths. Sedberry and Musick (1978) concluded that mesopelagic prey are important to the diet of many demersal fishes, including the abyssal macrourid, *Coryphaenoides (Nematonurus) armatus*. They suggested that vertical excursions off the bottom by predators as well as the impingement of the mesopelagic fauna on the bottom along the continental slope create opportunities for predation on pelagic prey by demersal predators. Haedrich and Henderson (1974) also reported evidence of pelagic feeding by *C. armatus* and suggested that feeding occurs off bottom. Pearcy and Ambler (1974) found pelagic prey in abyssal macrourids and suggested that scavenging along with vertical migrations off the bottom may occur. Direct evidence of off-bottom excursions by demersal deep-sea fishes was reported by Haedrich (1974), who captured 49 specimens of the demersal macrourid, *Coryphaenoides rupestris*, in midwater trawls fished from 270 to 1440 m above the bottom. Haedrich's collections demonstrated the potential of benthopelagic species to travel considerable distances above the bottom.

In contrast, Pereyra *et al.* (1969) and Marshall and Merrett (1977) suggested that demersal foraging on pelagic items reflects the abundance of pelagic taxa near the bottom over continental shelf regions and seamounts.

Hopkins *et al.* (1981) suggested that the rapid landward disappearance of oceanic micronekton (swimming organisms whose adult size generally ranges from 1 to 30 cm) in the Gulf of Mexico was the result of vertical migrations bringing these taxa into contact with demersal predators.

2. PELAGIC SPECIES

Among the pelagic fishes, the diets of mesopelagic species, especially of the Myctophidae, have been the most intensively studied. Sufficient information is available, however, on enough representatives of other midwater fish families to indicate that there are three major dietary guilds applicable to meso- and bathypelagic fishes: zooplanktivores, micronektonivores (includes piscivores and cephalopod mollusk predators), and generalists. We define "generalists" as fishes whose diets include significant components (>10% frequency in stomachs) of a broad array of unrelated taxa (e.g., crustaceans, gelatinous organisms, and fishes).

The zooplanktivores can be subdivided into several subguilds. Crustacean zooplanktivores constitute the majority of deep-sea pelagic fish species and families examined. Less common are predators that primarily ingest soft-bodied or gelatinous zooplankton, gastropod mollusks, and polychaete worms. These categories of predators are generally represented by a few individual species within different families.

Among the crustacean zooplanktivores, the primary prey types are calanoid copepods, followed by ostracods, euphausiids, and decapod crustaceans. In many species, the incorporation of euphausiids into the diet represents an ontogenetic shift (Gjösaeter, 1973; Hopkins and Baird, 1973, 1985a; Kinzer, 1977; Gorelova, 1981; Clarke, 1982; Hopkins and Gartner, 1992).

Most midwater fishes, particularly mesoopelagic species, feed primarily on copepods, which is correlated with the abundance of copepods in oceanic waters. Although in early research this was interpreted as evidence of nonselective feeding, in many cases feeding selectivity has been demonstrated for a variety of fishes, including selection of copepod prey (Hopkins and Gartner, 1992) (see Section II,B). Among prey groups, ostracods in particular seem to be preyed on selectively because the abundance of ostracods as prey items is disproportionate to their natural abundance (Merrett and Roe, 1974; Hopkins and Baird, 1985a).

Fishes that feed mainly on soft-bodied prey, especially cnidarian medusae and members of the subphylum Urochordata (salps and larvaceans) are not well represented among midwater groups. This is surprising because of the high abundance of gelatinous plankton in oceanic environments (e.g., Wiebe *et al.*, 1979). However, gelatinous prey are major prey items in the Bathylagidae (Cailliet, 1972; Mauchline and Gordon, 1983b; Gorelova and

Kobylyanskiy, 1985; Balanov *et al.*, 1995; Hopkins *et al.*, 1997), *Opisthoproctus* (family Opisthoproctidae) (Cohen, 1964; Mauchline and Gordon, 1983b), and some melamphaids (Gartner and Musick, 1989; Hopkins *et al.*, 1997). Some authors suggest that gelatinous organisms lack much nutritional value because of the binding of various lipid, carbohydrate, and protein components into indigestible forms (e.g., Madin *et al.*, 1981; Gorelova and Kobylyanskiy, 1985), but this may be dependent on the type of organisms ingested. Salps have a digestible stomach that may constitute 20% of the body weight, and are often found in enormous numbers, particularly over continental slope regions (Wiebe *et al.*, 1979; Kashkina, 1986). Gelatinous organisms probably go unrecognized quite often, particularly if well digested, and their contribution is thus underestimated in diet analyses. Mauchline and Gordon (1984a) suggested this very possibility when reporting "unidentifiable soft tissue" from a large number of stomachs of *Scopelogadus beanii*, a species later reported to feed mainly on gelatinous plankton (Gartner and Musick, 1989).

One other pattern of feeding on soft-bodied plankton has been reported. An extremely selective feeding habit is observed in the myctophid genus *Centrobranchus*, which feeds solely on gastropod mollusks (mainly pteropods and some heteropods) (Gorelova, 1977; Hopkins and Gartner, 1992).

Two groups of nektonic organisms, fishes and cephalopods, serve as dominant prey items for various pelagic predators. Piscivory, or predation mainly on fishes, is the common form of predation among larger bodied meso- and bathypelagic species (Table II). For many piscivores, myctophids seem to be a predominant prey item (Borodulina, 1972), but in an extensive review of piscivorous mesopelagic fishes, Hopkins *et al.* (1997) found that six predominantly or exclusively piscivorous mesopelagic fish families fed on prey items from five different families (Table II). Only one of the prey taxa were shared prey items in two of the piscivorous groups, suggesting a high degree of selectivity among piscivores.

Reports of diets composed mainly of cephalopods are rare among meso- or bathypelagic fish species, probably owing to the relatively larger sizes and faster locomotory speeds attained by many deep-sea squid species (Roper *et al.*, 1984). Hopkins *et al.* (1997) noted cephalopods as the principal diet component in three species from three different families of mesopelagic fishes (Table II).

Surprisingly, despite the prevailing concept that fishes in energy-poor deep-ocean waters should be opportunistic predators with a broad array of prey (see Ebeling and Cailliet, 1974), true generalists that eat a wide variety of unrelated taxa are rare. Some exceptions are the mesopelagic lanternfish, *Ceratoscopelus warmingii*, many of the stomiid species of the genus *Astronesthes*, and *Echiostoma barbatum* (Sutton and Hopkins, 1996) and the bathypelagic eurypharyngid eel *Eurypharynx pelecanoides* (Böhlke, 1966; J. V. Gartner, unpublished data). The myctophid *C. warmingii* (Gorelova, 1978; Kinzer and Schulz, 1985; Duka, 1987; Hopkins and Gartner, 1992; Hopkins *et al.*, 1997) appears not only to be a generalist, but a true omnivore. Robison (1984) reported significant amounts of diatoms in the diet of North Pacific *C. warmingii*.

B. Indirect Evidence

1. BENTHIC AND DEMERSAL SPECIES

Indirect evidence of food habits is sometimes provided by parasites, sediment, rocks, and other items found in fish stomachs, even in the absence of food. Fishes are often the secondary or definitive hosts (the organism in or on which the parasite reaches maturity) for parasites with complex life cycles. The incidence of particular parasite taxa for which the intermediate fish or invertebrate hosts are known may reveal prey specificity among demersal fish species. Thus, Mauchline and Gordon (1984b) found that two macrourids, *Coryphaenoides brevibarbis* and *Coelorinchus coelorinchus*, known to feed extensively on mysids, were also the most heavily infested with nematodes among ~40 species of deep-living bottomfishes examined from the Rockall Trough region. Mysids may be intermediate hosts of nematodes. Thus, it might reasonably be hypothesized that other fishes found to be heavily infested with nematodes also feed on mysids. Similarly, based on intense acanthocephalan infections in *Dicrolene intronigra*, Campbell *et al.* (1980) hypothesized that this ophidiid fish feeds extensively on amphipods, the known acanthocephalan intermediate hosts. Crabtree *et al.* (1991) subsequently provided support for this hypothesis, determining that amphipods were one of several crustacean taxa comprising the stomach contents of *D. intronigra*. Campbell *et al.* (1980) further hypothesized that the large abyssal skate *Bathyraja richardsoni* feeds on the common abyssal macrourid *Coryphaenoides* (*Nematonurus*) *armatus*. This skate is the definitive host for the trypanorhynch cestode (tapeworm) *Grillotia rowei*, a juvenile stage (pleurocercus) of which occur in macrourids, particularly in *C. armatus*. Comparative study of the parasite faunas of sympatric demersal fishes provides a means of assessing the degree of trophic generalization versus specialization within species and overlay among species, a topic also explored in Campbell *et al.* (1980).

The absence of parasites can also prove instructive in deciphering trophic behavior. Various studies have found that midwater fishes in the deep sea have very low incidence of platyhelminth (flatworm) and nematode (roundworm) parasites (Noble and Collard, 1970; Campbell *et al.*, 1980; Mauchline and Gordon, 1984c; Gartner and Zwerner, 1989). Accordingly, a diet consisting primarily of midwater fishes could leave a demersal piscivore relatively free of parasites as well. Thus, low parasite load may be indicative of a predominantly pelagic diet.

Sediment is a common constituent of stomach contents in certain taxa (Bright, 1970; Ribbink, 1971; Sedberry and Musick, 1978; Merrett and Marshall, 1980; Mauchline and Gordon, 1984d, Crabtree and Sulak, 1986). The regular presence of substantial amounts of sediment can be interpreted in at least two ways. The first explanation is that the fish feeds on buried benthic prey, ingesting sediment along with infauna. The alternative explanation is that the fish feeds on prey that themselves contain sediment at the time of ingestion (e.g., deposit-feeding brittle stars, polychaetes, or holothurians) (Smith *et al.*, 1979). As with interpretation of parasite load, the absence of sediment can also be instructive. Thus, the absence of sediment in the stomach contents of notacanths indicates feeding behavior that differs substantially from that of their benthivorous (predation on animals living within the bottom sediments) halosaur relatives. Halosaurs ingest sediment while feeding on infaunal prey (Sedberry and Musick, 1978; Crabtree *et al.*, 1991), whereas notacanths selectively crop epifauna (animals living on the surface of the sediments) from the sediment surface (Crabtree *et al.*, 1985). The absence of sediment from the stomachs of *Nezumia* and *Coelorinchus* in the Rockall Trough (Mauchline and Gordon, 1984b) contradicts the inference from head morphology that macrourids with projecting snouts are specifically adapted to grub in the sediment for prey (McLellan, 1977).

Despite the foregoing example, morphology can often provide another indirect indication of food habits. Thus, jaws equipped with numerous long, sharp, depressible teeth coordinate with ambush predation on fishes and other mobile nekton. Gill arches with a low number of short rakers often coordinate with piscivory, and numerous, long, closely set gill rakers coordinate with retention of small prey. A simple stomach, absence of pyloric caecae (accessory digestive pouches), and short, straight gut indicate predation on large prey ingested whole. However, fishes often display surprising plasticity in the use of their feeding apparatus. They may also display facultative feeding or prey switching to different prey types when their primary prey is scarce, or when secondary prey types are particularly abundant. Adaptive "specialization" deduced from morphology is best evaluated in the light of direct evidence from food habits analysis.

A final indirect method for food habits analysis utilizes an antigen–antibody reaction (Feller, 1979, 1985). This antisera methodology is particularly useful when prey is reduced to an unrecognizable mass during mastication, or is too rapidly digested to be identified. Even in conventional food habits analyses, considerable stomach content material is typically unidentifiable. The antisera method involves preparation of whole-organism extracts of potential prey species. Extracts are injected into small mammals, leading to production of antibodies specific to individual prey species. Prey species present in gut contents are identified in analyses consisting of antigen–antibody reactions. The method holds promise for defining trophic pathways and food webs for species that defy conventional stomach content analysis. However, a limitation of the method is that serological identification is nonquantitative, resulting in a list of prey taxa consumed, but without a measure of relative importance.

Baited camera arrays are another source of information on the feeding habits of deep-sea fishes. The importance of scavenging in the deep sea has been inferred from the presence of shallow-water prey in the guts of abyssal forms and from the quick response of abyssal species to the presence of baited cameras at abyssal depths. Clarke and Merrett (1972) reported cetacean remains from the stomachs of the shark, *Centroscymnus coelolepsis*—clear evidence of scavenging. Other researchers have suggested scavenging in a variety of species based on apparently scavenged prey in the guts of demersal fishes (Pearcy and Ambler, 1974; Merrett and Domanski, 1985).

Abyssal macrourids and other species are attracted by baited camera arrays and have been reported to arrive at baits within 10 min of the arrival of baits at the bottom (Wilson and Smith, 1984; Armstrong *et al.*, 1992) (Fig. 1). Mahaut *et al.* (1990) found that most species known to be present in the Bay of Biscay from trawl and visual surveys are not seen at baits; only sharks, chimaeras, macrourids, morids, and synaphobranchids are attracted to baits. Patterns of arrival, times of first arrival at baits, and numbers of animals at baits have been used to estimate abundance and distance of attraction for several species of deep-sea fishes. Models used by Sainte-Marie and Hargrave (1987) suggest that scavengers are not abundant in the deep sea and that the distances from which scavengers are attracted to baits are greater in the deep sea than in shallow waters. Wilson and Smith (1984) suggested that abyssal macrourids use olfaction to locate baits and usually arrive from a down-current location. They proposed that a "wait" rather than a "search" strategy is used by deep-sea fishes to locate baits. Sainte-Marie and Hargrave (1987) reached a similar conclusion and suggested that tidal currents are important in transporting odors.

Recent studies have deployed ultrasonic transmitters in conjunction with baited camera and hydrophone arrays. Transmitters designed to be swallowed by fish were deployed on the bottom, and those ingested were then tracked for various periods of time. Priede *et al.* (1990) placed baited arrays at abyssal depths in the Pacific and tracked two macrourid species, *Coryphaenoides yaquinae* and *Coryphaenoides* (*Nematonurus*) *armatus*. They concluded that both species are active foragers and not sit-and-wait predators. Residence times of fish at baits correspond with optimal foraging

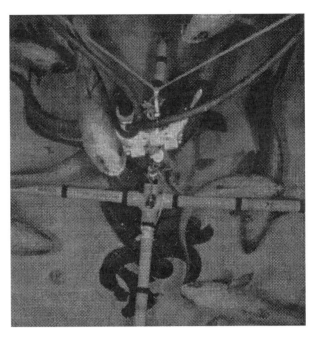

Fig. 1. Demersal macrourids, *Coryphaenoides* (*Nematonurus*) *armatus*, attracted to bait in the North Atlantic Ocean. Black marks represent 20-cm intervals. From Armstrong *et al.* (1992), by permission of Springer-Verlag.

theory; at deeper stations with lower food availability, fish remain near baits longer than at shallower stations. Armstrong *et al.* (1992) placed baited arrays at two locations at depths of 4800 and 4900 m in the North Atlantic. Four species were seen at the arrays: *C. armatus, Spectrunculus grandis, Synaphobranchus bathybius,* and *Barathrites* species. *Coryphaenoides armatus* dominated activities around the baits and was the only species observed to eat a transmitter. Tagged fish dispersed rapidly from the baits, suggesting active swimming; not a sit-and-wait foraging strategy. Over 60% of the fish tagged moved to altitudes of over 15 m above bottom at some point during the tracking period. Priede *et al.* (1994a) attracted four species to baits: *C. armatus, Synaphobranchus kaupi, Antimora rostrata,* and *Centroscymnus coelolepis.* They concluded that though several species will scavenge, the fraction of the diet derived from scavenging is probably small.

2. Pelagic Species

Observations made from submersibles of "feeding activities" of meso- or bathypelagic fish species are mainly anecdotal behavioral notes, because often the activity presumed to be feeding involves "prey" that are too small to be seen clearly, and the fishes are usually of small size (<300 mm) (J. V. Gartner and K. J. Sulak, personal observations). Adding to the difficulties of *in situ* work is the question of behavioral alterations (positive or negative tropisms) caused by brightly lit, generally noisy submersibles or by the artifically large food concentrations in baited trap arrays.

Observations from submersibles suggest that vertically migrating zooplanktivores, such as myctophids and sternoptychids, are more active at depth, even during nonmigration periods, than are nonmigratory zooplanktivores, such as the gonostomatid genus *Cyclothone,* and certain piscivorous stomiids such as *Chauliodus* and *Stomias* (Jannsen *et al.,* 1986; J. V. Gartner and K. J Sulak, personal observations). This suggests that migratory piscivores generally make more restricted diel vertical migrations than do the zooplanktivores (Sutton and Hopkins, 1997). In addition, nonmigratory species may conserve energy to some degree by adopting an ambush predation strategy (Borodulina, 1972; DeWitt and Cailliet, 1972).

As with benthic and demersal species, parasitological examinations of pelagic species may prove useful in elucidating predator–prey relationships. Although pelagic species in general have lower incidences of infection than do their benthic and demersal counterparts (Noble and Collard, 1970; Noble, 1973; Campbell, 1983; Gartner and Zwerner, 1989), some pelagic eel species (*Nessorhamphus ingolfianus, Nemichthys scolopaceus,* and *Eurypharynx pelecanoides*) are second intermediate and definitive hosts for digenetic trematode, nematode, and cestode parasites (Campbell and Gartner, 1982; Gartner and Zwerner, 1989). Gartner and Zwerner (1989) suggested links between prey types and parasite incidences for several genera of parasitic nematode and cestode life history stages.

Long-line gear and baited traps have not been useful tools for examination of most midwater fish taxa. Most midwater fishes are not large enough to be taken on longlines. Furthermore, most midwater fishes have been shown to be "swallowers," i.e., the food is ingested intact without much chewing—the jaws of these fish are adapted for pelagic feeding. It is unlikely that these fishes would use baited traps that are usually set on or near the bottom, and none have ever been observed to do so.

Behavioral data based on laboratory maintenance and observations are few because deep-sea pelagic fishes are poor candidates for confinement in aquaria (Robison, 1973). One mesopelagic zoarcid, *Melanostigma pammelas,* has been maintained with success for extended periods and its feeding habits have been observed by Belman and Anderson (1979). They reported from anatomical studies and examination of stomach contents that *M. pammelas* has a small mouth and is probably best adapted for feeding on small crustacean zooplankton, which it primarily locates visually. These suppositions were borne out by their laboratory observations.

The feeding behavior of live *Anoplogaster cornuta,* a cosmopolitan lower mesopelagic/bathypelagic species, in shipboard aquaria has also been described (Childress and Meek, 1973). Unlike *Melanostigma,* tactile and chemical stimuli, when applied to the head, appeared to elicit the primary feeding responses, but produced a flight response when applied to posterior regions. Childress and Meek (1973) concluded that such responses may augment visual stimuli for acquiring food in the deeper, darker zones of the mesopelagial.

Tchernavin (1953) and Pietsch (1978) used anatomical and physiological modeling to study the feeding behaviors of mesopelagic piscivore *Chauliodus sloani* and the unusual mesopelagic copepod predator *Stylephorus chordatus,* respectively. Their examinations of jaw and skull articulations coupled with the probable physiology of jaw manipulation and visual fields (in *Stylephorus*) led to accurate descriptions of feeding behaviors in these species. In similar fashion, comparisons of jaw and branchial basket construction among meso- and bathypelagic fish species allowed Ebeling and Cailliet (1974) to draw some general conclusions about expected meso- and bathypelagic feeding patterns. They noted that bathypelagic fishes that have mesopelagic relatives (e.g., Melamphaidae) have larger mouths but similar pharyngeal baskets, even though their overall body size is not appreciably different. They hypothesized that these adaptations of the mouth and pharyngeal basket enabled bathypelagic predators to ingest a broader array of prey taxa and prey sizes successfully, without additional energetic costs needed to maintain a larger body.

III. PATTERNS IN THE DIETS OF DEEP-SEA FISHES

A. Categories of Trophic Specialization

1. BENTHIC AND DEMERSAL SPECIES

Demersal and benthic deep-sea fishes have traditionally been viewed as generalized opportunists. In part this has resulted from particular attention to the demersal family Macrouridae (Fig. 2), a taxon dominant in the fauna of the North Atlantic and northeastern Pacific, where research on deep-sea fishes was first initiated and has been most concentrated. Since the beginning of oceanic exploration, macrourids have been more readily available for life history investigations than have other taxa, and studies of the more abundant species provided the first evidence of generalized

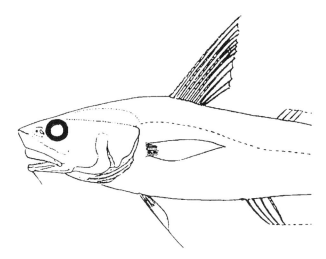

Fig. 2. Line drawing of the demersal macrourid *Coryphaenoides (Nematonurus) armatus*. Drawn by N. B. Marshall; modified from Marshall (1973).

food habits among deep-living demersal fishes (Haedrich and Henderson, 1974; Pearcy and Ambler, 1974). This evidence matched theoretical predictions of nonselective foraging under conditions of low food availability (Schoener, 1971; Dayton and Hessler, 1972), and led to a general model of nonselective predation for mobile deep-sea bottom fishes. This model, which continues to be advanced (e.g., Campbell *et al.*, 1980), could readily be rationalized in view of the scarcity of prey in the deep sea. However, it lacks rigor and general applicability across various taxa. Recent investigations of sympatric macrourids, halosaurs, and other demersal fishes have revealed differences in diets, suggesting that different trophic strategies are at play in partitioning food resources and structuring demersal fish communities (Sedberry and Musick, 1978; Macpherson, 1979, 1981; Merrett and Marshall, 1980; Mauchline and Gordon, 1986; Blaber and Bulman, 1987).

New analytical techniques are beginning to supplement conventional dietary analyses in an effort to explore quantitatively trophic specialization and resource partitioning and differential trophic strategies among demersal fishes. Mauchline and Gordon (1985) have applied rarefaction diversity methodology in an effort to quantify comparatively prey electivity by analyzing dietary diversity. Subsequently Mauchline and Gordon (1986) have utilized quantitative comparison of multiple incidences of prey in stomachs to evaluate species-specific patch exploitation. Macpherson (1981) has applied measures of niche breadth to bring greater rigor to analyses of resource partitioning. Mattson (1981) has devised a novel method for determining adequate sample size in assessment of dietary breadth, based on the cumulative number of first records of prey species in stomachs. In an effort to more precisely determine the relative energetic importance of prey species, Blaber and Bulman (1987) used caloric content of prey in lieu of conventional percent frequency, percent abundance, and percent biomass types of data.

Many macrourid species are indeed broadly euryphagous taxonomically, but remain selective regarding prey type and prey size, particularly when sympatric congeners are compared (Macpherson, 1979; Mauchline and Gordon, 1984b). Furthering the early model of broad, nonselective predation in the Macrouridae and other demersal taxa have been feeding habit studies employing taxonomic categories rather than functional or behavioral prey guilds, and others in which distinct juvenile and adult diets were not differentiated. Thus, a fish that specializes on infaunal polychaetes and amphipods would go undistinguished from one that specializes on epibenthic/benthopelagic polychaetes and amphipods, and another that is a benthos specialist as a juvenile and a nekton specialist as an adult would be classed a broad generalist. Prey size is another parameter that often goes undifferentiated when taxonomic prey categorization is the primary method of feeding habits classification. Overall, it appears that macrourids are fundamentally no more or less generalized in feeding habits than their typically more shallow-dwelling gadid relatives. Indeed, most gadiform fishes appear to be broadly euryphagous and opportunistic. Thus, euryphagy in the model deep-sea taxon Macrouridae may have less to do with adaptation to the energetic exigencies of the deep sea than to phylogenetic affinity.

Among the several hundred demersal fish species inhabiting the ocean floor, a number of common themes are evident in terms of prey selection. Mauchline and Gordon (1986) advanced four types of feeding strategies based on degree of generalized (opportunistic) versus specialized (locked-on to a given prey type) feeding. A synthesis of available food habits information for demersal deep-sea fishes results in our identification of 10 major guilds, or groups of species with similar feeding habits, of trophic specialization, categorized as follows:

    Trophic Guild 1: Piscivores
      Sit-and-wait ambush predator subguild
      Active forager subguild
    Trophic Guild 2: Macronekton foragers
    Trophic Guild 3: Micronekton/epibenthos predators
    Trophic Guild 4: Benthivorous infaunal predators
      Durophagous subguild
    Trophic Guild 5: Microphagous epifaunal browsers

Trophic Guild 6: Megafaunal croppers and browsers
Trophic Guild 7: Macroplanktonivores
Trophic Guild 8: Specialist necrophages
Trophic Guild 9: Necrophagivores
Trophic Guild 10: Detritivores

2. Pelagic Species

There are three main guilds of predators: zooplanktivores, nektonivores, and generalists. Based on morphological specializations coupled with feeding data, two of the three guilds can be subdivided in a manner similar to our benthic and demersal subguild designations. The divisions are as follows:

Trophic Guild 1: Micronektonivores
  Piscivorous subguild
  Cephalopod predator subguild
  Sit-and-wait ambush predator subguild
  Active forager subguild
Trophic Guild 2: Zooplanktivores
  Hard-bodied (crustacean) subguild
    Copepod predator subguild
    Penaeidean/caridean predator subguild
  Soft-bodied zooplanktivores
Trophic Guild 3: Generalists

B. Morphological and Behavioral Specializations among Trophic Guilds of Deep-Sea Fishes

1. Benthic and Demersal Species

Each trophic guild listed in Sections III,A,1 and III,A,2 typically comprises disparate taxa that have converged on similar diets, often taking very different morphological and behavioral routes in adaptation to arrive at the same functional end point. Each guild is discussed in the following sections with regard to constituent taxa and common themes in morphology and behavior. Some fishes defy ready classification, at times functioning as members of more than one of the guilds. Prominent examples include the eel, *Synaphobranchus kaupii* (mobile piscivore, macronekton forager, and scavenger), and the gadoid *Phycis chesteri* (which facultatively switches behavior from a benthivore to a mobile nekton forager). Many species switch categories ontogenetically (e.g., *Hoplostethus atlanticus*) (Bulman and Koslow, 1992), or under different conditions of ecology or prey availability (Crabtree *et al.*, 1991). A very small number of species, such as the abyssal macrourid *Coryphaenoides* (*Nematonurus*) *armatus*, are truly euryphagous, ingesting both live and dead pelagic and benthic animals, along with plant debris and human refuse (Sedberry and Musick, 1978).

   a. Trophic Guild 1: Piscivores. A number of demersal taxa are primarily piscivorous, although many also consume benthopelagic invertebrates behaviorally analogous to fishes (e.g., shrimps, cephalopods). Within the category of piscivores, two primary feeding strategies are used, resulting in two divergent trophic subguilds. The first is the sit-and-wait ambush strategy displayed by sedentary benthic fishes. Prey is attacked in a sudden short-range strike from a stationary position. Morphological characteristics of sit-and-wait piscivores include large body size, large gape, long sharp depressible teeth, heavy body musculature, absence of a gas bladder, and large eyes. Typical behavioral attributes are displayed by the deep-sea lizardfish *Bathysaurus* (Fig. 3). It rests motionless on the bottom, perched on its pelvic fins, tactile pectoral filaments curved out to its sides, alligator-like jaws ready to snatch up passing fishes (Sulak, 1977; Sulak *et al.*, 1985). Prey pinioned in the jaws are typically ingested whole, often ratcheted into the pharynx by their own struggles by the one-way action of hinged depressible teeth. *Bathysaurus* consumes primarily benthopelagic demersal fishes (Sedberry and Musick, 1978; Campbell *et al.*, 1980; Sulak *et al.*, 1985). Large body size enables this piscivore to select large demersal prey such

Fig. 3. *Bathysaurus ferox*, a benthic ambush piscivore (benthic/demersal trophic guild 1) in normal resting stance on the bottom at 2000 m depth. From Sulak *et al.* (1985).

as the cutthroat eel *Synaphobranchus kaupii*, the halosaur *Halosauropsis macrochir*, and rattails of the genus *Coryphaenoides*. Smaller demersal fishes are also eaten, along with midwater fishes (e.g., *Gonostoma* and myctophids) occurring near the bottom, and occasional decapod crustaceans and cephalopods. Prey strikes are probably limited to very short horizontal or oblique lunges; it is indeed very unlikely that the heavy-bodied *Bathysaurus* ever rises more than 0.5 m off the bottom.

The merlucciid of the upper slope, *Merluccius albidus*, closely mimics the morphology and solitary, bottom-perching behavior of *Bathysaurus*, and even displays blotchy lateral markings similar to those of lizardfish. However, *M. albidus* has a gas bladder and can hover off bottom after a short feeding lunge or predator escape maneuver. Other sedentary ambush piscivores are the goosefish *Lophius*, the scorpaenids *Cottunculus* and *Helicolenus*, and the pleuronectid *Reinhardtius*. A special case is the ambush piscivore, *Thaumatichthys*, a singular ceratioid anglerfish adapted for demersal existence (Bertelsen and Struhsaker, 1977) (Fig. 4). Although it does not "sit" on the substrate, *Thaumatichthys* employs an analogous behavior, hovering just above the substrate, its luminous lure dangling from the interior of an overshot mouth. Equipped with long hooked marginal teeth, its capacious jaws with stretched ligaments are set to spring shut on prey like a Venus fly-trap.

Most sit-and-wait predators are solitary animals of limited mobility. Maintaining sufficient mutual spacing among stationary apex predators is important to success of the species in a food-poor environment. Also important is the ability to survive for long intervals without feeding. In *Bathysaurus*, the large, lipid-rich liver probably does not function as a buoyancy control device as previously suggested (Marshall and Merrett, 1977), but as an important energy store (Savvatimskii, 1969; Hurcau, 1970; Stein and Pearcy, 1982), sustaining metabolism and growth between sporadic feeding episodes (Smith, 1978). Consistent with a sporadic macrophagy plus energy storage hypothesis, Sulak *et al.* (1985) found that over 50% of *Bathysaurus*

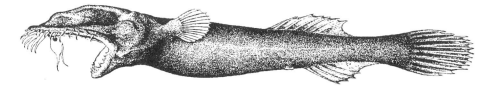

Fig. 4. Drawing of the unusual demersal anglerfish *Thaumatichthys axeli*, a sit-and-wait ambush piscivore (benthic/demersal trophic guild 1), showing the dentition and position of the luminescent "lure" inside the mouth. Drawn by P. H. Winther; from Bertelsen and Struhsaker (1977).

stomachs contained no food, and that liver size (as percent total body weight) varied dramatically among large fish. Presumably, large fish with small livers have not fed for a long time. In this regard, Smith (1978) has determined that the abyssal macrourid *C. armatus* carries a lipid/glycogen energy reserve sufficient to sustain its energetic requirements for an estimated 186 days.

Other smaller benthic sit-and-wait predators, such as the aulopiform genera *Bathysauropsis* and *Bathytyphlops,* the scorpaenid, *Helicolenus dactylopterus,* and large species of the zoarcid *Lycodes* (Fig. 5), have smaller teeth and probably depend on a mixed diet of fishes and crustaceans. The trophic morphology of most sit-and-wait piscivores reflects dependency primarily on large prey. For example, most have nonfunctional gill rakers (reduced in size and number) that allow small items to pass easily through the buccal cavity and out the gill apertures. Sedentary ambush piscivores are typically a minor component of the demersal deep-sea fish fauna, limited mostly to the upper and middle slope, where large eyes can function most effectively in prey detection.

An alternative feeding strategy displayed by a second subguild of demersal piscivores is mobile benthopelagic foraging. This active search strategy is much more widely used than the sit-and-wait strategy, and is common at all depths. Examples include some of the most abundant and familiar demersal deep-sea fishes, such as the synaphobranchid eels *Synaphobranchus* (Saldanha, 1980; Merrett and Marshall, 1980; Merrett and Domanski, 1985) and *Diastobranchus,* the morid *Antimora* (Mauchline and Gordon, 1984e), the gadid *Molva,* large macrourids of the genus *Coryphaenoides* (Priede *et al.,* 1990), the trachichthyid *Hoplostethus atlanticus* (Gordon and Duncan, 1987; Bulman and Koslow, 1992), and selachians, including squaloid and scyliorhinid sharks, and skates. Other active foragers that may also be primarily piscivorous include the alepocephalid *Narcetes,* and the ophidiids *Spectrunculus* (Fig. 6) and possibly *Apagesoma.*

Features common to most benthopelagic piscivores include large body size, large gape, and robust form. Large size is important to accommodate

Fig. 5. Illustration of *Lycodes lavalae,* a large benthic piscivorous eelpout (family Zoarcidae, benthic/demersal trophic guild 1). Drawn by P. MacWhirter.

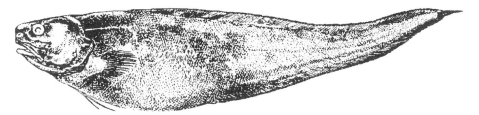

Fig. 6. Drawing of *Spectrunculus grandis*, an active benthopelagic piscivore (family Ophidiidae, benthic/demersal trophic guild 1). From Hureau and Nielsen (1981).

large prey, to avoid predation by other large piscivores, to achieve sufficient energy storage capacity between unpredictable feeding events (Smith, 1978; Dahl, 1979; Sulak *et al.*, 1985), and to enable energetically efficient low-speed cruising while foraging. Except in the sharks and skates, dentition consists of either rows of small sharp teeth or bands of minute teeth. Most have large terminal or subterminal mouths, capable of accommodating large prey. Except in synaphobranchids, for which a lunging strike is probable, most mobile piscivores probably engulf prey via suction feeding, combined with rapid forward locomotion.

In teleosts, mobile piscivory is more prevalent among the large species of the lower slope, rise, and abyss, where benthic prey becomes scarce relative to nektonic fishes. Synaphobranchid eels and large macrourids are the dominant mobile piscivores of the slope and rise in most parts of the world ocean. Sharks are important as well, particularly on the upper and middle slope. Mauchline and Gordon (1983a) found that the dominant, broadly distributed squalids *Centroscymnus coelolepis, Centroscyllium fabricii, Deania calceus,* and *Lepidorhinus squamosus* feed primarily on both demersal and midwater fishes, supplemented with squid and pelagic crustaceans. Morid and gadoid fishes were notable prey items (Mauchline and Gordon, 1983a). At greater depths skates such as *Bathyraja richardsoni* assume the role of apex piscivores, feeding on large teleosts, including *Coryphaenoides* (*Nematonurus*) *armatus* (Campbell *et al.*, 1980).

Active piscivory may be supplemented by opportunistic scavenging among mobile foragers (Pearcy and Ambler, 1974). Live and dead prey location is probably dependent on chemical (Wilson and Smith, 1984) and lateral-line senses. Priede *et al.* 1990) found that abyssal macrourids always arrived at baited free vehicles from down current. Vision is less important, particularly at abyssal depths, except in species that select bioluminescent prey. Submersible and baited camera observations indicate that although mobile piscivores do not form organized aggregations, they may often occur in high abundance, and may opportunistically mass on food falls (Jannasch and Wirsen, 1977; Jannasch, 1978; Priede *et al.*, 1990). Such behavior contrasts markedly from that of sedentary piscivores, which appear never to mass on prey. Among mobile piscivores, food discovery and exploitation is facilitated by denser spatial packing of individuals. Mobile foragers include some of the most numerically dominant fishes on the ocean floor, such as the eel *Synaphobranchus kaupii*, the rattail *C. armatus*, the morid *Antimora rostrata*, and the squalid sharks *Centroscymnus coelolepis* and *Centroscyllium fabricii*. Although large mobile abyssal piscivores have occasionally been captured far above the substrate (Pearcy, 1976; Smith *et al.*, 1979), such vertical excursions out of the benthic boundary layer are probably rare events (Priede *et al.*, 1990). In addition to scavenging, some species in this subguild may facultatively switch to small prey when necessary. Accordingly, the gill rakers in mobile foragers may be reduced in number and length, but are rarely obsolete as in sedentary ambush predators.

The largest predators in the deep sea are sharks. Herdendorf and Berra (1995) reported a Greenland Shark, *Somniosus microcephalus*, estimated to be 6 m long from a photograph taken at a depth of 2200 m off the coast of North Carolina (Fig. 7). The diets of such large specimens have not been

Fig. 7. The demersal Greenland shark, *Somniosus microcephalus*, photographed swimming over the shipwreck of the SS Central America in the Atlantic Ocean at 2200 m depth. Lines are part of scales from original photograph. From Herdendorf and Berra (1995).

studied; however, smaller species have been investigated. Most feed on variety of fish, decapods, and cephalopods (Crabtree *et al.*, 1991; Mauchline and Gordon, 1983a; Sedberry and Musick, 1978).

b. Trophic Guild 2: Macronekton Foragers. Fishes comprising this guild display a mixed diet of pelagic prey, including nektonic crustaceans (mysids, euphausiids, decapods), cephalopods, chaetognaths, and midwater fishes. Epibenthic invertebrates may be consumed facultatively, as in the macrourids *Trachyrinchus trachyrinchus* (Merrett and Marshall, 1980) and *Coelorinchus* sp. (Blaber and Bulman, 1987). Although broadly euryphagous, such fishes are nonetheless selective in targeting primarily small, schooling, off-bottom prey. Midwater fishes may be the predominant prey in many regions (Sedberry and Musick, 1978; Blaber and Bulman, 1987). Various taxa of nektonic crustacea may form the preferred prey of individual fish species. For example, the catshark *Apristurus* specializes on sergestid shrimps (Mauchline and Gordon, 1986).

Macronekton foragers are most prevalent on the upper and midde slopes, where prey concentrations facilitate predation on large numbers of small individuals. Prominent macronekton specialists include many numerically dominant species, such as the macrourids, *Coryphaenoides rupestris* (Podrazhanskaya, 1971; Geistdoerfer, 1979a), *Coryphaenoides guentheri* (Mauchline and Gordon, 1984b), *Coryphaenoides pectoralis* and *Macrourus berglax;* the gadoid *Gadiculus argenteus* (Macpherson, 1981; Mattson, 1981; Mauchline and Gordon, 1984e); certain slickheads of the genus *Alepocephalus* and other genera; the long-nose eel *Venefica procera;* the morid *Phycis chesteri;* a number of medium-size ophidiids such as species of *Bassozetus* (Crabtree *et al.*, 1991); and the beryciform fishes *Cyttus, Neocyttus,* and *Hoplostethus* (Fig. 8).

Macronekton foragers frequently occur in dense feeding aggregations and may engage in organized schooling. Some may also engage in off-bottom forays, following pelagic prey into midwater (Haedrich, 1974; Pearcy, 1976; Blaber and Bulman, 1987). Most are active swimmers of moderate to large size with terminal or subterminal mouths, moderate gapes, well-developed gill rakers, and large eyes. Bioluminescence may be exploited to locate and maintain contact with mobile schooling prey. Some species may regularly switch from nektonic to epibenthic prey. Included here are the gadoids *Phycis chesteri* and *Phycis blennoides.* A few species appear to be selective for a preferred prey type. For example, Merrett and Marshall (1980) reported that 51% of the macrourid *Bathygadus melanobranchus* examined had fed on the large mysid *Gnathophausia zoea,* in addition to copepods and chaetognaths. The preference for *G. zoea* was invariant across fish of all sizes.

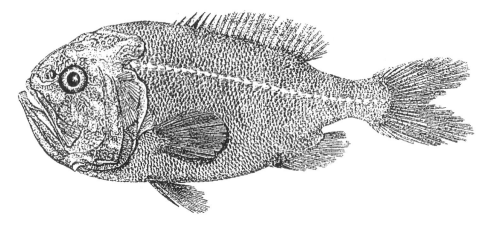

Fig. 8. Drawing of *Hoplostethus atlanticus*, an example of a demersal beryciform macronekton forager (benthic/demersal trophic guild 2). Drawn by M. A. Holloway; from Woods and Sonoda (1973).

c. Trophic Guild 3: Micronekton/Epibenthos Predators. Most demersal deep-sea fish species are microphagous, preying on small benthopelagic and epibenthic invertebrates, especially crustaceans (e.g., mysids, amphipods, isopods, tanaids, copepods). Micronekton/epibenthos specialists are most prevalent on the upper and middle slopes, where the densest populations of small prey are found. Examples include morids such as *Lepidion eques* (Mauchline and Gordon, 1980) and *Halargyreus johnsonii* (Mauchline and Gordon, 1984e), macrourids such as *Coelorinchus coelorinchus* (Du Buit, 1978; Mauchline and Gordon, 1984b) and four small species of the macrourid genus *Nezumia* (Merrett and Marshall, 1980; Crabtree *et al.*, 1991), halosaurs such as *Aldrovandia gracilis* and *Aldrovandia affinis* (Crabtree *et al.*, 1991), and ophidiids such as *Porogadus silus* and *Porogadus catena* (Crabtree *et al.*, 1991). However, this trophic guild is evident at all depths. The deepest dwelling fish, the ophidiid *Abyssobrotula galathea* (Fig. 9A), feeds on epibenthic polychaetes, isopods, and amphipods (Nielsen, 1977). Other abyssal fishes of small size or limited mobility, including the chlorophthalmids *Bathymicrops* and *Discoverichthys*, the alepocephalid *Rinoctes nasutus*, the macrourid *Echinomacrurus mollis*, and many species of Aphyonidae (Fig. 9B), appear to survive on a microphagous diet of copepods and other tiny crustacea (Nielsen, 1969; Merrett, 1987). The eyes of such species are often reduced, and vision is generally less important in micronekton/epibenthos specialists. Prey location may depend more on olfaction, the tactile sense, and the lateral-line sense.

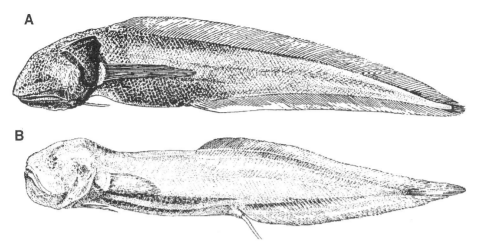

Fig. 9. Demersal micronekton/epibenthos predators (benthic/demersal trophic guild 3). (A) Illustration of *Abyssobrotula galatheae* (family Ophidiidae), the deepest dwelling fish species, an epibenthic predator. Note the relatively large mouth and reduced eyes. From Nielsen (1977). (B) *Barathronus bicolor* (family Aphyonidae), a micronekton predator. Note the similarities in gape and eye size to those features in *Abyssobrotula*. From Nielsen (1969).

Like their macrophagous counterparts, micronekton feeders have two alternative feeding strategies. The tripodfishes, genus *Bathypterois* (Fig. 10A), and many related chlorophthalmid genera such as *Ipnops, Bathymicrops* (Fig. 10B), and *Discoverichthys* are sit-and-wait predators. Tripodfish mimic stalked planktivorous invertebrates, facing into the current perched on elongate pelvic and caudal fin rays, feeding on small nektonic prey that approach too closely. The eyes are minute and probably unimportant in prey detection. Instead, an umbrella of delicate elongate pectoral rays surrounds the head like a forward looking satellite dish antenna, probably pinpointing the position of the slightest nearby disturbance. The lateral line is also well developed in *Bathypterois* and related genera, especially the superficial organs arrayed on the head (Marshall and Staiger, 1975). Prey ingestion is facilitated by a buccal cavity that opens to capacious dimensions and is sealed off posteriorly by a network of numerous long closely set gill rakers. Although rarely abundant, tripodfishes are ubiquitous in the deep sea between 200 and 6000 m. The largest species, *Bathypterois grallator*, stands on fins reaching over 0.5 m in length. It feeds on micronekton and on larger nektonic prey, including midwater fishes (Crabtree *et al.*, 1991).

More common among micronekton predators is the alternative strategy of continuous foraging while slowly moving along just above the substrate.

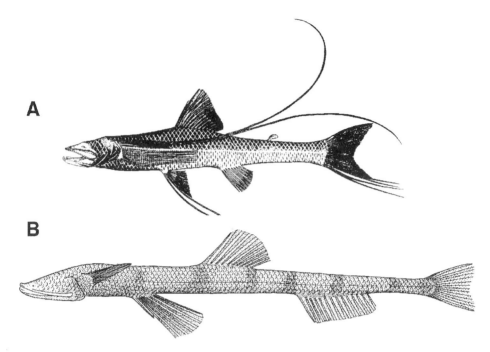

Fig. 10. Benthic sit-and-wait micronekton predators (benthic/demersal trophic guild 3). (A) Drawing of *Bathypterois andriashevi* (family Bathypteroidae). (B) The benthic chlorophthalmid *Bathymicrops regis*.

This mode is used by many macrourids, ophidiids, notacanths, morids, small ophidiids, and other taxa. Micronekton predators are often found in loose aggregations, a habit that may improve foraging efficiency by facilitating location and exploitation of prey patches. Most micronekton predators are small to medium-size fishes with small subterminal to inferior mouths. They consume a broad range of taxa, but select primarily small pelagic and epibenthic prey. They seldom probe the substrate or engulf sediment to extract infaunal prey. Most appear to favor crustaceans and polychaetes, avoiding small refractory megafauna (e.g., brittle stars, mollusks) even when abundant. Thus, prey selectivity is not only defined by prey types consumed, but also by those not consumed even when readily available.

Within the guild of demersal microvores, most species display a very broad taxonomic range of prey; however, many species feed selectively. The small continental rise macrourid *Coryphaenoides carapinus* is selective for amphipods and the small brittle star *Ophiura ljungmani* (Haedrich and Polloni, 1976). Selectivity must be involved because frequencies of preferred prey in stomach contents vastly exceed relative frequencies in box-core samples. Similarly, Carter (1984) and Crabtree *et al.* (1991) found that various small to medium-size ophidiids from the Bahamas and Middle Atlantic Bight regions display considerable prey selectivity. Carter (1984) expressed comparative prey selection in terms of percent frequency occurrence in stomachs (F), percent numerical abundance among prey items consumed (N), and percent weight (W). Among microphagous species analyzed, *Xyelacyba myersi* is a specialist on isopods ($F = 57, N = 12, W = 48$); *Acanthonus armatus*, on polychaetes ($F = 80, N = 82, W = 34$) (Fig. 11); *Porogadus miles*, on gammarid amphipods ($F = 57, N = 81, W = 74$); *Porogadus silus*, on calanoid copepods ($F = 77, N = 64, W = 76$); and *Barathrodemus manatinus*, on tanaids ($F = 69, N = 30, W = 45$). Relatively few ophidiids distribute effort roughly equally over several prey groups. One exception is *Bathyonus pectoralis*, which feeds on calanoid copepods, mysids, isopods, and amphipods ($F = 38, 22, 24,$ and 20, respectively). *Dicrolene kanazawai* selects two prey categories, calanoid copepods and tubiculous polychaetes, in roughly equal amounts. However, its congener of nearly identical morphology and size, *Dicrolene intronigra*, is selective for isopods ($F = 57, N = 23, W = 48$) (versus *D. kanazawai*: $F = 11, N = 1, W = 3$).

Prey specialization is particularly evident within the Ophidiidae, a family more prevalent at tropical latitudes. Indeed, had the earliest deep sea sampling efforts concentrated on tropical areas populated by ophiidids, our initial trophic model for demersal deep-sea fishes might well have been one of specialization rather than generalization.

d. Trophic Guild 4: Benthivorous Infaunal Predators. In any given region relatively few demersal fishes appear to depend predominantly on

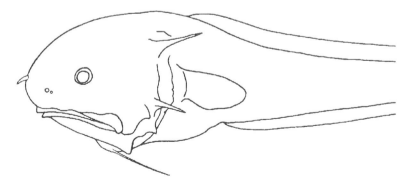

Fig. 11. Line drawing of *Acanthonus armatus*, a demersal microvore (benthic/demersal trophic guild 3) with selective feeding habits. Modified from Cohen and Nielsen (1978).

infaunal prey (Mauchline and Gordon, 1984d). Prominent among them are eelpouts of the family Zoarcidae. Important faunal components at temperate to arctic latitudes, eelpouts are most abundant in areas of high primary productivity. They sometimes occur in dense aggregations where benthic prey are particularly abundant (Sulak and Ross, 1993; Hecker, 1994) (Fig. 12A). They are benthic and often snake along the bottom instead of swimming. Predominant prey include mollusks, polychaetes, and other infaunal organisms, although large *Lycodes* species also feed partially on echinoderms, crabs, and fish. Typically, considerable sediment is ingested during feeding (Sedberry and Musick, 1978), but prey is probably manipu-

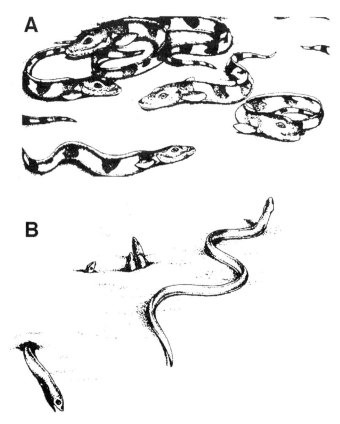

Fig. 12. Benthivorous infaunal predators (benthic/demersal trophic guild 4). (A) Sketch drawn from submersible videotapes showing aggregations of *Lycenchelys verrilli* (family Zoarcidae) on the bottom in the western North Atlantic Ocean. Drawn by P. MacWhirter. (B) Composite sketch from submersible videotapes showing the burrowing and snakelike swimming behaviors of the ophichthid eel, *Ophichthus cruentifer*. Drawn by P. MacWhirter.

lated from the substrate using the enlarged lips, rather than ingested indiscriminately in mouthfuls of sediment. Zoarcids have well-developed eyes, but these bottom-hugging fishes may employ tactile senses to detect buried prey.

Other slope fishes exploiting infaunal prey include the pleuronectid *Glyptocephalus cynoglossus,* the synaphobranchid eel *Ilyophis,* the burrowing snake eel *Ophichthus cruentifer* (Wenner, 1978) (Fig. 12B), the small morid *Laemonema barbatula,* the halosaur, *Halosauropsis macrochir* (Sedberry and Musick, 1978), the ogcocephalid, *Dibranchus atlanticus* (Crabtree *et al.,* 1991), and perhaps the ophidiid *Penopus macdonaldi* (Carter, 1984).

A special subguild of benthivores is adapted as durophages, capable of crushing thick-shelled mollusks and other armored invertebrates. These include the chimaeras, with heavily muscled beaklike jaws (Ribbink, 1971) and crushing palatine plates, together with species of skates having crushing molariform teeth. The peculiarly modified shovelnose chimaera, *Callorhynchus capensis* (Fig. 13), feeds on pelecypods, gastropods, crabs, and other crustacea dislodged from the substrate with the spadelike snout appendage (Ribbink, 1971). Mechanical excavation of prey from the sediment may be assisted by jets of water forcefully expelled through the mouth, a habit also employed by some skates. Hard-shelled prey are then crushed by the beak. Species of three other chimaerid genera also consume infauna, including burrowing tubiculous anemones, spatangoid urchins, ophiuroids, and polychaetes (Scott, 1911; Sedberry and Musick, 1978; Macpherson, 1980; Mauchline and Gordon, 1983a).

The durophagous habit may have been much more important in the geological past, when large, thick-shelled invertebrates dominated the ocean floor. Thus, despite the elegant morphological adaptation for duro-

Fig. 13. Illustration of the chimaera *Callorhinchus capensis,* a benthivorous infaunal predator (benthic/demersal trophic guild 4). Modified from Smith and Heemstra (1986), by permission of Springer-Verlag.

phagy, many present-day chimaeras depend as well on soft-bodied prey (Mauchline and Gordon, 1983a). Moreover, benthic foraging may be supplemented by opportunistic scavenging and piscivory. Chimaeras are readily attracted to food falls and baited traps and are among the suite of large fishes predictably captured on deep-baited lines (Forster, 1964, 1968, 1971, 1973). Mauchline and Gordon (1983a) attribute a cutthroat eel found in *Chimaera monstrosa* to net feeding, but opportunistic piscivory seems a more probable explanation.

e. Trophic Guild 5: Microphagous Epifaunal Browsers. A small group of demersal deep-sea fishes specialize in microcrustaceans (e.g., amphipods, isopods, cumaceans, tanaids, mysids) and polychaetes picked off the sediment surface. The best examples are species of the notacanthid genus *Polyacanthonotus* (Fig. 14). These fishes have slender pointed snouts and very small ventral mouths and feed on small benthic crustaceans and polychaetes (Crabtree *et al.*, 1985). Sediment is rarely ingested, in contrast to the habit of related halosaurs (Sedberry and Musick, 1978; Crabtree *et al.*, 1991). Two small, slender ophidiids, *Porogadus miles* and *Barathrodemus manatinus*, closely mimic notacanth head morphology. These species parallel spiny eels in feeding habits (Crabtree *et al.*, 1991), although *B. manatinus* is particularly selective for tanaids.

f. Trophic Guild 6: Megafaunal Croppers and Browsers. Relatively few demersal fishes feed on sessile megafaunal invertebrates. Perhaps this is because many such large epibenthic and/or burrowing animals are protected by spicules, spines, nematocysts, and tough integuments, and provide a relatively low caloric return per unit biomass ingested. However, at least a few demersal fishes overcome these obstacles to feed preferentially on large epibenthic invertebrates, including sponges, anemones, soft corals, sea pens, brittle stars, sea stars, sea urchins, and crinoids. The large bathyal notacanthid or spiny eel, *Notacanthus chemnitzi* preys selectively on anemones, corals, bryozoans, and colonial hydrozoans using special knife-edge

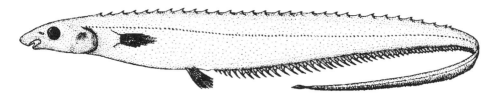

Fig. 14. Drawing of *Polyacanthonotus rissoanus* (family Notacanthidae), a microphagous epifaunal browser (benthic/demersal trophic guild 5). Drawn by P. Pebbles.

dentition to nip off the tenacles (Lozano Cabo, 1952; McDowell, 1973) (Fig. 15). The large abyssal chimaera, *Chimaera monstrosa*, also feeds heavily on anemones in the Rocktall Trough (Mauchline and Gordon, 1983a). Elsewhere, ophiuroids are the dominant prey of this fish (Macpherson, 1980). A small macrourid dominant on the continental rise, *Coryphaenoides carapinus*, preferentially exploits populations of the brittle star, *Ophiura ljungmani*, together with amphipods (Haedrich and Polloni, 1976). In the Rockall Trough area the witch flounder, *Glyptocephalus cynoglossus*, feeds on anemones and brittle stars (Mauchline and Gordon, 1984c), but elsewhere smaller fish feed primarily on smaller infaunal prey (Wenner, 1978). One of the most distinctive diets is that of the short-snouted ophidiid *Barathrites parri*, which preys on elasipod holothurians and tubiculous polychaetes. Most other small ophidiids specialize on various groups of crustaceans, all of which are singularly unimportant in the diet of *Barathrites* (Carter, 1984; Crabtree *et al.*, 1991). It appears that *B. parri* is also a browser by habit, nipping off pieces of its prey, rather than ingesting whole animals. Other megafaunal nippers found along rocky outcrops on the upper slope and along canyon walls may include morphologically specialized beryciform fishes such as *Neocyttus* and *Antigonia*. These unusual highly compressed, tubular-mouthed deep-water fishes resemble reef fishes. Their likely prey includes crinoids, sea pens, and other stalked cnidarians, taxa of very limited and patchy distribution in the deep sea.

g. Trophic Guild 7: Macroplanktonivores. Large slow-moving gelatinous animals form the primary food source for a group of demersal specialists. Gelatinous prey include jellyfish, comb jellies, salps, and ceratioid anglerfishes. Off Tasmania, Blaber and Bulman (1987) report that the colonial salp *Pyrosoma* is the main prey of the scorpaenid *Helicolenus percoides*, and is also important in the diet of *Neocyttus rhomboidalis*. In the zone of high primary production off West Africa, two well-studied

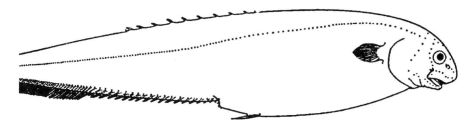

Fig. 15. Drawing of *Notacanthus chemnitzi* (family Notacanthidae), a megafaunal cropper/browser (benthic/demersal trophic guild 6). Modified from Tucker and Jones (1951), in McDowell (1973).

slickhead species, *Alepocephalus bairdi* (Fig. 16) and *Alepocephalus rostratus*, feed predominantly on medusae, ctenophores, and *Pyrosoma*, together accounting for the majority of food items consumed by frequency of occurrence (Golovan' and Pakhorukov, 1975, 1980). The remainder of the diet is composed of incidental pelagic prey, including midwater fishes, cephalopods, pteropods, and shrimps. Benthic prey items are essentially absent. In the Rockall Trough area the diet of *A. bairdi* is again dominated by gelatinous macroplankton (Mauchline and Gordon, 1983b).

Another slickhead, *Conocara fiolenti*, feeds predominantly on gelatinous salps, while its close congener *Conocara macropterum* feeds mainly on benthos, supplemented by salps (Crabtree and Sulak, 1986). Many slickhead species forage passively using a hover-and-wait strategy, drifting passively in unusual attitudes (head up, head down, upside down) (Sulak, 1977, 1982; Markle, 1978) between feeding bouts. Bioluminescence may be used in prey detection; alephocephalids and closely related platytroctids have very large eyes with a notable aphakic space, and deep retinal fovea (Lockett, 1971) indicating a capacity for low-level light detection and good distance-ranging capability.

The importance of jelly animals as energy resources in the deep sea has barely been recognized. However, it is probable that numerous species of Alepocephalidae, and other fish taxa, feed extensively on jelly animals, which contain about 10% utilizable protein. Other predators of pelagic prey, including squaloid sharks, are at least facultative consumers of jellyfish such as the abundant deep-water genus *Atolla* (Bigelow and Schroeder, 1948; Mauchline and Gordon, 1983a). Slickheads closely match their prey in basic composition, consisting of about 90% water (Golovan' and Pakhorukov, 1980; Crabtree, 1995). They are specifically adapted to process jellyfish. The posterior-most branchial arches and gill rakers are modified into a pair of triturating organs, called crumenal or epibranchial organs. These organs are analogous to the pharyngeal mills of stromateid fishes (Haedrich, 1967),

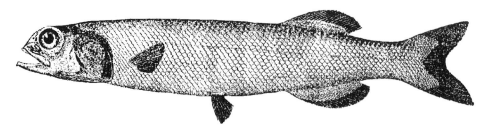

Fig. 16. Drawing of *Alepocephalus bairdii* (family Alepocephalidae), a macroplanktonivore on gelatinous prey (benthic/demersal trophic guild 7). From Goode and Bean (1895).

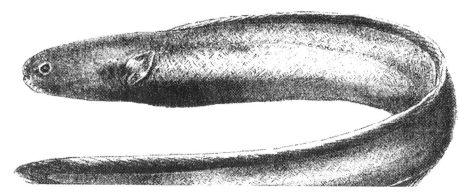

Fig. 17. Illustration of *Simenchelys parasitica* (family Synaphobranchidae), a specialist necrophage (benthic/demersal trophic guild 8). From Robins and Robins (1989).

also adapted to process jellyfish. Both structures appear to function in crushing gelatinous tissue and discharging coelenterate nematocysts to enable detoxification. In alepocephalids, a thick (125–140 μm), tough subepithelial layer of connective tissue lines the oral cavity, pharynx, and esophagus in all species studied (Veriginia, 1979; Veriginia and Golovan', 1978). This layer may shield internal organs against nematocyst discharge during the initial stage of mastication.

h. Trophic Guild 8: Specialist Necrophages. Particularly on the continental rise and on the abyssal plain, a number of large benthopelagic foragers such as *Synaphobranchus kaupii* (Crabtree *et al.,* 1991) and *Centroscymnus coelolepis* (Mauchline and Gordon, 1983a) use scavenging as an optional strategy. However, a few demersal fishes are necrophages. These include the hagfishes *Myxine* and *Eptatretus* and the snubnose cutthroat eel *Simenchelys parasitica* (Fig. 17). Both taxa are strongly attracted to baited traps (Solomon-Raju and Rosenblatt, 1971; Isaacs and Schwartzlose, 1975). Both attack injured fishes or carcasses, tearing off pieces of flesh. Hagfishes accomplish this with their unique rasping "tongue," whereas *Simenchelys* uses short, stout teeth and powerfully muscled jaws. *Simenchelys* never seems to be particularly abundant in any region, but hagfishes occur in very high densities in productive regions of the oceans. A uniquely specialized life-style of general near-torpor, punctuated by brief bouts of intense activity, enables the hagfish to maintain high standing population densities in the absence of constant energy resources. Between feeding episodes hagfish rest totally inactive, burrowed in the sediment with only the snout and sensory barbels protruding. When a local food fall appears they exit their burrows to mass on the carrion (Isaacs and Schwartzlose, 1975), exuding quantities of viscous, perhaps noxious, slime that effectively sequesters the windfall for hagfishes alone.

i. Trophic Guild 9: Necrophagivores. Large food falls in the deep ocean rapidly attract not only necrophagous fishes, but also scavenging amphipods (Isaacs and Schwartzlose, 1975) such as *Eurythenes gryllus*. It appears that at least one abyssal fish, *Paraliparis bathybius* (Fig. 18), exploits concentrations of scavenging amphipods at food falls as its primary prey (Lampitt *et al.*, 1983). This cyclopterid is a small gelatinous fish with limited mobility. It presumably hovers and drifts passively off bottom, descending in concert with deep tidal currents, to arrive just as amphipod abundance has peaked around a carcass. There are several very similar species of *Paraliparis* occurring at different depths and in different regions.

Another unrelated fish very similar in size, gelatinous body composition, and passive off-bottom hovering behavior (Wenner, 1978) is the peculiar zoarcid *Melanostigma*, and yet another unrelated taxon with many species of similar diminutive, gelatinous form is the neotenic ophidiiform family Aphyonidae (Nielsen, 1969), characterized as well by degenerate eyes. The limited data on feeding habits of other *Paraliparis* and *Melanostigma*

Fig. 18. Necrophagivores (benthic/demersal trophic guild 9). Photograph of *Paraliparis bathybius* (family Cyclopteridae) attracted to bait at 4009 m depth in the eastern North Atlantic Ocean. Adapted from Lampitt *et al.* (1983), by permission of Springer-Verlag.

indicate mixed pelagic crustacea including euphausiids, sergestids, mysids, and copepods (Wenner, 1978). Copepods and mostly unidentifiable crustacean remains have been found in the stomachs of aphyonids (Nielsen, 1969). Some aphyonids (e.g., *Leucochlamys*) have fangs at the tip of the jaws, suggesting specialized predation despite limited locomotory ability. Passive hovering and drifting, coupled with chemical attraction to carcasses, could explain how such tiny fishes of limited swimming ability and reduced sensory capabilities are able to locate and consume active crustacean prey. In company with *P. bathybius,* perhaps some of these peculiar gelatinous fishes are also specific necrophagivores, exploiting the exploiters of food falls to the ocean floor. Economies in body size and composition, together with passive locomotion via neutrally buoyant drifting, are energetic accommodations consistent with a trophic strategy dependent on rare, unpredictable food falls (Stockton and DeLaca, 1982).

j. Trophic Guild 10: Detritivores/Meiofaunal Predators. Making a living by gleaning the sediment surface for organic detritus and/or processing sediment for microscopic animals appears to be the particular forté of a host of megafaunal and infaunal invertebrates. Mobile, muscle-laden animals such as fishes have comparatively high metabolic requirements. In the deep sea, the energy expenditure required to process indiscriminately large amounts of sediment may exceed the energetic return gained from contained detritus and meiofauna. Sedberry and Musick (1978) suggest that meiofauna may assume greater dietary importance for fish at greater depths. However, meiofauna has rarely been identified in the stomach contents of demersal deep-sea fishes. It has been hypothesized (McDowell, 1973) that the peculiarly modified notacanthiform fish, *Lipogenys gilli,* is an indiscriminate "vacuum cleaner" of the sediment surface. This widespread but rare bottom fish lacks teeth, and has the mouth formed into a broad ventral "sucker." The digestive tract includes a distensible stomach and long, complex intestine. McDowell (1973) found mixed detritus, sediment, and unidentified matter in the few fish he examined. However, if *Lipogenys* is indeed well-adapted as a detritivore, its rarity seems curious relative to the ubiquity of soft sediment substrate along continental margins.

Certain alepocephalids also appear to ingest considerable sediment. Included here is *Conocara macropterum* (Crabtree and Sulak, 1986; Crabtree *et al.,* 1991), a passive hoverer/drifter that periodically descends to the bottom to ingest mouthfuls of substrate, often including foraminiferans. This behavior pattern is analogous to that of sediment-ingesting pelagic holothurians (Barnes *et al.,* 1976). However, it seems unlikely that *Conocara* ingests sediment nonselectively to obtain microscopic meiofauna, but rather targets sediment-living macrofaunal invertebrates.

Another alepocephalid, *Rouleina*, may indeed feed on organic detritus and tiny organisms obtained not from the sediment surface, but from the water column. Food collection in this genus may represent one of the most bizarre adaptations yet observed in any benthopelagic fish. In trawl-caught *Rouleina* the skin is always parted exactly along the ventral and dorsal midlines. In submersible observations, Markle (1978) noted sheets of shredded mucus hanging from the jaws and body of this hovering fish. Perhaps this is a device to trap suspended organic particles, which are then ingested by the fish along with the mucus. Such a habit would mimic somewhat the mucus feeding behavior of many surface deposit-feeding worms.

2. Pelagic Species

Our remarks introducing the topic of morphological and behavioral specializations for benthic and demersal fish species apply equally to pelagic species, as do the caveats. Data for many species are sparse for several reasons: the low numbers of animals actually examined, stomach content loss through expansion of the swim bladder in some species, or regurgitation of stomach contents (Borodulina, 1972; Nielsen and Bertelsen, 1984). Some species cannot be readily placed in a specific guild, and in some species, ontogenetic shifts in guilds or subguilds occur.

One behavioral characteristic of most species of mesopelagic fishes is that of diel vertical migration from daytime resident depths of usually >500 m to nighttime occupation of depths <200 m and, in some species, the actual surface layer (<5 m) (Gartner *et al.*, 1987, 1989). We will discuss this pattern in relation to feeding in the next section, but a few comments regarding general body construction and vertical migration should be provided (see also Chapter 3, this volume).

Vertically migrating mesopelagic fishes usually have a gas-filled swim bladder (Marshall, 1980). The Myctophidae are one of the most abundant of the migratory species and in some species the swim bladder either becomes invested with lipids or the body has a naturally high lipid content that secondarily provides buoyancy (Bone, 1973; Butler and Pearcy, 1972). In nonmigrating mesopelagic species, swim bladders are either regressed or absent, whereas bathypelagic species typically lack a swim bladder (Marshall, 1980). Because they affect movement patterns, these characteristics are also directly related to feeding behavior.

On the basis of diet and morphological and behavioral specializations, we have identified three guilds. Guilds 1 and 2 have several subguilds based on diet or activity patterns:

Trophic Guild 1. Micronektonivores
  Piscivorous subguild
  Cephalopod predator subguild

4. FEEDING AT DEPTH   151

   Sit-and-wait ambush predator subguild
   Active forager subguild
Trophic Guild 2. Zooplanktivores
   Hard-bodied (crustacean) zooplanktivores
      Small Crustacea (Copepoda, Ostracoda, Amphipoda,
Euphausiacea) subguild
         Large Crustacea (Penaeidea, Caridea) subguild
   Soft-bodied Zooplanktivores
      Gelatinous prey subguild
      Gastropod mollusk subguild
Trophic Guild 3. Generalists

a. Trophic Guild 1: Micronektonivores. This guild includes two subguilds based on taxa that feed primarily on fish versus those that feed on cephalopod mollusks, and two subguilds based on predatory activity levels.

The piscivorous subguild contains the largest number of species. In the mesopelagic zone, piscivorous predators include members of the families Stomiidae, Paralepididae (Fig. 19A), Evermannellidae (Fig. 19B), Scopelarchidae, Alepisauridae, Giganturidae, Melanocetidae, Chiasmodontidae, and Gempylidae (see Chapter 2, this volume, for systematic relationships). Bathypelagic piscivores include members of several families of the Ceratioidei and the Saccopharyngidae.

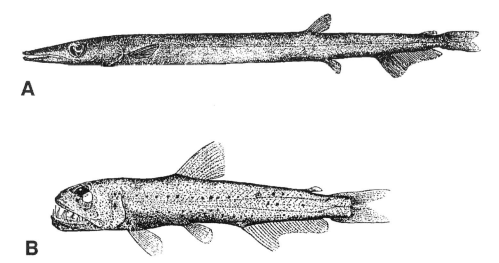

Fig. 19. Illustrations of pelagic active micronektonivores (pelagic trophic guild 1). (A) *Notolepis rissoi* (family Paralepididae). (B) *Evermannella indica* (family Evermannellidae). From Rofen (1966).

Characteristics of all piscivores are the presence of well-developed dentition (less so in bathypelagic species), a typically large gape (Marshall, 1980), and a relatively straight intestine (Fange and Grove, 1979). In some cases (e.g., many stomiids and ceratioid anglerfishes), the teeth are hinged and fold inward, allowing the prey to be drawn into the mouth while preventing the prey from withdrawing (e.g., *Eustomias*) (Morrow, 1964). The stomiid piscivore *Chauliodus* has a hinged skull that can rotate upward to allow large prey items to be ingested (Tchernavin, 1953).

Gut structure is somewhat variable, but all piscivores typically have large elongate stomachs (e.g., *Chauliodus*) (see Morrow, 1964). Expansible stomachs that allow for the ingestion of disproportionately large prey have been noted among the Saccopharyngidae (Nielsen and Bertelsen, 1984), Alepisauridae (Gibbs and Wilimovsky, 1966), Chiasmodontidae, and many ceratioid anglerfishes (Marshall, 1980).

Pigmentation of the gut walls is not unusual. Fishelson (1994) examined the construction of the gut in deep-sea benthic and pelagic eels. Varying degrees of melanization of the gut wall were observed (unless the eel possessed a black body, like *Eurypharynx*), and Fishelson concluded that one function was to hide any light produced by luminescent prey, which might expose the predator, from shining through the gut walls.

The eyes of many mesopelagic piscivores are often large relative to head size, suggesting that prey are visually tracked. In many piscivores, such as the Evermannellidae, Paralepididae, Scopelarchidae, and Giganturidae, the eyes are tubular, an accommodation to produce a binocular field and increase depth perception (see review in Marshall, 1980). One piscivorous species, *Scopelarchus analis,* has yellow eye lenses, which may produce a spectral shift such that the wavelengths of light from bioluminescence are more detectable (Muntz, 1976). This does not appears to be an adaptation solely for piscivory because zooplanktivores such as *Argyropelecus affinis* and *Malacosteus niger* and the generalist stomiid *Echiostoma barbatum* also possess yellow lenses (Somiya, 1976, 1979, 1982).

The cephalopod predator subguild includes members of the Evermannellidae, Omosudidae, and interestingly, the ceratioid anglerfish *Cryptopsaras couesi.* Sutton and Hopkins (1996) suggested that at least one stomiid species (*Heterophotus opisthoma*) may feed primarily on cephalopods. The general body characteristics of the taxa in this subguild are similar to those of the piscivorous subguild.

Two behavioral subguilds can be differentiated, an ambush predator subguild and an active forager subguild. Included in the first of these subguilds are most stomiids, the ceratioid anglerfishes (Fig. 20), and the saccopharyngid gulper eels. In ambush nektonivores, prey are generally thought to be lured by a luminescent device projecting from the lower jaw

4. FEEDING AT DEPTH

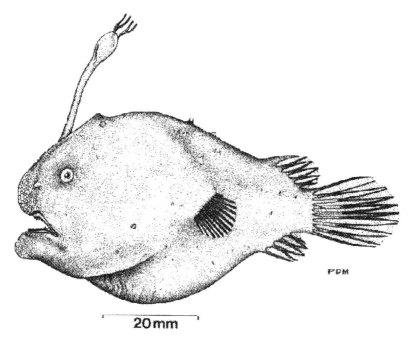

Fig. 20. Illustration of the pelagic ambush micronektonivore (pelagic trophic guild 1) *Himantolophus* species (cf. *albinares*), a ceratioid anglerfish. Note the specialized dorsal fin ray for luring prey. Drawn by P. MacWhirter.

(mental barbel, most stomiids), a modified dorsal fin [ceratioid anglerfishes (Bertelsen, 1951) and *Chauliodus* (Marshall, 1980)], or the tip of the caudal fin (Saccopharyngidae) (Nielsen and Bertelsen, 1984). Observations have been made from submersibles of the mesopelagic fish genera *Chauliodus* and *Stomias* (Fig. 21) by various researchers, including two of us (J. V. Gartner and K. J. Sulak). These fishes remain virtually motionless, with the elongate dorsal ray looped anteriorly over the mouth in the former and the mental barbel held outstretched and angled forward in the latter species. Bathypelagic ambush piscivores such as ceratioids and saccopharyngids, with eyes capable of light detection but lacking the ability to form images (Munk, 1984), are thought to be true "sit-and-wait" predators. However, some species such as *Chauliodus* and *Stomias* exhibit what Sutton and Hopkins (1996) term an asynchronous diel vertical migration. Nighttime distribution patterns for these genera clearly show that many individuals do not migrate and that the overall range of vertical migration is more limited than the range of their prey. The presumption is that these piscivores

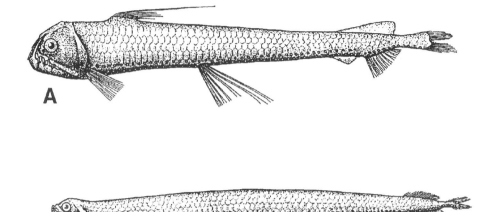

Fig. 21. Illustrations of pelagic ambush piscivores (pelagic trophic guild 1). (A) *Chauliodus sloani.* (B) *Stomias boa ferox.* Note the mental barbel for luring prey. From Morrow (1964).

position themselves at the main depths of upward and downward vertical migration and then wait for their prey to migrate through their locations.

In contrast, active foragers would include most "astronesthine" stomiids, and members of the Evermannellidae, Paralepididae, Scopelarchidae, Alepisauridae, Giganturidae, Chiasmodontidae, and Gempylidae. These fishes have well-muscled bodies, well-developed eyes, and strong dentition. Collection data for many of these families are sparse, presumably because of their ability to evade capture (Rofen, 1966).

b. Trophic Guild 2: Zooplanktivores. Based on the diversity of predator species, this is by far the largest overall pelagic feeding guild. The largest of the subguilds is the "hard-bodied" or crustacean zooplanktivores, including most species of the families Nemichthyidae, Derichthyidae, Gonostomatidae (Fig. 22A), Sternoptychidae (Fig. 22B), Phosichthyidae, Myctophidae (Fig. 22C), Bregmacerotidae, and some Melamphaidae, as well as many of the ceratioid anglerfish families. General characteristics for many of these groups include numerous fine teeth in the jaws and, excluding the Nemichthyidae and Derichthyidae, long and usually numerous gill rakers. A study of the feeding habits of the stomiids *Malacosteus niger* and *Photostomias guernei* by Sutton and Hopkins (1996) also places these two species in this guild.

Subdivisions of this would include small crustacean predators, overwhelmingly dominated by copepods as the primary prey items [see Hopkins

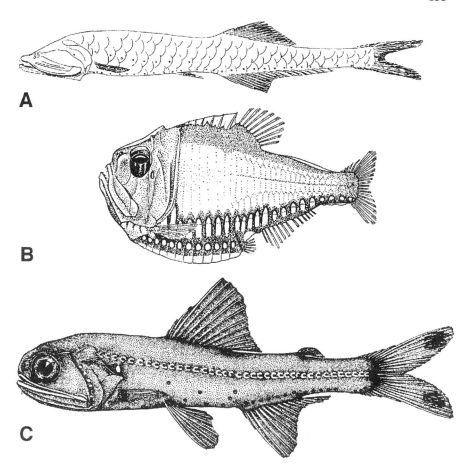

Fig. 22. Illustrations of representative pelagic zooplanktivores (pelagic trophic guild 2). (A) *Cyclothone microdon*, a nonmigratory bristlemouth in the family Gonostomatidae. Note the reduced eyes. From Grey (1964). (B) *Argyropelecus gigas*, a migratory hatchetfish in the family Sternoptychidae. Note the upwardly directed tubular eyes. From Schultz (1964). (C) *Diaphus dumerilii*, a vertically migrating lanternfish of the family Myctophidae. From Nafpaktitis *et al.* (1977).

and Baird (1977) for pertinent references] and large crustacean predators feeding on pelagic penaeidean and caridean shrimps. Major structural differences between the two subguilds would include differences in gape and dentition. Small crustacean predators include the afore-mentioned groups except for the Nemichthyidae and *Photostomias guernei*. Although morphology is usually closely correlated with the diest of these subguilds, there are some exceptions. The stomiid *Malacosteus niger* lacks gill rakers, has a head that hinges upward like *Chauliodus*, has enlarged teeth, and lacks a floor in the mouth to allow for jaw expansion. Such a morphology is suggestive of predation on larger prey items, although *M. niger* feeds mainly on copepods (Sutton and Hopkins, 1996). The other large crustacean predator, *Photostomias guernei*, has physical characteristics similar to those of *M. niger*, but feeds mainly on penaeidean shrimps (Sutton and Hopkins, 1996; J. V. Gartner, unpublished data, 1980).

Nemichthyid eels (Fig. 23A) feed primarily on penaeidean shrimps, especially sergestids (Nielsen and Smith, 1978; Gartner, unpublished data, 1980). An early hypothesis suggested by Meade and Earle (1970) was that these eels, which have large recurved jaws studded with tiny, inward-pointing teeth, are ambush predators using their jaws to entangle the antennae of their prey as the shrimps swim nearby. Observations we have made from submersibles (J. V. Gartner and K. J. Sulak) suggest that these are active predators that chase down their prey.

The derichthyid eels *Derichthys serpentinus* (Fig. 23B) and *Nessorhamphus ingolfianus* (Fig. 23C) also ingest crustaceans. Although morphologically similar, with overlapping vertical distributions in the western North Atlantic, these species show evidence of resource partitioning among individuals of similar body size. *Derichthys serpentinus* feeds primarily on sergestid shrimps whereas *N. ingolfianus* feeds on large euphausiids. These

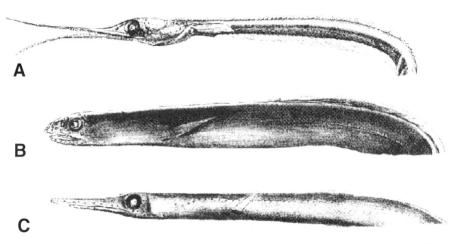

Fig. 23. Drawings of the heads of pelagic zooplanktivorous eels (pelagic trophic guild 2). (A) *Nemichthys scolopaceus* (family Nemichthyidae). Modified from Smith (1989). (B) *Derichthys serpentinus* (family Derichthyidae). Modified from Robins (1989). (C) *Nessorhamphus ingolfianus* (family Derichthyidae). Modified from Robins (1989).

differences appear to be related to the width of the gape, with adult *Nessorhamphus* showing a distinctly smaller gape compared to *Derichthys* of similar length (Gartner, unpublished data, 1980).

Predators on soft-bodied zooplankton are divided into two subguilds, one that feeds on various gelatinous prey, including various cnidarians, as well as thaliacean and larvacean chordates, and a second subguild that eats gastropod mollusks.

The gelatinous predator subguild is characterized by members of the mesopelagic families Bathylagidae (Fig. 24A), Opisthoproctidae (Fig. 24B), and some members of the Melamphaidae. All three groups share the common characteristics of relatively small gape, very fine teeth in the jaws, and a longer, more coiled intestine than is found in members of other guilds (Balanov *et al.*, 1995). Interestingly, opisthoproctids and bathylagids are closely related to the demersal Alepocephalidae, many of which are gelatinous plankton predators (see Section III,B,1,g).

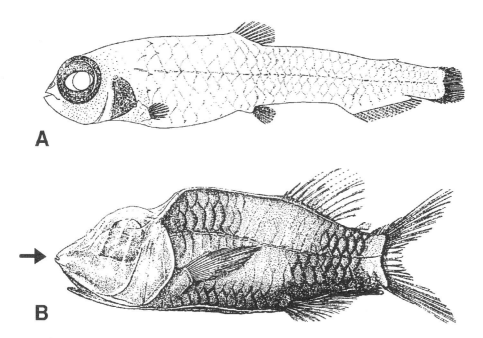

Fig. 24. Illustrations of pelagic zooplanktivores feeding on soft-bodied prey (pelagic trophic guild 2). (A) *Bathylagus euryops* (family Bathylagidae). (B) *Opisthoproctus soleatus* (family Opisthoproctidae). The arrow indicates position of the mouth. Modified from Cohen (1964).

Predation on gastropod mollusks has thus far been observed in one myctophid species, *Centrobranchus nigroocellatus* (Fig. 25), a species that feeds solely on pteropod and heteropod molluscs (Gorelova, 1977; Hopkins and Gartner, 1992). The species differs in morphology and behavior from almost all other myctophids, possessing no gill rakers, a subterminal mouth, and a very narrow caudal peduncle. At night it occupies surface waters and is commonly collected in neuston nets (Gartner *et al.*, 1989).

c. Trophic Guild 3: Generalists. Some pelagic fishes appear to feed on a wide range of prey, which is much more in keeping with the idealized notion of broad-spectrum opportunism in feeding in deep-ocean fishes. Most of these species come from taxa mentioned in the previous guilds (e.g., the stomiid *Echiostoma barbatum*), and there are few distinctive morphological or behavioral differences that have been reported to separate them from their related taxa. Robison (1984) noted a longer more convoluted intestine in the myctophid *Ceratoscopelus warmingii* and presented this as evidence supporting occasional herbivory by this species on algal mats. Others have reported this species to be a broad-spectrum feeder (Clarke, 1980; Duka, 1987; Hopkins and Gartner, 1992).

One additional family that can be included in this guild is the monotypic bathypelagic gulper eel family Eurypharyngidae (*Eurypharynx pelecanoides*) (Fig. 26). It possesses a very large gape, weak jaws, and an extremely flaccid body and is presumed to behave like a living net, engulfing prey by slowly swimming over them with its mouth open (Böhlke, 1966). Its prey include caridean shrimps, fishes, and copepods. Surprisingly, benthic prey

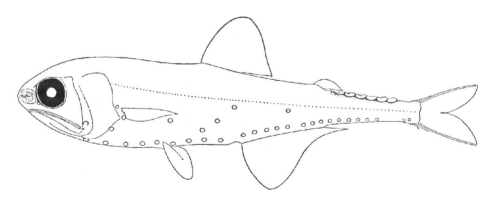

Fig. 25. Drawing of the zooplanktiovorous specialist *Centrobranchus nigroocellatus* (family Myctophidae). From Nafpaktitis *et al.* (1977).

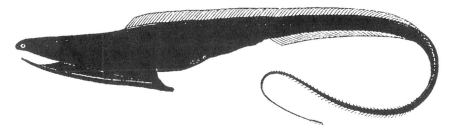

Fig. 26. The bathypelagic generalist (pelagic trophic guild 3) *Eurypharynx pelecanoides* (family Eurypharyngidae). Note the enormous gape and reduced eyes. From Böhlke (1966).

have also been recovered from the stomachs of *E. pelecanoides* (Bertin, 1934; Gartner, unpublished data, 1980).

C. Congruent Patterns in Morphological Specialization among Benthic and Demersal Fish Species: Common Themes on the Shelf and in the Deep Sea

Relatively few studies of morphological feeding specialization have been undertaken for deep-sea fishes, and fewer still have been substantiated by feeding habits data. An overview of available information suggests that demersal fishes of the open slope and abyss present a range of dietary and morphological specialization fairly comparable to that seen in their soft-substrate counterparts of the continental shelf and estuaries. Of course, there are a few notable exclusions due to limiting conditions in the deep sea. Thus, herbivores are absent, along with plankton filterers. Also rare in the deep sea are trophic specialists equivalent to certain fishes found on coral reefs, and in other structured habitats. This is due to the general rarity of such hard-substrate live-bottom habitats in the deep sea beneath high-productivity surface waters (where thick sediments tend to prevail).

Congruence in trophic morphology may be drawn between various dominant demersal deep-sea taxa and apparent shallow-water counterparts. For example, the Macrouridae present a level of trophic adaptation and diversity equivalent to the very remotely related Sciaenidae of coastal and estuarine waters. Both families include mobile epibenthic/benthopelagic foragers of small to large size (0.1–50 kg) and varied food habits. Analogy in feeding morphology (i.e., snout shape, mouth position, type of dentition) is evident between respective sciaenid (Chao and Music, 1977) and macrourid (McLellan, 1977) genera with equivalent food habits. A composite figure of these adaptations is presented in Fig. 27. Macrourid (*Bathygadus* and *Gadomus*) and sciaenid (*Bairdiella* and *Larimus*) nekton specialists closely parallel one another (terminal mouth, long jaws, blunt snout, restricted jaw protrusibility, small teeth, and large round gape). The primarily benthivorous macrourid *Nezumia* closely parallels the sciaenid *Menticirrhus* (subterminal mouth, moderate jaws, short pointed snout, moderate jaw protrusibility, small teeth, and reduced gape). Both even have a single short barbel in the same position. The macrourid genus *Coelorinchus* parallels the sciaenid *Leiostomus* (inferior mouth, short jaws, high jaw protrusibility, small to obsolete teeth, and limited gape), although the *Leiostomus* lacks the elongate snout of its deep-sea counterpart. Both are small fishes that feed on small benthic prey and adopt the same inclined body attitude when foraging, reverting to a horizontal attitude when swimming. Large abyssal species such as *Coryphaenoides leptolepis* and *C. armatus* round out the Macrouridae/Sciaenidae analogy, offering an approximate trophic sequel to large sciaenids such as *Sciaenops* and *Cynoscion* (terminal to subterminal mouths, long jaws well equipped with teeth, limited protrusibility, large gapes, and broad feeding habits centering on fishes and large decapod crustaceans).

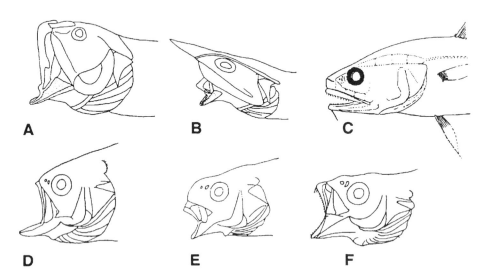

Fig. 27. A composite of line drawings showing congruence in trophic morphology between shallow-water and deep-sea fishes. Deep-water macrourids: (A) *Bathygadus;* (B) *Coelorinchus.* From McLellan (1977). (C) *Chalinura leptolepis.* From Marshall (1973). Shallow-water sciaenids [from Chao and Musick (1977)]: (D) *Larimus fasciatus;* (E) *Leiostomus xanthurus;* (F) *Cynoscion regalis.*

4. FEEDING AT DEPTH

D. Inferences from the Morphology of Deep-Sea Fishes: Trophic Strategies and Prey Selection

1. BENTHIC AND DEMERSAL SPECIES

Morphology provides fundamental insight into trophic behavior and suggests functional parallels, as just discussed. However, there is danger in trying to draw too direct an interpretation of feeding habits from morphology. The Macrouridae is the most speciose, and perhaps trophically most diverse, demersal fish family in the deep sea. McLellan (1977) investigated trophic morphology in this family, relating mouth size, mouth position, jaw protrusibility, jaw musculature, and snout development among macrourids to possible feeding strategies. She recognized two extremes in specialization as represented by the genera *Bathygadus* (subfamily Bathygadinae) and *Coelorinchus* (subfamily Macrourinae). The blunt snout, terminal jaws, and voluminous gape in *Bathygadus* correspond with macrophagous predation on nektonic prey (fishes and decapod crustaceans). Relative to upper jaw length, the jaws are only moderately protrusible. Protrusibility is limited by an ascending process much shorter than premaxilla length. But when the mouth is opened during a feeding strike, the volume of the buccal cavity is suddenly greatly expanded, and prey is engulfed via oral suction much as in the largemouth bass (Nyberg, 1971). The jaws in bathygadine macrourids bear bands of tiny teeth, poorly adapted for piercing, holding, or manipulating prey. Moreover, the adductor mandibulae (McLellan, 1977) is configured to emphasize gape enlargement rather than holding power. Teeth and musculature both suggest that prey is swallowed whole with little manipulation or mastication.

In *Coelorinchus* the long bony rostrum, inferiorly positioned jaws, and limited gape correspond with benthic foraging on small prey (e.g., euphausiids, amphipods, polychaetes) removed from the sediment or picked off the substrate surface. Relative to upper jaw length, the jaws are quite protrusible in a downward direction. Protrusibility is enabled by an ascending process longer than premaxilla length. Mouth opening results in limited volume expansion of the buccal cavity. Prey is apparently ingested via jaw manipulation, together with buccal suction. Although jaw teeth in *Coelorinchus* and many other macrourine macrourids are tiny, prey is probably manipulated orally before swallowing such that sediment can be ejected out the gill openings or mouth. The superficial and inner sections of the adductor mandibulae (Geistdoerfer, 1977; McLellan, 1977) are appropriately configured to facilitate dorsal/ventral movement and strong retraction of the lower jaw. When foraging, *Coelorinchus* and related genera such as *Nezumia* move along, snout to the substrate, with the tail elevated at a steep angle. In this body attitude the open mouth protrudes directly toward the substrate.

Marshall and Bourne (1964) and McLellan (1977) have speculated that macrourines use the elongate snout to probe the substrate for prey. However, such probing behavior has not been reported during submersible observations. Remote camera observations of burst swimming into the sediment [termed "explosive sediment diving" by Isaacs and Schwartzlose (1975)], followed by sediment expulsion among macrourids, probably represents a startle response to intense illumination rather than feeding. Although the lateral ridges of the rostrum and snout tip are bony and protected by special hard stellate scales in some genera, the underside of the snout is typically soft, naked, and unprotected from abrasion. The true function of the elongate snouts of benthophagous macrourids (and similarly halosaurs, chimaeras, and other taxa) is probably enlargement of surface area of the sensory apparatus used to detect small prey (McDowell, 1973). A tactile and/or chemical sensory function facilitating prey detection would be consistent with the snout oriented to the substrate plus continuous slow swimming behavior routinely observed in genera such as *Nezumia*.

Sturgeons may provide an approximate model for benthic feeding in macrourids equipped with long snouts and protrusible ventral mouths (e.g., *Coelorinchus, Trachyrinchus*). The projecting snout of sturgeons is not used to dig for prey, but as a rostrum to suspend sensory barbels ahead of the mouth. Prey is detected as the fish swims along the substrate and is instantaneously sucked into the oral cavity as the mouth is protruded. Sediment and debris are ejected through the mouth and gill apertures. A similar sensory function for the elongate snout of the abyssal chimaera *Harriotta raleighana* (Fig. 28A) has been suggested by Sedberry and Musick (1978), and for the projecting snouts of halosaurs (Fig. 28B) by McDowell (1973).

While instructive, the dichotomy between the extremes of bathygadine and macrourine feeding specializations is imprecise. Indeed, many bathyal genera display intermediate morphologies and body size, suggesting varied or optional feeding habits. Nor is morphology a definitive predictor of actual prey selection. Thus, a ventral mouth position does not prevent *Nezumia* from feeding extensively on benthopelagic prey. Another macrourine with a highly protrusible inferior mouth, *Coryphaenoides rupestris*, feeds predominantly not on benthic prey, but on benthopelagic and pelagic prey (Savvatimskii, 1969; Podrazhanskaya, 1971; Geistdoerfer, 1979a). Absolute body size and mouth size are important factors in the ultimate equation of trophic adaptation in a given species. Thus, small-bodied, small-mouthed macrourids of all subfamilies are limited to small prey, whereas large-bodied, large-mouthed macrourids (e.g., *Bathygadus, Coryphaenoides*) can accommodate both large and small prey (Geistdoerfer, 1979a; Merrett and Marshall, 1980). Moreover, diet may shift ontogenetically, such

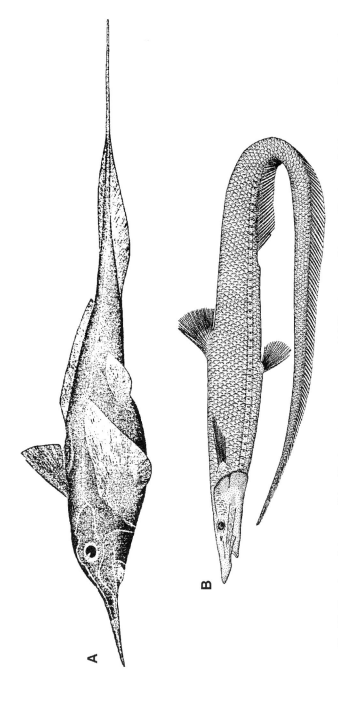

Fig. 28. Drawings of deep-sea fish with elongate sensory snouts. (A) The deep-sea chimaera *Harriotta raleighana*. From Goode and Bean (1895). (B). *Aldrovandia affinis* (family Halosauridae). Drawn by P. Pebbles.

that body size overrides specialized morphology, especially in macrourine species that attain large size. Thus, small *Macrourus berglax* use their ventral mouth to feed predominantly on benthic amphipods and polychaetes, but large *M. berglax* (>50 cm) switch to a diet of benthic fishes, shrimps, and ophiuroids (Geistdoerfer, 1979b). Large size is typical of abyssal rattails, which probably depend heavily on more occasional large prey and scavenged food falls. Such large abyssal species have larger, less numerous teeth set in rows, enabling holding, manipulating, and tearing (based on baited-camera evidence) of large food items.

2. PELAGIC SPECIES

a. Feeding Selectivity. Determination of selectivity toward prey (either positive or negative) is an aspect of feeding ecology of deep-sea pelagic fishes that has historically proved ambiguous. This is mostly with respect to inclusion of specific prey taxa; it is easier to determine that certain prey available in the environment are *not* being selected than to determine if certain taxa are more often eaten. A central assumption that has long been held is that in deep-ocean environments, opportunistic generalists should be favored owing to decreased food availability, and that this pattern should become more apparent with increasing depth (Ebeling and Cailliet, 1974; see also review in Marshall, 1980). Many earlier studies of feeding habits and stomach contents routinely supported this assumption. However, more recent studies suggest this assumption may be erroneous and possibly even invalid. By examining stomach contents in greater detail, a number of researchers suggest that in fact there is a high degree of inter- and intraspecies selectivity among mesopelagic fishes at least (e.g., Borodulina, 1972; Merrett and Roe, 1974; Clarke, 1980, 1982; Hopkins and Gartner, 1992; Sutton and Hopkins, 1996). Unlike the difficulties in assessing specializations or lack thereof among benthic and demersal fishes (see Section III,A,1), most studies conducted on the feeding habits of pelagic fishes are accompanied by concomitant sampling to ascertain prey availability (e.g., Clarke, 1980; Hopkins and Gartner, 1992).

Several factors render determination of prey selectivity a complex task, not the least of which is how one determines selectivity. Selectivity can be related to the taxa, sizes, and distributions of both predator and prey species [see discussions in Hopkins and Baird (1977) and Hopkins and Gartner (1992)]. Selectivity may be made on the basis of prey taxa being taken by many individuals of a predator species in disproportionate numbers to their environmental abundance (e.g., Merrett and Roe, 1974; Clarke, 1980; Hopkins and Baird, 1985a). Specific size ranges of a single prey taxon may be ingested in disproportionate numbers (e.g., Clarke, 1980; Hopkins and Gartner, 1992). Similarly, specific size ranges of the predator may ingest a certain size range of different prey items *or* a specifically sized prey taxon (e.g., Gorelova, 1981; Lancraft *et al.*, 1988).

Resource partitioning and selectivity among predatory species on the basis of size, taxa, and predator distribution patterns can often be ascertained if stomach contents are identified to the lowest possible taxa and prey sizes are determined as much as possible. Use of broad taxonomic categories (e.g., copepods, euphausiids, ostracods) has been done in many diet studies (Collard, 1970; Gjösaeter, 1973; Kinzer, 1977; Gorelova, 1978; Kinzer and Schulz, 1985; Duka, 1987; Dalpadado and Gjösaeter, 1988; Gorelova and Krasil'nikova, 1990) and provide an excellent base line for more detailed examinations of diet, but by their very nature often cannot provide clear evidence of selection for particular taxa or size groups within taxa.

Interpretation of the evidence from such papers may lead to the assumption, for example, that myctophids typically opportunistically forage on copepods, at least small-sized ones, and may ontogenetically switch to larger euphausiid taxa because the actual selectivities have been obscured by the use of broad categories. Studies that have divided the data into more numerous, smaller categories such as predator and prey sizes and prey taxa have usually concluded that at least some species clearly exhibit some form of prey selection. In truth, it would appear that this more often the rule than the exception (e.g., Borodulina, 1972; Clarke, 1980; Gorelova, 1981; Hopkins and Baird, 1977, 1985a; Young and Blaber, 1986; Lancraft *et al.*, 1988; Gartner and Musick, 1989; Balanov *et al.*, 1995; Sutton and Hopkins, 1996).

b. Daily Ration Estimates and Gut Evacuation Rates. Some studies have estimated the daily ration of various pelagic deep-sea fish (Table II). The data in many cases may not be directly comparable because of differing methods used, but the agreement among studies is striking, ranging from approximately 1.0 to 4.5% of the dry weight of the prey to the dry body weight of the predators per day.

The types and amounts of food (in biomass and calories) needed to meet metabolic demands presumably vary with species, diel and seasonal periods, age, and sexual maturity of individuals. Only a few studies have addressed aspects of the energetic needs of mesopelagic fishes (Baird and Hopkins, 1981b; Hopkins and Baird, 1977, 1985b; Gartner, 1993) and none has integrated all of these variables into a single model for feeding dynamics of any species.

There are few experimental data on gut evacuation rates. Most studies that include estimates have based their calculations on shallow-water fishes

Table II
Estimated Daily Rations of Mesopelagic Fishes

Family/species	Daily ration (estimation method)	Reference
Bathylagidae		
Leuroglossus ochotensis	1.5–2.1%[a]	Gorbatenko and Il'inskii (1992)
Leuroglossus stilbius	2.0%[a]	Gorbatenko and Il'inskii (1992)
Gonostomatidae		
Gonostoma elongatum	2–4%[b]	Lancraft *et al.* (1988)
Myctophidae		
Diaphus taaningi	0.8%[c]	Baird *et al.* (1975)
Hygophum proximum	5.75%[d]	Clarke (1978)
Lampanyctus alatus	2–4%[b]	Hopkins and Baird (1985b)
Stenobrachius leucopsarus	1.1%[a]	Gorbatenko and Il'inskii (1992)
Stenobrachius nannochir	1.0%[a]	Gorbatenko and Il'inskii (1992)
Benthosema pterotum	4.5%[e]	Dalpadado and Gjösaeter (1988)
Phosichthyidae		
Vinciguerria nimbaria	6–18%[d]	Clarke (1978)
Sternoptychidae		
Danaphos oculatus	1.9%[d]	Clarke (1978)
Valenciennellus tripunctulatus	4–7%[d]	Clarke (1978)
Stomiidae		
Piscivores	2–4%[e]	Sutton and Hopkins (1996)
Crustacean zooplanktivores	0.6–1.5%[e]	Sutton and Hopkins (1996)

[a] Based on digestion rate and amount of prey in stomach relative to abundance of prey in environment.
[b] Mean values of stomach ash-free dry weight (AFDW) as percentage of maximum full stomach AFDW.
[c] Percentage of body weight as food.
[d] Changes in stomach fullness over time period.
[e] Average dry weight (DW) of prey as percentage DW of predator.

of similar morphologies and temperature regimes (e.g., Hopkins and Baird, 1977, 1985b; Sutton and Hopkins, 1996).

In the cases of evident diel periodicity in feeding (see Section III,E), evacuation rates have been assumed to be less than 24 h for zooplanktivores, and some authors have suggested the same rates for mesopelagic piscivores (Legand and Rivaton, 1969; Merrett and Roe, 1974). Clarke (1982), examining piscivorous stomioids in Hawaiian waters, presented trophic impact models based on digestion/evacuation ranges of between 1 and 4 days. Sutton and Hopkins (1996), examining the most abundant stomiid piscivores in the Gulf of Mexico, suggested that, in these species, gut evacuations are considerably slower than in zooplanktivores and that the piscivores exhibit "snake-like feeding," i.e., acquiring large meals asynchronously and digesting meals slowly, perhaps on the order of 5 days or more.

c. Predation Effects on Prey Populations. Most assessments of the effects of predation by deep-sea pelagic fishes on prey populations have provided data mainly for the abundant mesopelagic fish groups Myctophidae, Gonostomatidae, Sternoptychidae, and zooplanktivorous Stomiidae. In all cases, the conclusions were that these groups at minimum estimates are the primary source of removal of the herbivorous zooplankton biomass in epipelagic waters (Table III).

Despite differences in regions and methods of sampling and methods of estimating the effects of predation on prey populations, the studies all indicate that, with current sampling technologies and lack of direct observations, mesopelagic fishes are capable of removing two to three times the total annual standing stock (secondary production integrated over a 1-year period) of herbivorous zooplankton. Sutton and Hopkins (1996) suggested a similar importance of the mesopelagic family Stomiidae as predators on these mesopelagic fish groups. Although these studies all concur that this is not ecologically possible, it is clear that turnover and replacement of prey populations are rapid. Available evidence from these

Table III
Estimated Removal of Annual Standing Stock of Prey Removed by Pelagic Fish Taxa

Taxa	Standing stock removed (%)	Reference
Nekton-eating stomiids[a]	57.5–230[b]	Clarke (1982)
Benthosema pterotum	300	Dalpadado and Gjösaeter (1988)
Gonostoma elongatum	<30[c]	Lancraft *et al.* (1988)
Bathylagidae, Myctophidae[d]	300	Gorbatenko and Il'inskii (1992)
Myctophidae[e]	300	Hopkins and Gartner (1992)
Piscivorous stomiids[f]	140–240[b]	Sutton and Hopkins (1996)
Zooplanktivorous stomiids[g]	31	Sutton and Hopkins (1996)

[a] Includes seven families (six now incorporated as Stomiidae) and 28 species.
[b] Percentage ranges based on estimated rates of gut evacuation.
[c] Calculated only for copepod genus *Pleuromamma*, a principal diet component.
[d] Includes two species in each family.
[e] Includes 17 species.
[f] Includes 49 species.
[g] Includes eight species.

studies suggest that mesopelagic fishes have an important and possibly critical role in causing these rapid changes in prey populations, as well as being pivotal in energy flow from surface waters to the deep ocean.

Hopkins *et al.* (1997) have done the most comprehensive work to date on the subject of trophic interactions in the mesopelagic zone, examining the predation effects of virtually the entire mesopelagic fish assemblage (16 families) in the eastern Gulf of Mexico, a system similar to the open-ocean central gyres. Their findings reinforce the concept of the crucial role in energy flow played by the dominant zooplanktivore families Myctophidae, Sternoptychidae, and Gonostomatidae (in decreasing order of importance) and piscivorous members of the Stomiidae.

E. Diel and Seasonal Feeding Patterns

1. BENTHIC AND DEMERSAL SPECIES

Many deep-sea fishes show diel feeding patterns, often related to the diel vertical migrations of mesopelagic prey. Bulman and Koslow (1992) reported that the diet of orange roughy, *Hoplostethus atlanticus*, collected at depths of 700–1200 m, changes over a diel cycle, and Golovan' and Pakhorukov (1975) reported that *Alepocephalus bairdi* feeds most intensively during the morning. Blaber and Bulman (1987) reported diel feeding patterns for three species collected from 420 to 550 m. Dudochkin (1988) reported that catches of *Macrourus holotrachys* are lowest at night and suggested that this macrourid leaves the bottom at night and follows its prey on a diel vertical migration toward the surface.

In addition to diel patterns, seasonal changes in diet have been reported for some species. Blaber and Bulman (1987) reported seasonal changes in diet and studied several species collected off Tasmania that were related to seasonal changes in prey abundance. Gordon (1979) reported seasonal changes in the diet of *Coryphaenoides rupestris*. Seasonal phenomena have also been reported from abyssal depths. Armstrong *et al.* (1991) reported that the staying time of abyssal macrourids around baited-camera arrays appeared to vary seasonally, perhaps reflecting seasonal changes in prey abundance. Similarly Priede *et al.* (1994b) found that the activity levels of macrourids that had swallowed ultrasonic transmitters were seasonally variable. They suggested that the activity pattern of fish could be coupled to the seasonal cycle of surface production. They also reported seasonal changes in the abundance of fishes, with lower densities during February.

2. PELAGIC SPECIES

The literature on the chronology of feeding activities is replete with correlations between migrations and feeding. Most studies clearly indicate that feeding among vertical migrators occurs most intensively or exclusively during nighttime hours (Table IV, see Hopkins and Baird, 1977 and Marshall, 1980 for reviews).

One prevalent concept is that diel migration is a foraging strategy moving predator populations from a region where food is sparse to where it is much more concentrated, i.e., migrations are food driven (see Marshall, 1980). In contrast to this is the idea that because deep-water migratory fish species are derived from ancestral stocks driven from shallow waters by competition and predation pressures, vertical migrations are actually downward migrations during the day, to avoid being eaten, with a return to ancestral shallow waters during the night, to avoid the primarily visual, diurnal, epipelagic predators.

Regardless of which of these strategies drives migration, most migrators show a synchronized patterns of feeding on a diel basis. In contrast, many weakly migratory or nonmigratory species have no clear temporal feeding pattern [e.g., the melamphaid *Scopelogadus beanii* (Gartner and Musick, 1989) and *Sternoptyx diaphana* and *Sternoptyx pseudobscura* (Hopkins and Baird, 1985a)]. In the cases of some limited migrators or nonmigrators such as *Danaphos oculatus* and *Valenciennellus tripuctulatus,* which are resident in the upper mesopelagic zone (<350 m), a diurnal feeding pattern is observed (Clarke, 1978; Hopkins and Baird, 1981).

Many stomiids, despite the fact that they are migratory, have often been suggested to be asynchronous feeders with respect to a diel cycle, based on the supposition that prey items are ingested and digested within a 24-h period (e.g., Merrett and Roe, 1974; Clarke, 1978). More recently, Sutton and Hopkins (1996) presented evidence to suggest that many of these stomiid migrators actually feed at night, but that they may take several days to digest the prey contents. Thus there is synchronization in terms of the time during a diel period when they will feed, but the pattern will not repeat on a nightly basis.

A number of conditions, such as ontogenetic stage or lunar period, have been shown to affect the range of vertical migration among active migrators, and these conditions may also affect feeding. Among myctophids, very young juvenile stages often show little or no vertical migration for a period of time, however, no information has been published on these stages specifically comparing their diet to that of migratory members of the population (Clarke, 1973; Badcock and Merrett, 1976; Willis and Pearcy, 1980; Gartner *et al.,* 1987; Karnella, 1987). Among migrators, older stages are generally found deeper than younger stages and may exhibit a reduced migration range or cease migrations altogether (Clarke and Wagner, 1976; Nafpaktitis *et al.,* 1977; Lancraft *et al.,* 1988).

Table IV
Diel and Seasonal Feeding Periodicity Reports for Various Midwater Fish Taxa

Taxa[a]	Chronology[b]				Reference
	N	C	D	S	
Benthosema glaciale (MC)	+	−	−	+	Gjösaeter (1973)
Myctophidae (3 spp.)	+	−	−	ND	Merrett and Roe (1974)
Valenciennellus tripunctulatus (SE)	−	−	+	ND	Merrett and Roe (1974)
Argyropelecus aculeatus (SE)	−	DK	−	ND	Merrett and Roe (1974)
Argyropelecus hemigymnus (SE)	−	DK	−	ND	Merrett and Roe (1974)
Chauliodus sloani (SO)	−	−	−	ND	Merrett and Roe (1974)
Diaphus taaningi (MC)	+	−	−	ND	Baird *et al.* (1975)
Benthosema glaciale (MC)	+	−	−	ND	Kinzer (1977)
Vinciguerria nimbaria (P)	+	−	−	ND	Ozawa *et al.* (1977)
Gonostoma (3 spp.) (G)	+	−	−	ND	Gorelova (1981)
Argyropelecus aculeatus (SE)	+	−	−	ND	Hopkins and Baird (1985a)
Argyropelecus hemigymnus (SE)	−	DK	−	ND	Hopkins and Baird (1985a)
Sternoptyx (2 spp.) (SE)	−	−	−	ND	Hopkins and Baird (1985a)
Lampanyctus alatus (MC)	+	−	−	ND	Hopkins and Baird (1985b)
Myctophidae (7 spp.)	+	−	−	ND	Kinzer and Schulz (1985)
Diaphus danae (MC)	−	−	−	ND	Young and Blaber (1986)
Lampanyctodes hectoris (MC)	−	−	−	ND	Young and Blaber (1986)
Maurolicus muelleri (SE)	+	−	−	ND	Young and Blaber (1986)
Ceratoscopelus warmingii (MC)	+	−	−	ND	Duka (1987)
Benthosema pterotum (MC)	+	−	−	−	Dalpadado and Gjösaeter (1988)
Gonostoma elongatum (G)	+	−	−	−	Lancraft *et al.* (1988)
Scopelogadus beanii (ML)	−	−	−	+	Gartner and Musick (1989)
Maurolicus muelleri (SE)	+	−	−	+	Gorelova and Krasil'nikova (1990)
Bathylagidae (2 spp.)	ND	ND	ND	+	Balanov *et al.* (1995)
Myctophidae (2 spp.)	ND	ND	ND	+	Balanov *et al.* (1995)
Myctophidae (6 spp.)	+	−	−	ND	Kinzer *et al.* (1993)
Stomiidae (3 abundant spp.)	−	−	−	ND	Sutton and Hopkins (1996)
Derichthys serpentinus (D)	−	−	+	−	Gartner, unpublished data, 1980

(*continues*)

Table IV Continued

Taxa[a]	Chronology[b]				Reference
	N	C	D	S	
Nemichthys scolopaceus (N)	−	−	−	−	Gartner, unpublished data, 1980
Nessorhamphus ingolfianus (D)	−	−	+	+	Gartner, unpublished data, 1980
Serrivomer beanii (SV)	−	−	−	−	Gartner, unpublished data, 1980

[a] Key: D, Derichthyidae; G, Gonostomatidae; MC, Myctophidae; ML, Melamphaidae; N, Nemichthyidae; P, Phosichthyidae; SE, Sternoptychidae; SO, Stomiidae; SV, Serrivomeridae.

[b] Key: N, night; C, crepuscular (DA, dawn; DK, dusk); D, day; S, seasonal; +, positive periodicity; −, negative periodicity; ND, no data.

The phase of the moon has been shown to affect the migration patterns of some lanternfishes. Migratory responses in mesopelagic fishes are correlated with ambient light intensities (Boden and Kampa, 1967). Several studies on migratory myctophids have noted a reduction in or cessation of vertical migration based on lunar periodicity (Clarke, 1973; Gartner et al., 1987; Linkowski, 1996). The extent to which these alterations affect feeding is currently unknown and needs to be explored.

That reductions in the range of vertical migration affect physiology is evident. Torres et al. (1979) and Donnelly and Torres (1988) have demonstrated decreasing oxygen consumption rates with increasing depth of occurrence in midwater fishes. Gartner (1991a) and Linkowski (1991, 1996) examined otolith microstructure in various actively migratory myctophid species, and both authors noted regions of reduced calcification in the otoliths, which they correlated with a reduction or cessation of diel migration in these animals. McLaren (1963) was the first to suggest that the migration into deep waters during the daytime produced the metabolic benefit of lowering activity levels. Meals taken during restricted periods of high activity were thought to provide enough energy not only to offset the costs of vertical migration, but also to meet basic metabolic needs and provide a surplus for growth, reproduction, etc. The physiological nature of vertical migration is still poorly understood and so the energetic cost/benefit analyses are at best approximations. At a population level, elucidating the energetics of vertical migration and feeding is not only complicated by the reduction or cessation of migratory activities by certain life history stages, as previously mentioned, but also by the fact that we do not know if all members of the actively migrating fraction of the population do so during each 24-h cycle.

Alteration of feeding patterns synchronized with diel vertical migration has been demonstrated by Pearcy et al. (1979) for the myctophid species *Stenobrachius leucopsarus*, which is dominant in eastern Pacific waters. They showed that this species exhibited a nighttime bimodal distribution peak, and an analysis of stomach contents indicated that the shallow migrators fed in shallow waters at night, whereas deep nonmigrators mainly fed at depth, apparently during the day.

Few studies have quantified seasonal patterns in feeding among midwater fish species, because few long-term seasonal collections both of fishes and of their prey have been made in a uniform and simultaneous fashion. As a result, many studies that reference such patterns are equivocal in summarizing their findings.

Published reports that demonstrate seasonality of feeding are primarily from studies on the abundant mesopelagic fish families Myctophidae and Gonostomatidae (Table IV; Gjösaeter, 1973; Dalpadado and Gjösaeter, 1988; Lancraft et al., 1988; Gartner and Musick, 1989; Gorelova and Efremenko, 1989; Balanov et al., 1995). All suggest that there are shifts in prey composition based on shifts in abundances of prey taxa, but that cross-taxa shifts are rare. Thus, a species feeding predominantly on copepods will continue to do so, but may seasonally shift among whatever copepods are most abundant.

IV. SOURCES OF FOOD IN THE DEEP SEA

A. Marine Snow and Foodfalls

Of paramount interest to deep-sea ecologists is the downward cycling of energy from the epipelagic zone to the deep benthos. Examination of the trophodynamics of deep-sea fishes is an exceptionally complex risk, owing to the variability of prey in both time and space. Patchiness of prey items in the pelagic environment has long been a source of discussion among oceanic biologists, and the relatively energy depauperate benthos, particularly at lower continental slope and rise and abyssal depths, is well known (see Marshall, 1980). In addition to evaluating feeding habits of deep-sea fishes, determinations of the sources of food in oceanic environments and its distribution and transfer among trophic levels and habitats are vital to evaluating deep-ocean feeding ecology.

Because of the general paucity of available food items in deep-ocean ecosystems, and in order to attempt to define vertical coupling of energy

4. FEEDING AT DEPTH   173

flow into the deep ocean from the surface, it is important to consider all aspects of what might constitute available energy to the various trophic guilds of fishes we have mentioned (see Section III,A). Accordingly, we examine here additional direct and indirect sources of potential energy to pelagic and demersal fishes.

Other than the animals living in the environments, there are other potentially large sources of energy available to pelagic, demersal, and benthic deep-sea fishes. These include such things as marine snow and large foodfalls, which may be of terrigenous origin (e.g., plant remains).

The use of marine snow as an energy source is being most intensively investigated as it pertains to the plankton (e.g., Alldredge, 1972; Alldredge and Silver, 1988; Lampitt *et al.*, 1993), and as a direct source is probably most heavily exploited by grazing plankton. Marine snow, generally defined as settling particles >0.5 mm in diameter (Lampitt *et al.*, 1993), can encompass an enormous variety of organic and inorganic source materials. Certain types of marine snow, such as larvacean houses, have been shown to serve as food for various zooplankton that aggregated around the discarded houses (Alldredge, 1972). Such aggregations can conceivably attract the attention of various pelagic zooplanktonivores and thus serve as a richer food source, both in the pelagic and benthic environments, when materials settle to the bottom. It is quite possible that marine snow aggregates could serve as an important food source for the demersal/benthic detritivore feeding guild.

Suspension of marine snow in larger quantities in density layers may also serve as sites of increased overall plankton prey density. A benthic boundary, or "nepheloid," layer has been observed in various continental slope regions and may attract pelagic species, drawing them close enough to the bottom (<10 m) that they become prey for various demersal or benthic predators (Sedberry and Musick, 1978).

Sinking rates of various marine snow components are quite variable (<100 to >1000 m/day (Wiebe *et al.*, 1979; Riemann, 1978; Robison and Bailey, 1981; Alldredge and Gottschalk, 1988; Alldredge and Silver, 1988; Lampitt *et al.*, 1993). Particles with slow sinking rates and high residence times in midwater probably are of most use to pelagic fishes by aggregating zooplankton, but many of these particles may be too small and too dispersed to produce many animal aggregations. The best likely indirect sources for pelagic animals seem to be the remains of gelatinous animals and perhaps some types of fecal pellets (Lampitt *et al.*, 1993). Fish fecal pellets, gelatinous aggregates, and even certain phytoplankton have extremely rapid settling rates of >100 m/day [1000 m/day (Robison and Bailey, 1981) for myctophid fecal pellets] and thus will serve as an enrichment source primarily or entirely for inhabitants of the benthos (Robison and Bailey, 1981; Riemann, 1989).

Foodfalls of large items, such as carcasses of various fishes, squids, and marine mammals, and large inputs of anthropogenic materials may periodically augment the diets of benthic and demersal fishes (see Section III,B,1), but only a few species are adapted to feeding solely on carrion (trophic guild 9). The body morphology and physiology of many broad-ranging demersal fishes such as the macrourids may be adaptations to allow for the rapid exploitation of foodfalls that are variable in both space and time (Stockton and DeLaca, 1982). Wilson and Smith (1984) suggested that macrourids may utilize a waiting rather than searching strategy in order to save energy in the food-poor deep-sea environment, and on olfactory stimulation following the arrival of a foodfall, move rapidly to the site following downstream scent trails. Given the relatively high degree of morphological and behavioral feeding specializations among various demersal and benthic species (see Section III,B), both of these hypotheses seem unlikely. It is more probable that large foodfalls are energy bonuses that are rapidly exploited by a variety of organisms, including fishes. In fact, Priede *et al.* (1991), using radio transmitters placed in bait, found that macrourids are active foragers that can rapidly home in on large foodfalls. They then disperse these foodfalls over great distances as fecal deposits.

B. Benthopelagic Interface

One aspect of the coupling of energy transport that has slowly emerged as an important link between epipelagic waters and the deep benthos is the direct interaction between demersal and pelagic organisms. These interactions appear to be especially pronounced at the interfaces between submerged bottom features such as islands, seamounts, or continental slope regions (Marshall and Merrett, 1977). These bottom features typically span depths that bring the lower limits of diel vertical migration ranges near the bottom (<10 m). This allows demersal and even benthic predators (see Sections II,A and III,A) to prey on meso- or bathypelagic fishes. Pereyra *et al.* (1969) showed that off Oregon, yellowtail rockfish (*Sebastodes flavidus*), a benthic shelf-edge fish, concentrated in locations where they could feed on aggregations of mesopelagic animals, especially myctophid fishes, that came into contact with bottom. Based on feeding chronologies, Mauchline and Gordon (1991) similarly noted that incidences of pelagic prey in demersal (referred to as benthopelagic) fishes in the northeast Atlantic were directly attributable to the movement of the midwater species near the bottom during their downward vertical migrations.

However, not all interactions at the benthic/pelagic interface are necessarily accidents of migratory impingement. From continued sampling in such regions, plus increasing numbers of reports made by observers from submersibles, such benthopelagic aggregations of midwater animals are quite common and can often be quite dense. For example, in several such near-bottom aggregations of the myctophid species *Diaphus dumerilii* and *Ceratoscopelus maderensis* in the continental slope region near Cape Hatteras, North Carolina, two of us (J. V. Gartner and K. J. Sulak, unpublished) estimated densities of >20 individuals/m^3. Captures of huge numbers of mesopelagic fishes, especially myctophids, have been reported in bottom trawls fishing various continental slope regions (see Nafpaktitis *et al.*, 1977).

Midwater fishes may approach the bottom not only because their migratory range incorporates those depths, but because of the concentration of prey in such regions, which greatly reduces the amount of search volume and hence time and energy needed for prey location. That the benthopelagic interface represents a significant energy resource to midwater fishes is apparent in the fact that there are at least two assemblages of midwater fishes that are associated with these benthopelagic environments. Within these midwater assemblages, it may be that their primary food sources are demersal rather than pelagic.

In the first assemblage, it appears that either specific taxa or ontogenetic stages of various taxa that are predominantly or solely midwater groups occupy a benthopelagic habitat as a normal habitat. There are certain species of midwater fishes, such as the myctophids *Diaphus adenomus* and *Diaphus watasei* (Clarke, 1973) or the gonostomatids *Yarella blackfordi* and *Polymetme corythaeola* (Grey, 1964), that are collected only in bottom trawls, whereas in other taxa, such as the myctophid genera *Lampanyctus* and *Lampadena*, large individuals are observed and collected only from near the bottom (Marshall and Merrett, 1977; Nafpaktitis *et al.*, 1977). Thus, it would appear that we have members of these families evolving toward a near-bottom existence. Unfortunately, we cannot compare how these animals might compare in feeding behavior, diet, etc., because no detailed morphological or diet analysis studies have yet been published.

A second well-defined community of primarily mesopelagic fish taxa has now been identified by various studies in different regions. This community has been termed either "pseudoceanic" (see Hulley and Lutjeharms, 1989) or the "mesopelagic boundary community" (Reid *et al.*, 1991). Either term refers to an assemblage of mesopelagic fish and other micronekton species that are found within a very narrow horizontal distance from the position where the neritic environment gives way to the oceanic—for example, islands and continental shelf breaks. Such fishes often are collected or observed in close proximity to the bottom [e.g., *Diaphus dumerilii* (J. V.

Gartner and K. J. Sulak, personal observations], however, they are still primarily captured by midwater trawls fishing off the bottom. Such communities are tightly linked to specific isobaths between about 400 and 1000 m and so are restricted to narrow horizontal ranges of only a few kilometers width (Hulley and Lutjeharms, 1989; Reid et al., 1991).

Like the "benthopelagic" midwater fish assemblage, this boundary community has thus far been characterized on the basis of distribution, and no focused feeding studies have yet been forthcoming. However, both of these groups should be closely examined for feeding to strengthen our understanding of vertical energy coupling among fishes between the shallow and deep-ocean zones.

V. DEEP-SEA ENERGETICS RELATED TO FEEDING

A. Chemical Composition Data

Compositional analyses have proved useful in understanding the adaptations of deep-sea fishes to their environment. Most studies of body composition have considered only pelagic species that generally occur at depths of less than 1000 m (Childress and Nygaard, 1973; Childress and Somero, 1979; Torres et al., 1979; Bailey and Robison, 1986; Childress et al., 1980; Donnelly et al., 1990). Only a few studies have considered deep-sea demersal species (Siebenaller et al., 1982; Steimle and Terranova, 1988; Crabtree, 1995). Chemical composition has been suggested to vary as a function of both depth (Childress and Nygaard, 1973; Stickney and Torres, 1989; Crabtree, 1995) and regional productivity (Bailey and Robison, 1986; Crabtree, 1995). Depth and productivity both affect food availability and thus influence chemical composition. Deep-sea species appear to have adapted to low food availability by substituting low-density body fluids for organic matter, thereby approaching neutral buoyancy and reducing the energy required for growth (Childress et al., 1980).

The water content of deep-sea demersal fishes ranges from about 73 to 92% of wet weight (Crabtree, 1995). Among dominant families, the alepocephalids and ophidiids have the highest overall water contents, contrasting with zoarcids and chlorophthalmids, which have much lower water contents. The energy content of demersal deep-sea species ranges from 110 to 666 kJ per 100 g wet weight (Steimle and Terranova, 1988; Crabtree, 1995) and decreases as a function of depth of occurrence for benthopelagic species with swim bladders (Crabtree, 1995).

Ranges in compositional parameters of midwater fishes are in general similar to those of demersal fishes. In addition, the increase in water content and corresponding decreases in carbon, nitrogen, and energy content as a function of depth observed for benthopelagic species with swim bladders are similar to those reported for midwater fishes. The increase in the water content of benthopelagic species with swim bladders with increasing depth of occurrence reported by Crabtree (1995) is also similar to that reported for midwater fishes by Childress and Nygaard (1973) off southern California and by Stickney and Torres (1989) in the eastern Gulf of Mexico. As water content increases, corresponding decreases occur in carbon, nitrogen, and energy content as a percentage of wet weight. These authors suggested that such trends presumably result in a greater growth efficiency at depth by reducing the energy input needed to produce a given body size, and could be a response to decreasing food availability as a function of increasing depth.

Chemical composition of demersal fishes is also correlated with buoyancy mechanisms. Benthic and benthopelagic species with swim bladders have lower water contents and higher skeletal ash, nitrogen, carbon and energy contents than do benthopelagic species without swim bladders (Crabtree, 1995). Similar results were reported for midwater fishes by Childress and Nygaard (1973), who found higher water contents and lower protein and skeletal ash contents in midwater fishes without swim bladders than in those with swim bladders. Benthopelagic species without swim bladders also have low nitrogen contents in the body tissues, indicating low protein levels. Because protein content is proportional to muscle content, these fishes probably show limited swimming capabilities (Crabtree, 1995). Species that achieve neutral buoyancy through low-density body fluids are probably relatively inactive "float-and-wait" predators, and could be among the most energy efficient of deep-sea fishes.

Food availability may have a considerable influence on the chemical composition of deep-sea fishes. Bailey and Robison (1986) reported on chemical composition of midwater fishes across a geographical productivity gradient and found consistent trends that were correlated with food availability. Water content is higher and lipid and energy content lower in fishes from areas with low surface productivity. Crabtree (1995) suggested that trends in the chemical composition of Middle Atlantic Bight and Bahamian demersal fishes also appear to reflect food availability, and are consistent with the ideas of Sulak (1982), Anderson *et al.* (1985), and Crabtree *et al.* (1991), who proposed that successful Middle Atlantic Bight species have high energy requirements, as evidenced by more active feeding modes, in contrast to less active Bahamian species. Families characterized by lower energy contents, such as the Ophidiidae and Alepocephalidae, are more prominent in terms of numbers of species and individuals in the Bahamas than in the Middle Atlantic Bight (Sulak, 1982).

Chemical composition and enzyme activity levels have been related to swimming capabilities (Siebenaller *et al.*, 1982; Childress *et al.*, 1990). Siebenaller *et al.* examined enzyme activities of four macrourid species as well as several other deep-sea demersal species and found that activity levels are not a function of depth, but rather reflect feeding habits. Of the fishes they examined, *Coryphaenoides (Nematonurus) armatus,* among the deepest living species, has the highest enzyme activity levels and thus the highest potential for active swimming.

Childress *et al.* (1990) suggested that changes in visual predator–prey interactions with depth, rather than food availability, could be critical in allowing the evolution of lower metabolic rates and reduced locomotor capabilities in deeper living midwater fishes. Childress *et al.* (1990) argued that as light intensity decreases with increasing depth, visual interactions decrease in importance. As the visual field of organisms decreases, selection pressures for strong swimming abilities diminish accordingly because predators and prey need move only a short distance to detect prey or escape predation. The diminishing selection pressures for strong swimming abilities are reflected in the trends in chemical compositional of fish observed with increasing depth. However, Crabtree (1995) pointed out that the data on demersal species, including fishes from much greater depths than examined by Childress *et al.* (1990), are not entirely consistent with this hypothesis. The existence of a significant relationship between water content and depth of occurrence at bathyal and abyssal depths, where light levels are presumably insignificant, suggests that factors other than visual interactions affect the chemical composition of demersal fishes.

B. Energetics

Construction of meaningful bioenergetics equations or models for deep-sea fish species, based on accurate quantification of physiological data, is still lacking and probably will continue to be so for some time, owing to the many problems inherent in data acquisition, as we have previously mentioned. As a result, there have been few attempts to construct bioenergetics equations for deep-sea fishes. No such attempts have been made for demersal or benthic species, although a recent publication (Moser *et al.*, 1997) has provided some of the first respiratory data for deep-sea benthic fishes.

A basic equation developed from examination of freshwater fishes has been suggested in several studies as applicable to midwater fishes (Hopkins and Baird, 1977; Baird and Hopkins, 1981b). The equation is

$$Q_c = Q_g + Q_w + Q_d + Q_s + Q_a,$$

where Q_c is energy of the ingested ration; Q_g is increased potential energy through growth; Q_w is energy loss through waste (feces, urine, various secretions); Q_d is cost of digestion, assimilation, and storage of energy; Q_s is cost of basal metabolism (resting); and Q_a is cost of activity (swimming).

The various problems mentioned previously in this chapter have allowed few direct measurements of the variables in the Q_c bioenergetic equation. As a result, most of these parameters are extrapolations from freshwater fish data or deductions based on what we know about mesopelagic fishes. The potential variability of each parameter based on what are essentially a series of assumptions is probably so large as to render such exercises almost futile, especially because many assumptions, based on animals collected by trawls, are now being challenged or discarded based on direct observations made by observers in submersibles.

Other assumed values may be erroneous based on faulty interpretation or inadequacy of data. For example, in general, tropical–subtropical myctophids have always been thought to be short-lived, fast-growing species (see review in Gjösaeter and Kawaguchi, 1980). However, Gartner (1991b) showed that even though such species may live less than a year, they grew no faster than epipelagic and inshore counterparts, thus assumptions of energy conversion for fast growth would skew an energetics model.

Another assumption in energetics has been that regarding the caloric value of gonads versus the body weight of the fish. As the assumption goes, the gonads of a species possess a certain caloric value, which is a fraction of the total caloric content of the body. In order to reproduce successfully then, a species would need to obtain from its diet that percentage of the body caloric value. Although such a calculation might be appropriate for species that spawn in a restricted annual period, many tropical–subtropical myctophid species are serial batch spawners that may release batches as frequently as every day for 4 to 6 months (Gartner, 1993).

Gartner (1993) has shown that for the myctophid *Lepidophanes guentheri*, which at maximum spawning intensity releases a batch every 4 days, the use of the gonad caloric value to body value would suggest that the fish needed to convert only about 5% of its energy to reproduction (63.94 calories mean weight of gonads versus 1271.20 calories for mean body weight). In fact, in order to spawn every fourth day, *L. guentheri* needed to convert about 30% of its *daily* caloric intake to oocyte production. Thus, the gonad weight to body weight percentage calculation for bioenergetics of this species would be grossly inaccurate.

More intensive investigations are needed to enable a realistic attempt to model patterns of energy transfer via feeding in fishes of the deep oceans.

VI. FUTURE DIRECTIONS IN DEEP-SEA FISH RESEARCH

The round-the-world expedition of HMS Challenger (1872–1876) sparked an extended period of deep-sea exploration. During this period of wide-ranging oceanographic expeditions, which explored both the pelagic and the benthic environments of the deep oceans, attention focused on descriptions of new species, genera, and families obtained from great depths and from previously unexplored regions. Knowledge of the deep-sea fauna increased dramatically in the wake of Challenger. The period of nationally sponsored expeditions culminated in the Danish Galathea expedition (1950–1952). Subsequently deep-sea research turned toward more intensive, localized, and long-term studies of the faunas of particular regions. In the late 1950s and 1960s attention was focused on evaluation of pelagic organisms forming layers that reflected sonar, the so-called deep-scattering layers (DSLs) typically found between the surface and 800 to 900 m depth. Analysis of these layers revealed that many DSL organisms were fishes and also revealed the vertically migrating nature of many of these fish species. Thus, with these and other active research programs, by the 1960s the focus of research on deep-sea fishes shifted from qualitative faunal inventory to quantitative community structure and ecology, combined with in-depth studies of the life histories of individual dominant species characteristic of the regional fauna. Among fish groups, the pelagic faunas have perhaps been better studies than their benthic and demersals counterparts because of the relative ease of collection in midwater versus bottom habitats.

Despite these advances, it would still be quite correct to say that our knowledge of deep-sea fishes and other deep-sea organisms remains in its infancy. It should be apparent from our presentation in this chapter that there are many aspects of deep-sea fish feeding that remain entirely speculative. It is also clear that research into feeding habits and ecology of pelagic fishes is perhaps more advanced than that of benthic and demersal fishes, but that work on both groups lags far behind research in shallow waters. Unfortunately, current trends suggest that this gap between our knowledge of shallow and deep-sea fish feeding habits and physiology will continue to grow. Oceanographic sampling has declined precipitously since the mid-1970s due to a progressively more stringent funding climate in the West and the simultaneous collapse of the Soviet Union, formerly a major player in exploration of remote regions of the world oceans. The fleet of oceanographic research vessels has diminished substantially over the past decade, and many ships and submersibles that remain "on-line" are currently inac-

4. FEEDING AT DEPTH 181

tive. The United States National Undersea Research Program has been at a virtual standstill over recent years, with no dedicated funding for submersible operations. Despite diminished sampling, however, deep-sea biologists continue to discover new taxa with regularity, even in the best sampled regions. Discoveries such as the megamouth shark reveal the inadequacy of our knowledge of the deep-ocean fauna. Thus, we remain in the empirical descriptive stage of inquiry. Moreover, fundamental knowledge of species composition of the deep ocean is heavily biased to the northern hemisphere, particularly the North Atlantic. Sampling has also been concentrated primarily on the continental slope. The continental rise and the vast abyssal areas remain very poorly sampled. Many apparently cosmopolitan species are represented by a handful of specimens from widely disparate localities. Although the outlines of faunal structure are now available for a few select study areas, our understanding of processes underlying that structure is very limited. Knowledge of the deep-sea bottom fish fauna rests primarily on trawl samples from soft-substrate, low-relief biotopes. Using submersibles and remotely operated vehicles (ROVs), we have only recently begun to explore hard-bottom and rough-topography biotopes, resulting in astounding discoveries of unique communities of unusual organisms associated with thermal vents and other novel rough-bottom biotopes. Evidence from *in situ* observations also reveals an array of behaviors previously unknown, as well as patterns of abundance among fishes and their potential prey groups that cannot be delineated from net captures. Experimentation in the deep ocean is difficult and expensive, and we have barely begun to test hypotheses concerning faunal structure and function, species life histories, and physiology. Sadly, funding for such observation platforms and experimentation continues to dwindle at a time when ever more powerful technology and analytical techniques are becoming available. Some speculative conclusions, such as the apparently cosmopolitan nature of a number of deep-sea fish species, can perhaps now be answered by genetic analyses. Such analyses, however, usually require special tissue handling and preparation, so specimens caught by older expeditions may not be useful samples; new material needs to be collected.

Future directions in deep-sea research, not only for fishes but for other organisms as well, should include a renewed effort to define fundamental taxonomic composition of the fauna, with emphasis on the southern hemisphere and abyssal midocean areas. This will require renewed funding for remote sampling from surface research vessels, supplemented by video transects accomplished from submersibles or ROVs. Only when this first stage of faunal exploration is complete will the broad patterns in worldwide faunal composition emerge. Second, additional focused intensive sampling efforts should be undertaken to define quantitatively regional faunal structure in areas outside the North Atlantic. Again, renewed availability of surface research vessels is critical to this stage of deep-sea faunal research. Developing data bases from such areas will enable a comparative approach to testing hypotheses of faunal organization. Additionally, a dedicated effort should be made to increase the availability to deep-sea biologists of submersibles, ROVs, acoustic arrays, and other high-tech tools to enable direct *in situ* observation, quantification, and experimentation in the deep ocean. Particular emphasis should be given to rough-topography biotopes, which cannot be sampled from surface vessels. It is essential to undertake comprehensive laboratory, shipboard, and *in situ* life history studies of individual species. This requires continued sampling to provide adequate numbers of specimens for analyses of morphology, feeding, reproduction, and physiology. Some deep-sea fish species lacking gas bladders can be captured quiescently by submersibles, retrieved to the surface in insulated containers, and successfully maintained for study in shipboard aquaria. Such species are providing insights into the unique physiological capabilities of fishes specifically adapted to life in the deep ocean (Moser *et al.,* 1997). Much more live animal research needs to be undertaken to answer specific questions about the physiological capabilities and limitations of deep-sea fishes.

A final important area of research that needs to be developed and evaluated comprehensively is the transfer of energy from surface waters to the deep ocean by fishes, either through production of fecal material or direct movement of pelagic fishes from near-surface waters to near bottom. There is growing evidence to show that many pelagic deep-sea fishes (and other organisms) regularly approach the bottom, and demersal counterparts may often rise well off the bottom to forage. Research on deep-ocean energy transfer is an attractive approach to returning to deep-sea studies because it is an integrated approach in which a number of projects, including taxonomic composition, life history, and physiology studies, can be conducted simultaneously for both pelagic and demersal/benthic organisms.

It is clear that a great deal of work still remains, even at very basic levels, to elucidate feeding habits and physiology, as well as most other aspects of physiology in deep-sea fishes. We can only hope that at some point such basic research attains a renewed emphasis among agencies funding marine research.

REFERENCES

Alldredge, A. A. (1972). Abandoned larvacean houses: A unique food source in the pelagic environment. *Science* **177,** 885–887.

Alldredge, A. A., and Gotschalk, C. C. (1988). *In situ* settling behavior of marine snow. *Limnol. Oceanogr.* **33**, 339–351.

Alldredge, A. A., and Silver, M. W. (1988). Characteristics, dynamics and significance of marine snow. *Prog. Oceanogr.* **20**, 41–82.

Anderson, M. E., Crabtree, R. E., Carter, H. J., Sulak, K. J., and Richardson, M. D. (1985). Distribution of bottom fishes of the Caribbean Sea found below 2,000 meters. *Bull. Mar. Sci.* **37**, 794–807.

Armstrong, J. D., Priede, L. G., and Smith, J. K. L. (1991). Temporal change in foraging behavior of the fish *Coryphaenoides* (*Nematonurus*) *yaquinae* in the central North Pacific. *Mar. Ecol. Prog. Ser.* **76**, 195–199.

Armstrong, J. D., Bagley, P. M., and Priede, I. G. (1992). Photographic and acoustic tracking observations of the behaviour of the grenadier *Coryphaenoides* (*Nematonurus*) *armatus*, the eel *Synaphobranchus bathybius*, and other demersal fish in the North Atlantic Ocean. *Mar. Biol.* **112**, 535–544.

Badcock, J., and Merrett, N. R. (1976). Midwater fishes in the eastern North Atlantic. I. Vertical distribution and associated biology in 30°N, 23°W, with developmental notes on certain myctophids. *Prog. Oceanogr.* **7**, 3–58.

Bailey, T. G., and Robison, B. H. (1986). Food availability as a selective factor on the chemical compositions of midwater fishes in the eastern North Pacific. *Mar. Biol.* **91**, 131–141.

Baird, R. C., and Hopkins, T. L. (1981a). Trophodynamics of the fish *Valenciennellus tripunctulatus*. II. Selectivity, grazing rates and resource utilization. *Mar. Ecol. Prog. Ser.* **5**, 11–19.

Baird, R. C., and Hopkins, T. L. (1981b). Trophodynamics of the fish *Valenciennellus tripunctulatus*. III. Energetics, resources and feeding strategy. *Mar. Ecol. Prog. Ser.* **5**, 21–28.

Baird, R. C., Hopkins, T. L., and Wilson, D. F. (1975). Diet and feeding chronology of *Diaphus taaningi* (Myctophidae) in the Cariaco Trench. *Copeia*, 356–364.

Balanov, A. A., Gorbatenko, K. M., and Efimkin, A. Ya. (1995). Foraging dynamics of mesopelagic fishes in the Bering Sea during summer and autumn. *J. Ichthy.* **34**, 65–77.

Barnes, A. T., Quetin, L. B., Childress, J. J., and Pawson, D. L. (1976). Deep-sea macroplanktonic sea cucumbers: Suspended sediment feeders captured from deep submergence vehicle. *Science* **194**, 1083–1085.

Belman, B. W., and Anderson, M. E. (1979). Aquarium observations on feeding by *Melanostigma pammelas* (Pisces: Zoarcidae). *Copeia*, 366–369.

Bertelsen, E. (1951). The ceratioid fishes. *Dana Report* **39**, 1–276.

Bertelsen, E., and Struhsaker, P. J. (1977). The ceratioid fishes of the genus *Thaumatichthys*, osteology, relationships, distribution and biology. *Galathea Report* **14**, 7–40.

Bertin, L. (1934). Les poissons apodes appartenant au sous-ordres des lyomeres. *Dana Report* **3**, 1–56.

Bigelow, H. B., and Schroeder, W. C. (1948). Sharks. *In* "Fishes of the Western North Atlantic," Memoir I, Part 1, pp. 59–546. Sears Foundation for Marine Research, Yale University, New Haven, Connecticut.

Blaber, S. J. M., and Bulman, C. M. (1987). Diets of fishes of the upper continental slope of eastern Tasmania: Content, calorific values, dietary overlap and trophic relationships. *Mar. Biol.* **95**, 345–356.

Boden, B. P., and Kampa, E. M. (1967). The influence of natural light on the vertical migrations of an animal community in the sea. *Symp. Zool. Soc. London* **19**, 15–26.

Böhlke, J. E. (1966). Family Eurypharyngidae. *In* "Fishes of the Western North Atlantic" (G. W. Mead, ed.-in-chief), Part 5, pp. 610–616. Sears Foundation for Marine Research, New Haven, Connecticut.

Bone, Q. (1973). A note on the buoyancy of some lantern-fishes (Myctophoidei). *J. Mar. Biol. Assoc. U.K.* **53**, 619–633.

Borodulina, O. D. (1972). The feeding of mesopelagic predatory fish in the open ocean. *J. Ichthy.* **12**, 692–702.

Bright, T. J. (1970). Food of the deep-sea bottom fishes. In "Contributions to the Biology of the Gulf of Mexico" (W. E. Pequegnat and F. A. Chace, eds.), Vol. 1, pp. 245–252. Texas A&M University, College Station.

Bulman, C. M., and Koslow, J. A. (1992). Diet and food consumption of a deep-sea fish, orange roughy *Hoplostethus atlanticus* (Pisces: Trachichthyidae), off southeastern Australia. *Mar. Ecol. Prog. Ser.* **82**, 115–129.

Butler, J. L., and Pearcy, W. G. (1972). Swimbladder morphology and specific gravity of myctophids off Oregon. *J. Fish. Res. Bd. Can.* **29**, 1145–1150.

Cailliet, G. M. (1972). The study of feeding habits of two marine fishes in relation to plankton ecology. *Trans. Am. Microsc. Soc.* **91**, 88–89.

Campbell, R. A. (1983). Parasitism in the deep sea. In "The Sea" (G. T. Rowe, ed.), Vol. 8, pp. 473–552, Wiley, NY.

Campbell, R. A., and Gartner, Jr., J. V. (1982). *Pistana eurypharyngis* gen et sp. n. (Cestoda: Pseudophyllidea) from the bathypelagic gulper eel, *Eurypharynx pelecanoides* Vaillant, 1882, with comments on host and parasite ecology. *Proc. Helminthol. Soc. Wash.* **49**, 218–225.

Campbell, R. A., Haedrich, R. L., and Munroe, T. A. (1980). Parasitism and ecological relationships among deep-sea fishes. *Mar. Biol.* **57**, 301–313.

Carter, H. J. (1984). Feeding strategies and functional morphology of demersal deep-sea ophidiid fish. Ph.D. Thesis, College of William and Mary in Virginia, Gloucester Point.

Chao, L. N., and Musick, J. A. (1977). Life history, feeding habits, and functional morphology of juvenile sciaenid fishes in the York River Estuary, Virginia. *Fish. Bull. U.S.* **75**, 657–702.

Childress, J. J., and Meek, R. P. (1973). Observations on the feeding behavior of a mesopelagic fish (*Anoplogaster cornuta*: Beryciformes). *Copeia*, 602–603.

Childress, J. J., and Nygaard, M. H. (1973). The chemical composition of midwater fishes as a function of depth of occurrence off southern California. *Deep-Sea Res.* **20**, 1093–1109.

Childress, J. J., and Somero, G. N. (1979). Depth-related enzymic activities in muscle, brain and heart of deep-living pelagic marine teleosts. *Mar. Biol.* **52**, 273–283.

Childress, J. J., Taylor, S. M., Cailliet, G. M., and Price, M. H. (1980). Patterns of growth, energy utilization and reproduction in some meso- and bathypelagic fishes off Southern California. *Mar. Biol.* **61**, 27–40.

Childress, J. J., Price, M. H., Favuzzi, J., and Cowles, D. (1990). Chemical composition of midwater fishes as a function of depth of occurrence off the Hawaiian Islands: Food availability as a selective factor? *Mar. Biol.* **105**, 235–246.

Clarke, M. R., and Merrett, N. R. (1972). The significance of squid, whale and other remains from the stomachs of bottom-living deep-sea fish. *J. Mar. Biol. Assoc. U.K.* **52**, 596–603.

Clarke, T. A. (1973). Some aspects of the ecology of lanternfishes (Myctophidae) in the Pacific Ocean near Hawaii. *Fish. Bull. U.S.* **71**, 401–434.

Clarke, T. A. (1978). Diel feeding patterns of 16 species of mesopelagic fishes from Hawaiian waters. *Fish. Bull. U.S.* **76**, 495–513.

Clarke, T. A. (1980). Diets of fourteen species of vertically migrating mesopelagic fishes in Hawaiian waters. *Fish. Bull. U.S.* **78**, 619–640.

Clarke, T. A. (1982). Feeding habits of stomiatoid fishes from Hawaiian waters. *Fish. Bull. U.S.* **80**, 287–304.

Clarke, T. A., and Wagner, P. J. (1976). Vertical distribution and other aspects of the ecology of certain mesopelagic fishes taken near Hawaii. *Fish. Bull. U.S.* **74**, 635–645.

Cohen, D. M. (1964). Families Bathylagidae and Opisthoproctidae. In "Fishes of the Western North Atlantic" (H. B. Bigelow, ed.-in-chief), Part 4, pp. 34–69. Sears Foundation for Marine Research, New Haven, Connecticut.

Cohen, D. M., and Nielsen, J. G. (1978). Guide to the identification of genera of the fish order Ophidiiformes with a tentative classification of the order. *NOAA Tech. Rep. NMFS Cir.* **417.**
Collard, S. B. (1970). Forage of some eastern Pacific midwater fishes. *Copeia,* 348–354.
Crabtree, R. E. (1995). Chemical composition and energy content of deep-sea demersal fishes from tropical and temperate regions of the western north Atlantic. *Bull. Mar. Sci.* **56,** 434–449.
Crabtree, R. E., and Sulak, K. J. (1986). A contribution to the life history and distribution of Atlantic species of the deep-sea fish genus *Conocara* (Alepocephalidae). *Deep-Sea Res.* **33,** 1183–1201.
Crabtree, R. E., Sulak, K. J., and Musick, J. A. (1985). Biology and distribution of species of *Polyacanthonotus* (Pisces: Notacanthiformes) in the western North Atlantic. *Bull. Mar. Sci.* **36,** 235–248.
Crabtree, R. E., Carter, H. J., and Musick, J. A. (1991). The comparative feeding ecology of temperature and tropical deep-sea fishes from the western North Atlantic. *Deep-Sea Res.* **38,** 1277–1298.
Dahl, E. (1979). Deep-sea carrion feeding amphipods; evolutionary patterns in niche adaptation. *Oikos* **33,** 167–175.
Dalpadado, P., and Gjösaeter, J. (1988). Feeding ecology of the lanternfish *Benthosema pterotum* from the Indian Ocean. *Mar. Biol.* **99,** 555–567.
Dayton, P. K., and Hessler, R. R. (1972). Role of biological disturbance in maintaining diversity in the deep sea. *Deep-Sea Res.* **19,** 199–208.
DeWitt, F. A., Jr., and Cailliet, G. M. (1972). Feeding habits of two bristlemouth fishes, *Cyclothone accllinidens* and *C. signata* (Gonostomatidae). *Copeia,* 868–871.
Donnelly, J., and Torres, J. J. (1988). Oxygen consumption of midwater fishes and crustaceans from the eastern Gulf of Mexico. *Mar. Biol.* **97,** 483–494.
Donnelly, J., Torres, J. J., Hopkins, T. L., and Lancraft, T. M. (1990). Proximate composition of Antartic mesopelagic fishes. *Mar. Biol.* **106,** 13–23.
Du Buit, M.-H. (1978). Alimentation de quelques poissons téléostéens de profondeur dans la zone du seuil de Wyville Thomson. *Oceanol. Acta* **1,** 129–134.
Dudochkin, A. S. (1988). The food of the grenadier, *Macrourus holotrachys,* in the Southwestern Atlantic. *Vopr. Ikhtiol.* **3,** 421–425.
Duka, L. A. (1987). Feeding of *Ceratoscopelus warmingii* (Myctophidae) in the tropical Atlantic. *J. Ichthy.* **28,** 89–95.
Ebeling, A. W., and Cailliet, G. M. (1974). Mouth size and predator strategy of midwater fishes. *Deep-Sea Res.* **21,** 959–968.
Fange, R., and Grove, D. (1979). Digestion. *In* "Fish Physiology" (W. S. Hoar, D. J. Randall, and J. R. Brett, ed.), pp. 161–260. Academic Press, New York.
Feller, R. J. (1979). The Analysis of "Gorp," or how to know the guts of recalcitrant predators. (C. A. Simenstad and S. Lipovsky, eds.), Proceedings of the Second Pacific Fish Food Habits Conference, University of Washington, Seattle.
Feller, R. J. (1985). Deep-sea food web analysis using cross-reacting antisera. *Deep-Sea Res.* **32,** 485–497.
Fishelson, L. (1994). Comparative internal morphology of deep-sea eels, with particular emphasis on gonads and gut structure. *J. Fish. Biol.* **44,** 75–101.
Forster, G. R. (1964). Line-fishing on the continental slope. *J. Mar. Biol. Assoc. U.K.* **44,** 277–284.
Foster, G. R. (1968). Line-fishing on the continental slope. II. *J. Mar. Biol. Assoc. U.K.* **48,** 479–483.

Forster, G. R. (1971). Line-fishing on the continental slope. III. Mid-water fishing with vertical lines. *J. Mar. Biol. Assoc. U.K.* **51,** 73–77.

Forster, G. R. (1971). Line-fishing on the continental slope. The selective effect of different hook patterns. *J. Mar. Biol. Assoc. U.K.* **53,** 73–77.

Gartner, J. V., Jr. (1991a). Life histories of three species of lanternfishes (Pisces: Myctophidae) from the eastern Gulf of Mexico. I. Morphological and microstructural analysis of sagittal otoliths. *Mar. Biol.* **111,** 11–20.

Gartner, J. V., Jr. (1991b). Life histories of three species of lanternfishes (Pisces: Myctophidae) from the eastern Gulf of Mexico. II. Age and growth patterns. *Mar. Biol.* **111,** 21–27.

Gartner, J. V., Jr. (1993). Patterns of reproduction in the dominant lanternfish species (Pisces: Myctophidae) of the eastern Gulf of Mexico, with a review of reproduction among tropical–subtropical Myctophidae. *Bull. Mar. Sci.* **52,** 721–750.

Gartner, J. V., Jr., and Musick, J. A. (1989). Feeding habits of the deep-sea fish, *Scopelogadus beanii* (Pisces: Melamphaidae), in the western North Atlantic. *Deep-Sea Res.* **36,** 1457–1469.

Gartner, J. V., Jr., and Zwerner, D. E. (1989). The parasite faunas of meso- and bathypelagic fishes of Norfolk submarine canyon, western North Atlantic. *J. Fish. Biol.* **34,** 79–95.

Gartner, J. V., Jr., Hopkins, T. L., Baird, R. C., and Milliken, D. M. (1987). The lanternfishes (Pisces: Myctophidae) of the eastern Gulf of Mexico. *Fish. Bull. U.S.* **85,** 81–98.

Gartner, J. V., Jr., Conley, W. J., and Hopkins, T. L. (1989). Escapement by fishes from midwater trawls: A case study using lanternfishes (Pisces: Myctophidae). *Fish. Bull. U.S.* **87,** 213–222.

Geistdoerfer, P. (1977). Étude biomecanique du mouvement de ferméture des mâchoires chez *Ventrifossa occidentalis* et *Coelorhynchus coelorynchus. Bull. Mus. Nat. Hist. Nat. Ser. 3,* **481**(Zoologie 338), 993–1020.

Geisdoerfer, P. (1979a). Alimentation du grenadier, *Coryphaenoides rupestris,* dans l'Atlantique nord-est. *J. Cons. Int. pour l'Exploration de la Mer,* Comité des poissons de fond. C.M. 1979/G:31.

Geistdoerfer, P. (1979b). Recherches sur l'alimentation de *Macrourus berglax* Lacépède 1801 (Macrouridae, Gadiformes). *Ann. Inst. Océanogr.,* **55,** 135–144.

Gerking, S. D. (1994). "Feeding Ecology of Fish," pp. 1–416. Academic Press, San Diego.

Gibbs, R. H., Jr., and Wilimovsky, N. J. (1966). Family Alepisauridae. "Fishes of the Western North Atlantic" (H. B. Bigelow, ed.-in-chief), Part 5, pp. 482–496. Sears Foundation for Marine Research, New Haven, Connecticut.

Gjösaeter, J. (1973). The food of the myctophid fish *Benthosema glaciale* (Reinhardt), from western Norway. *Sarsia* **52,** 53–58.

Gjösaeter, J. and Kawaguchi, K. (1980). A review of the world resources of mesopelagic fish. *FAO Fish. Tech. Paper* **193,** 1–151.

Golovan', G. A., and Pakhorukov, N. P. (1975). Some data on the morphology and ecology of *Aleopocephalus bairdi* (Alepocephalidae) of the Central and Eastern Atlantic. *J. Ichthyol.* **15,** 44–50.

Golovan', G. A., and Pakhorukov, N. P. (1980). New data on the ecology and morphometry of *Alepocephalus rostratus* (Alepocephalidae). *Vopr. Ikhtiol.* **20,** 77–83.

Goode, G. B., and Bean, T. H. (1895). Oceanic Ichthyology. *U.S. Natl. Mus. Spec. Bull.* **2,** 1–553.

Gorbatenko, K. M., and Il'inskii, E. N. (1992). Feeding behavior of the most common mesopelagic fishes in the Bering sea. *J. Ichthy.* **32,** 52–60.

Gordon, J. D. M. (1979). Lifestyle and phenology in deep sea anacanthine teleosts. *Symp. Zool. Soc. London* **44,** 327–359.

Gordon, J. D. M., and Duncan, J. A. R. (1987). Aspects of the biology of *Hoplostethus atlanticus* and *H. mediterraneus* (Pisces: Berycomorphi) from the slopes of the Rockall

Trough and the Porcupine Sea Bight (North-Eastern Atlantic). *J. Mar. Biol. Assoc. U.K.* **67,** 119–133.

Gorelova, T. A. (1973). Zooplankton from the stomachs of juvenile lantern fish of the family Myctophidae. *Oceanology* **14,** 575–580.

Gorelova, T. A. (1977). Some characteristics of the nutrition of the young of nictoepipelagic and mesopelagic lantern fish (Pisces, Myctophidae). *Oceanology* **17,** 220–222.

Gorelova, T. A. (1978). The feeding of lanternfishes, *Ceratoscopelus warmingi* and *Bolinichthys longipes,* of the family Myctophidae in the western equatorial part of the Pacific Ocean. *J. Ichthy.* **18,** 588–598.

Gorelova, T. A. (1981). Notes on feeding and gonad condition in three species of the genus *Gonostoma* (Gonostomatidae). *J. Ichthy.* **21,** 82–92.

Gorelova, T. A., and Efremenko, V. N. (1989). On the food composition of the larvae of two species of lantern anchovies (Myctophidae) from the Scotia Sea. *J. Ichthy.* **29,** 106–109.

Gorelova, T. A., and Kobylyanskiy, S. G. (1985). Feeding of deepsea fishes of the family Bathylagidae. *J. Ichthy.* **25,** 89–100.

Gorelova, T. A., and Krasil'nikova, N. A. (1990). On the diet of *Maurolicus muelleri* in the vicinity of seamounts Discovery, Nasca, and Mt. Africana. *J. Ichthy.* **30,** 42–52.

Grey, M. (1964). Family Gonostomatidae. *In* "Fishes of the Western North Atlantic" (H. B. Bigelow, ed.-in-chief), Part 4, pp. 78–240. Sears Foundation for Marine Research, New Haven, Connecticut.

Haedrich, R. L. (1967). The stromateoid fishes: Systematics and a classification. *Bull. Mus. Comp. Zoo., Harvard* **135,** 31–139.

Haedrich, R. L. (1974). Pelagic capture of the epibenthic rattail *Coryphaenoides rupestris. Deep-Sea Res.* **21,** 977–979.

Haedrich, R. L., and Henderson, N. R. (1974). Pelagic food of *Coryphaenoides armatus,* a deep benthic rattail. *Deep-Sea Res.* **21,** 739–744.

Haedrich, R. L., and Polloni, P. T. (1976). A contribution to the life history of a small rattail fish, *Coryphaenoides carapinus. Bull. S. Calif. Acad. Sci.* **75,** 203–211.

Hecker, B. (1994). Unusual megafaunal assemblages on the continental slope off Cape Hatteras. *Deep-Sea Res.* **41,** 809–834.

Herdendorf, C. E., and Berra, T. M. (1995). A Greenland shark from the wreck of the SS *Central America* at 2,200 meters. *Trans. Am. Fish Soc.* **124,** 950–953.

Hopkins, T. L., and Baird, R. C. (1973). Diet of the hatchetfish, *Sternoptyx diaphana. Mar. Biol.* **21,** 34–46.

Hopkins, T. L., and Baird, R. C. (1977). Aspects of the feeding ecology of oceanic midwater fishes. *In* "Oceanic Sound Scattering Prediction" (N. R. Andersen and B. J. Zahuranec, eds.), pp. 325–360. Plenum, New York.

Hopkins, T. L., and Baird, R. C. (1981). Trophodynamics of the fish *Valenciennellus tripunctulatus.* I. Vertical distribution, diet and feeding chronology. *Mar. Ecol. Prog. Ser.* **5,** 1–10.

Hopkins, T. L., and Baird, R. C. (1985a). Feeding ecology of four hatchetfishes (Sternoptychidae) in the eastern Gulf of Mexico. *Bull. Mar. Sci.* **36,** 260–277.

Hopkins, T. L., and Baird, R. C. (1985b). Aspects of the trophic ecology of the mesopelagic fish *Lampanyctus alatus* (Family Myctophidae) in the eastern Gulf of Mexico. *Biol. Oceanogr.* **3,** 285–313.

Hopkins, T. L., and Gartner. J. V., Jr. (1992). Resource-partitioning and predation impact of a low-latitude myctophid community. *Mar. Biol.* **114,** 185–197.

Hopkins, T. L., Milliken, D. M., Bell, L. M., McMichael, E. J., Heffernan, J. J., and Cano, R. V. (1981). The landward distribution of oceanic plankton and micronekton over the west Florida continental shelf as related to their vertical distribution. *J. Plank. Res.* **3,** 645–658.

Hopkins, T. L., Sutton, T. T., and Lancraft, T. M. (1997). The trophic structure and predation impact of a low latitude midwater fish assemblage. *Prog. Oceanogr.* **38,** 205–239.
Hulley, P. A., and Lutjeharms, J. R. E. (1989). Lanternfishes of the southern Benguela region. Part 3. The pseudoceanic-oceanic interface. *Ann. S. Afr. Mus.* **98,** 409–435.
Hureau, J.-C. (1970). Biologie comparée de quelques poissons antarctiques (Nototheniidae). *Bull. Institut Oceanogr. (Monaco)* **68,** 1–244.
Hureau, J.-C., and Nielsen, J. G. (1981). Les poissons Ophidiiformes des campagnes du N.O. "Jean Charcot" dans l'Atlantique et la Mediterranée. *Cybium* **5,** 3–27.
Isaacs, J. D., and Schwartzlose, R. (1975). Active animals of the deep-sea floor. *Sci. Am.* **233,** 84–91.
Jannasch, H. W. (1978). Experiments in deep-sea microbiology. *Oceanus* **21,** 50–57.
Jannasch, H. W., and Wirsen, C. O. (1977). Microbial life in the deep sea. *Sci. Am.* **236**(6), 42–52.
Janssen, J., Harbison, G. R., and Craddock, J. E. (1986). Hatchetfishes hold horizontal attitudes during diagonal descents. *J. Mar. Biol. Assoc., U.K.* **66,** 825–833.
Karnella, C. (1987). Family Myctophidae. Biology of midwater fishes of the Bermuda Ocean Acre (R. H. Gibbs, Jr., and W. H. Krueger, eds.). Smithson. Contrib. Zool. **452,** 51–168.
Kashkina, A. A. (1986). Feeding of fishes on salps (Tunicata, Thaliacea). *J. Ichthy.* **26,** 57–64.
Kinzer, J. (1977). Observations on the feeding habits of the mesopelagic fish *Benthosema glaciale* (Myctophidae) off NW Africa. *In* "Oceanic Sound Scattering Prediction" (N. R. Andersen and B. J. Zahuranec, eds.), pp. 381–392. Plenum, New York.
Kinzer, J., and Schultz, K. (1985). Vertical distribution and feeding patterns of midwater fish in the central equatorial Atlantic. I. Myctophidae. *Mar. Biol.* **85,** 313–322.
Kinzer, J., Böttger-Schnack, R., Schulz, K. (1993). Aspects of horizontal distribution and diet of myctophid fish in the Arabian Sea with reference to the deep water oxygen deficiency. *Deep-Sea Res.* **40,** 783–800.
Lampitt, R. S., Merrett, N. R., and Thurston, M. H. (1983). Inter-relations of necrophagous amphipods, a fish predator, and tidal currents in the deep sea. *Mar. Biol.* **74,** 73–78.
Lampitt, R. S., Wishner, K. F., Turley, C. M., and Angel, M. V. (1993). Marine snow studies in the northeast Atlantic Ocean: Distribution, composition and role as a food source for migrating plankton. *Mar. Biol.* **116,** 689–702.
Lancraft, T. M., Hopkins, T. L., and Torres, J. J. (1988). Aspects of the ecology of the mesopelagic fish *Gonostoma elongatum* (Gonostomatidae, Stomiiformes) in the Gulf of Mexico. *Mar. Ecol. Prog. Ser.* **49,** 27–40.
Legand, M., and Rivaton, J. (1969). Cycles biologique des poissons mésopélagiques dans l'est de l'Ocean Indien. Troisieme note: Action predatrice des poissons micronectoniques. *Cahiers: Off Recherche Sci. Tech. Outre-mer. Oceanogr.* **7,** 29–45.
Linkowski, T. B. (1991). Otolith microstructure and growth patterns during the early life history of lanternfishes (family Myctophidae). *Can. J. Zool.* **69,** 1777–1792.
Linkowski, T. B. (1996). Lunar rhythms of vertical migrations coded in otolith microstructure of North Atlantic lanternfishes, genus *Hygophum* (Myctophidae). *Mar. Biol.* **124,** 495–508.
Lockett, A. (1971). Retinal structure in *Platytroctes apus*, a deep-sea fish with a pure rod fovea. *J. Mar. Biol. Assoc. U.K.* **51,** 79–91.
Lozano Cabo, F. (1952). Estudio preliminar sobre la biometria, la biologia y la anatomia general de *Notacanthus bonapartei*. *Bol. Inst. Espanol Oceanogr.* **49,** 1–30.
McDowell, S. B. (1973). Order Heteromi (Notacanthiformes). *In* "Fishes of the Western North Atlantic" (D. M. Cohen, ed.-in-chief), Part 6, pp. 1–228. Sears Foundation for Marine Research, New Haven, Connecticut.
McLaren, I. A. (1963). Effects of temperature on growth of zooplankton and the adaptive value of vertical migration. *J. Fish. Res. Bd. Can.* **20,** 685–727.
McLellan, T. (1977). Feeding strategies of the macrourids. *Deep-Sea Res.* **24,** 1019–1036.

Macpherson, E. (1979). Ecological overlap between macrourids in the Western Mediterranean Sea. *Mar. Biol.* **53,** 149–159.

Macpherson, E. (1980). Food and feeding of *Chimaera monstrosa* in the western Mediterranean. *J. Cons. Int. Exploration Mer* **39,** 26–29.

Macpherson, E. (1981). Resource partitioning in a Mediterranean demersal fish community. *Mar. Ecol. Prog. Ser.* **4,** 183–193.

Madin, L. P., Cetta, C. M., and McAlister, V. L. (1981). Elemental and biochemical composition of salps (Tunicata: Thaliacea). *Mar. Biol.* **63,** 217–226.

Mahaut, M.-L., Geistdoerfer, P., and Sibuet, M. (1990). Trophic strategies in carnivorous fishes: Their significance in energy transfer in the deep-sea benthic ecosystem (Meriadzek Terrace—Bay of Biscay). *Prog. Oceanogr.* **24,** 223–237.

Markle, D. F. (1978). Taxonomy and distribution of *Rouleina attrita* and *Rouleina maderensis* (Pisces: Alepocephalidae). *Fish. Bull. U.S.* **76,** 79–87.

Marshall, N. B. (1971). "Explorations in the Life of Fishes." Harvard Univ. Press, Cambridge, Massachusetts.

Marshall, N. B. (1973). Family Macrouridae. *In* "Fishes of the Western North Atlantic" (D. M. Cohen, Jr., ed.-in-chief), Part 6, pp. 496–667. Sears Foundation for Marine Research, New Haven, Connecticut.

Marshall, N. B. (1980). "Developments in Deep-Sea Biology." Blandford, London.

Marshall, N. B., and Bourne, D. W. (1964). A photographic survey of benthic fishes in the Red Sea and the Gulf of Aden, with observations on their population density, diversity and habits. *Bull. Mus. Comp. Zool. Harvard* **132,** 223–244.

Marshall, N. B., and Merrett, N. R. (1977). The existence of a benthopelagic fauna in the deep-sea. A voyage of discovery. *Deep-Sea Res.* George Deacon 70th Anniversary Volume. (Suppl.) 483–497.

Marshall, N. B., and Staiger, J. C. (1975). Aspects of the structure, relationships, and biology of the deep-sea fish *Ipnops murrayi* Gunther (Family Bathypteroidae). *Bull. Mar. Sci.* **25,** 101–111.

Mattson, S. (1981). The food of *Gadus melastomus, Gadiculus argenteus thori, Trisopterus esmarkii, Rhinonemus cimbrius* and *Glyptocephalus cynoglossus* (Pisces) caught during the day with shrimp trawl in a west Norwegian fjord. *Sarsia* **66,** 109–127.

Mauchline, J., and Gordon, J. D. M. (1980). Food and feeding of the morid fish *Lepidion eques* (Gunther, 1887) in the Rockall Trough. *J. Mar. Biol. Assoc. U.K.* **60,** 1053–1059.

Mauchline, J., and Gordon, J. D. M. (1983a). Diets of sharks and chimaeroids of the Rockall Trough, northeastern Atlantic Ocean. *Mar. Biol.* **75,** 269–278.

Mauchline, J., and Gordon, J. D. M. (1983b). Diets of clupeoid, stomiatoid and salmonoid fish of the Rockall Trough, northeastern Atlantic Ocean. *Mar. Biol.* **77,** 67–78.

Mauchline, J., and Gordon, J. D. M. (1984a). Occurrence and feeding of berycomorphid and percomorphid fishes teleost fish in the Rockall Trough. *J. Cons. Int. Exploration Mer.* **41,** 239–247.

Mauchline, J., and Gordon, J. D. M. (1984b). Diets and bathymetric distributions of the macrourid fish of the Rockall Trough, northeastern Atlantic Ocean. *Mar. Biol.* **81,** 107–121.

Mauchline, J., and Gordon, J. D. M. (1984c). Incidence of parasitic worms in stomachs of pelagic and demersal fish of the Rockall Trough, northeastern Atlantic Ocean. *J. Fish Biol.* **24,** 281–285.

Mauchline, J., and Gordon, J. D. M. (1984d). Occurrence of stones, sediment and fish scales in stomach contents of demersal fish of the Rockall Trough. *J. Fish Biol.* **24,** 357–362.

Mauchline, J., and Gordon, J. D. M. (1984e). Feeding and bathymetric distribution of the gadoid and morid fish of the Rockall Trough. *J. Mar. Biol. Assoc. U.K.* **64,** 657–665.

Mauchline, J., and Gordon, J. D. M. (1985). Trophic diversity in deep-sea fish. *J. Fish Biol.* **26,** 527–535.
Mauchline, J., and Gordon, J. D. M. (1986). Foraging strategies of deep-sea fish. *Mar. Ecol. Prog. Ser.* **27,** 227–238.
Mauchline, J., and Gordon, J. D. M. (1991). Oceanic pelagic prey of benthopelagic fish in the benthic boundary layer of a marginal oceanic region. *Mar. Ecol. Prog. Ser.* **74,** 109–115.
Mead, G. W., and Earle, S. A. (1970). Notes on the natural history of snipe eels. *Proc. Calif. Acad. Sci.* **38,** 99–103.
Merrett, N. R. (1987). A zone of faunal change in assemblages of abyssal demersal fish in the eastern North Atlantic; a response to seasonality in production? *Biol. Oceanogr.* **9,** 185–244.
Merrett, N. R., and Domanski, P. A. (1985). Observations of the ecology of deep-sea bottom living fishes collected off northwest Africa: II. The Moroccan Slope (27°–34° N), with special reference to *Synaphobranchus kaupi. Biol. Oceanogr.* **3,** 349–399.
Merrett, N. R., and Marshall, N. B. (1980). Observations on the ecology of deep-sea bottom-living fishes collected off northwest Africa (08°–27° N). *Prog. Oceanogr.* **9,** 185–244.
Merrett, N. R., and Roe, H. S. J. (1974). Patterns and selectivity in the feeding of certain mesopelagic fishes. *Mar. Biol.* **28,** 115–126.
Morrow, J. E. (1964). Families Chauliodontidae and Stomiatidae. *In* "Fishes of the Western North Atlantic" (H. B. Bigelow, ed.-in-chief), Part 4, pp. 274–310. Sears Foundation for Marine Research, New Haven, Connecticut.
Moser, M. L., Ross, S. W., and Sulak, K. J. (1997). Metabolic responses to hypoxia of *Lycenchelys verrillii* (wolf eelpout) and *Glyptocephalus cynoglossus* (witch flounder): Sedentary bottom fishes of the Hatteras/Virginia middle slope. *Mar. Ecol. Prog. Ser.*
Munk, O. (1984). Non-spherical lenses in the eyes of some deep-sea teleosts. *Arch. Fisch. Wiss.* **34,** 145–153.
Muntz, W. R. A. (1976). On yellow lenses in mesopelagic animals. *J. Mar. Biol. Assoc. U.K.* **56,** 963–976.
Nafpaktitis, B. G., Backus, R. H., Craddock, J. E., Haedrich, R. L., Robison, B. H., and Karnella, C. (1977). Family Myctophidae. *In* "Fishes of the Western North Atlantic" (R. H. Gibbs, Jr., ed.-in-chief), Part 7, pp. 13–265. Sears Foundation for Marine Research, New Haven, Connecticut.
Nielsen, J. G. (1969). Systematics and biology of the Aphyonidae. *Galathea Report* **10,** 7–89.
Nielsen, J. G. (1977). The deepest living fish *Abyssobrotula galathea,* a new genus and species of oviparous ophidioids (Pisces, Brotulidae). *Galathea Report* **14,** 41–48.
Nielsen, J. G., and Bertelsen, E. (1984). The gulper-eel family Saccopharyngidae (Pisces, Anguilliformes). *Steenstrupia* **11,** 157–206.
Nielsen, J. G., and Smith, D. G. (1978). The eel family Nemichthyidae. *Dana Report* **88,** 1–71.
Noble, E. R. (1973). Parasites and fishes in a deep-sea environment. *Adv. Mar. Biol.* **11,** 121–195.
Noble, E. R., and Collard, S. B. (1970). The parasites of midwater fishes. *Am. Fish. Soc. Spec. Publ.* **5,** 57–68.
Nyberg, D. W. (1971). Prey capture in the largemouth bass. *Am. Mid. Nat.* **96,** 128–144.
Ozawa, T., Fujii, K., and Kawaguchi, K. (1977). Feeding chronology of the vertically migrating gonostomatid fish, *Vinciguerria nimbaria* (Jordan and Williams), off southern Japan. *J. Oceanogr. Soc. Japan* **33,** 320–327.
Pearcy, W. G. (1976). Pelagic capture of abyssobenthic macrourid fish. *Deep-Sea Res.* **23,** 1065–1066.
Pearcy, W. G., and Ambler, J. W. (1974). Food habits of deep-sea macrourid fishes off the Oregon coast. *Deep-Sea Res.* **21,** 745–759.

Pearcy, W. G., Lorz, H. V., and Peterson, W. (1979). Comparison of the feeding habits of migratory and non-migratory *Stenobrachius leucopsarus* (Myctophidae). *Mar. Biol.* **51,** 1–8.

Pereyra, W. T., Pearcy, W. G., and Carvey, F. E., Jr. (1969). *Sebastodes flavidus,* a shelf rockfish feeding on mesopelagic fauna, with consideration for ecological implications. *J. Fish. Res. Bd. Can.* **26,** 2211–2215.

Pietsch, T. W. (1978). The feeding mechanism of *Stylephorus chordatus* (Teleostei: Lampridiformes): Functional and ecological implications. *Copeia,* 255–262.

Podrazhanskaya, S. G. (1971). Feeding and migrations of the roundnose grenadier, *Macrourus rupestris,* in the northwest Atlantic and Iceland waters. *International Commission for the Northwest Atlantic Fisheries, Redbook,* Part 3, 115–123.

Priede, I. G., and Smith, K. L., Jr. (1986). Behaviour of the abyssal grenadier, *Coryphaenoides yaquinae,* monitored using ingestible acoustic transmitters in the Pacific Ocean. *J. Fish Biol.*(Suppl. A.), 199–206.

Priede, I. G., Smith, K. L., Jr., and Armstrong, J. D. (1990). Foraging behavior of abyssal grenadier fish: Inferences from acoustic tagging and tracking in the North Pacific Ocean. *Deep-Sea Res.* **37,** 81–101.

Priede, I. G., Bagley, P. M., Armstrong, J. D., Smith, K. L., Jr., and Merrett, N. R. (1991). Direct measurement of active dispersal of food-falls by deep-sea demersal fishes. *Nature* (*London*) **351,** 647–649.

Priede, I. G., Bagley, P. M., Smith, A., Creasey, S., and Merrett, N. R. (1994a). Scavenging deep demersal fishes of the Porcupine Seabight, North-east Atlantic: Observations by baited camera, trap and trawl. *J. Mar. Biol. Assoc. U.K.* **74,** 481–498.

Priede, I. G., Bagley, P. M., and Smith, J. K. L. (1994b). Seasonal change in activity of abyssal demersal scavenging grenadiers *Coryphaenoides* (*Nematonurus*) *armatus* in the eastern North Pacific Ocean. *Limnol. Oceanogr.* **39,** 279–285.

Reid, S. B., Hirota, J., Young, R. E., and Hallacher, L. E. (1991). Mesopelagic-boundary community in Hawaii: Micronekton at the interface between neritic and oceanic ecosystems. *Mar. Biol.* **109,** 427–440.

Ribbink, A. J. (1971). The jaw mechanism and feeding of the holocephalan, *Callorhynchus capensis* Dumeril. *Zool. Afr.* **6,** 45–73.

Riemann, F. (1989). Gelatinous phytoplankton detritus aggregates on the Atlantic deep-sea bed. Structure and mode of formation. *Mar. Biol.* **100,** 533–539.

Robins, C. H. (1989). Family Derichthyidae. *In* "Fishes of the Western North Atlantic" (A. E. Parr, Ed. emeritus), Part 9, Vol. 1, pp. 420–431. Sears Foundation for Marine Research, New Haven, Connecticut.

Robins, C. H., and Robins, C. R. (1989). Family Synaphobranchidae. *In* "Fishes of the Western North Atlantic" (A. E. Parr, ed. emeritus), Part 9, Vol. 1, pp. 247–253. Sears Foundation for Marine Research, New Haven, Connecticut.

Robison, B. H. (1973). A system for maintaining midwater fishes in captivity. *J. Fish. Res. Bd. Can.* **30,** 126–128.

Robison, B. H. (1984). Herbivory by the myctophid fish *Ceratoscopelus warmingii. Mar. Biol.* **84,** 119–123.

Robison, B. H., and Bailey, T. G. (1981). Sinking rates and dissolution of midwater fish fecal matter. *Mar. Biol.* **65,** 135–142.

Rofen, R. R. (1966). Families Paralepididae, Omosudidae, Evermannellidae and Scopelarchidae. *In* "Fishes of the Western North Atlantic" (G. W. Mead, ed.-in-chief), Part 5, pp. 205–602. Sears Foundation for Marine Research, New Haven, Connecticut.

Roper, C. F. E., Sweeney, M. J., and Nauen, C. E. (1984). Cephalopods of the world. An annotated and illustrated catalogue of species of interest to fisheries. *FAO Fish. Synop.* **125,** 1–277.

Sainte-Marie, B., and Hargrave, B. T. (1987). Estimation of scavenger abundance and distance of attraction to bait. *Mar. Biol.* **94,** 431–443.

Saldanha, L. (1980). Régime alimentaire de *Synaphobranchus kaupi* Johnson, 1862 (Pisces Synaphobranchidae) au large des côtes européenes. *Cybium (Ser. 3)* 91–98.

Sanders, H. L. (1968). Marine benthic diversity: A comparative study. *Am. Nat.* **102,** 243–282.

Savvatimskii, P. I. (1969). The grenadier of the North Atlantic. *Proc. Polar Res. Ins. Mar. Fish. Oceanogr.* 3–72. (*Fish. Res. Bd. Can. Trans. Ser.* No. 2879, 1974).

Schoener, T. (1971). Theory of feeding strategies. *Annu. Rev. Ecol. Syst.* **2,** 369–404.

Schultz, L. P. (1964). Family Sternoptychidae. *In* "Fishes of the Western North Atlantic" (H. B. Bigelow, ed.-in-chief), Part 4, pp. 241–273. Sears Foundation for Marine Research, New Haven, Connecticut.

Scott, T. (1911). On the food of the halibut, with notes on the food of *Scorpaena, Phycis blennoides*, the garpike and *Chimaera monstrosa*. *Sci. Inv. Fish. Bd. Scotland* **1909**(28), 24–37.

Sedberry, G. R., and Musick, J. A. (1978). Feeding strategies of some demersal fishes of the continental slope and rise off the Mid-Atlantic Coast of the USA. *Mar. Biol.* **44,** 357–375.

Siebenaller, J. F., Somero, G. N., and Haedrich, R. L. (1982). Biochemical characteristics of macrourid fishes differing in their depths of distribution. *Biol. Bull.* **163,** 240–249.

Smith, D. G. (1989). Family Nemichthyidae. *In* "Fishes of the Western North Atlantic" (A. E. Parr, ed. emeritus), Part 9, Vol. 1, pp. 441–459. Sears Foundation for Marine Research, New Haven, Connecticut.

Smith, K. L., Jr. (1978). Metabolism of the abyssopelagic rattail *Coryphaenoides armatus* measured *in situ*. *Nature (London)* **274,** 362–364.

Smith, K. L., Jr., White, G. A., Laver, M. B., McConnaughey, R. R., and Meador, J. P. (1979). Free vehicle capture of abyssopelagic animals. *Deep-Sea Res.* **26,** 57–64.

Smith, M. M., and Heemstra, P. C. (1986). "Smith's Sea Fishes." Macmillan, Johannesburg, South Africa.

Solomon-Raju, N., and Rosenblatt, R. H. (1971). New record of the parasitic eel, *Simenchelys parasiticus,* from the central North Pacific with notes on its metamorphic form. *Copeia*, 312–314.

Somiya, H. (1976). Functional significance of the yellow lens in *Argyropelecus affinis*. *Mar. Biol.* **34,** 93–99.

Somiya, H. (1979). 'Yellow lens' eyes and luminous organs of *Echiostoma barbatum* (Stomiatoidei, Melanostomiatidae). *Jpn. J. Ichthy.* **25,** 269–272.

Somiya, H. (1982). 'Yellow lens' eyes of a stomiatoid deep-sea fish, *Malacosteus niger*. *Proc. R. Soc. London* **B215,** 481–489.

Steimle, F. W. J., and Terranova, R. J. (1988). Energy contents of northwest Atlantic continental slope organisms. *Deep-Sea Res.* **35,** 415–423.

Stein, D. L., and Pearcy, W. G. (1982). Aspects of reproduction, early life history, and biology of macrourid fishes off Oregon, U.S.A., *Deep-Sea Res.* **29,** 1313–1329.

Stickney, D. G., and Torres, J. J. (1989). Proximate composition and energy content of mesopelagic fishes from the eastern Gulf of Mexico. *Mar. Biol.* **103,** 13–24.

Stockton, W. L., and DeLaca, T. E. (1982). Food falls in the deep-sea: Occurrence, quality, and significance. *Deep-Sea Res.* **29,** 157–169.

Sulak, K. J. (1977). ALVIN, Window in the Deep. *Sea Front.* **23,** 113–119.

Sulak, K. J. (1982). A comparative taxonomic and ecological analysis of temperate and tropical demersal deep-sea fish faunas in the western North Atlantic. Ph.D. Thesis, University of Miami, Miami, Florida.

Sulak, K. J., and Ross, S. W. (1993). Analysis of submersible videotapes for demersal fish faunas from the continental slope in "The Point" region. *In* "Benthic Processes in an

Unusual Area of the U.S. Atlantic Continental Slope off Cape Hatteras" (R. J. Diaz and J. A. Blake, eds.), U.S. Department of the Interior, Minerals Management Service, Washington, D.C., MMS Report 30672, December 1992.

Sulak, K. J., Wenner, C. A., Sedberry G. R., and Van Guelpen, L. (1985). Life history and systematics of deep-sea lizardfishes, genus *Bathysaurus* (Synodontidae). *Can. J. Zool.* **63,** 623–642.

Sutton, T. T., and Hopkins, T. L. (1996). Trophic ecology of the stomiid (Pisces, Stomiidae) assemblage of the eastern Gulf of Mexico: Strategies, selectivity and impact of a mesopelagic top predator group. *Mar. Biol.* **127,** 179–192.

Tchernavin, V. V. (1953). The feeding mechanisms of a deep-sea fish *Chauliodus sloanei* Schneider. *Br. Mus. Nat. Hist.* 1–101.

Torres, J. J., Belman, B. W., and Childress, J.-J. (1979). Oxygen consumption rates of midwater fishes as a function of depth of occurrence. *Deep-Sea Res.* **26,** 185–197.

Veriginia, I. A. (1979). Morphological anatomy of the digestive tract of the deep-sea fish *Bajacalifornia calcarata* (Weber) (Family Alepocephalidae, Salmoniformes). *J. Ichthy.* **19**(6), 164–168.

Veriginia, I. A., and Golovan', G. A. (1978). Structural peculiarities of fishes of family Alepocephalidae. *Vopr. Ikhtiol.* **18**(2), 277–279 (in Russian).

Wenner, C. A. (1978). Making a living on the continental slope and in the deep-sea: Life history of some dominant fishes of the Norfolk Canyon area. Ph.D. Thesis, College of William and Mary in Virginia, Gloucester Point.

Wiebe, P. H., Madin, L. P., Haury, L. R., Harbison, G. R., and Philbin, L. M. (1979). Diel vertical migration by *Salpa aspera* and its potential for large-scale particulate organic matter transport to the deep-sea. *Mar. Biol.* **53,** 249–255.

Willis, J. M., and Pearcy, W. G. (1980). Spatial and temporal variations in the population size structure of three lanternfishes (Myctophidae) off Oregon, USA. *Mar. Biol.* **57,** 181–191.

Wilson, R. R. J., and Smith, K. L. (1984). Effect of near-bottom currents on detection of bait by the abyssal grenadier fish *Coryphaenoides* spp., recorded *in situ* with a video camera on a free vehicle. *Mar. Biol.* **84,** 83–91.

Woods, L. P., and Sonoda, P. M. (1973). Family Trachichthyidae. *In* "Fishes of the Western North Atlantic" (D. M. Cohen, ed.-in-chief), Part 6, pp. 298–327. Sears Foundation for Marine Research, New Haven, Connecticut.

Young, J. W., and Blaber, S. J. M. (1986). Feeding ecology of three species of midwater fishes associated with the continental slope of eastern Tasmania, Australia. *Mar. Biol.* **93,** 147–156.

5

BUOYANCY AT DEPTH

BERND PELSTER

I. Introduction
II. The Problem of Buoyancy
   A. Hydrostatic Pressure
   B. Energy Costs of Neutral Buoyancy
III. Swim Bladder Function
   A. Morphology of the Swim Bladder
   B. Mechanisms of Gas Deposition
   C. Resorption of Gas
   D. Depth Limitations on the Utility of a Gas-Filled Swim Bladder
   E. Lipid-Filled Swim Bladders
IV. Lipid Accumulation
   A. Density of Lipids
   B. Lipid Droplets in Eggs and Larvae
   C. Lipid Accumulation in the Liver
   D. Bone Lipids
   E. Lipid Accumulation in Other Tissues
V. Watery Tissues
   A. Basic Principle
   B. Eggs and Larvae
   C. Reduction of Skeletal Density
   D. Watery Muscle
   E. Gelatinous Masses
VI. Hydrodynamic Lift
VII. Conclusions
   References

I. INTRODUCTION

Any submerged body or organism experiences upthrust, given by the weight of the displaced fluid medium. Thus, to stay on the bottom of a water column is very easy if the density of an organism is greater than that of water, usually given as 1.00 kg liter^{-1} for fresh water and 1.026 kg liter^{-1} for seawater. In fact, most animal tissues are denser than water. To stay in open water is thus a problem for any animal of density greater than that of the surrounding water. Therefore, organisms that have successfully invaded the pelagic space, irrespective of their systematic allocation, show special adaptations that allow a comfortable equilibrium between body weight and buoyancy plus hydrodynamic lift. For example, they have developed hydrofoils, providing hydrodynamic lift, and they have reduced tissue density or special very low-density tissues or spaces with densities lower than that of the surrounding water. With only few exceptions the density of lipids and fat is less than 1.00 kg liter^{-1} making these substances effective buoyancy aids. An even more effective buoyancy aid is a gas cavity; in the fish swim bladder, the density of gas is negligible at low or moderate hydrostatic pressure. The accumulation of fluid with a lower salinity than that of the surrounding water will also provide upthrust, because water density increases with increasing salinity. Another strategy is to reduce the density of typically "high-density tissues." A reduction in skeleton mineralization, for example, will not provide upthrust, but it will significantly reduce the density of the skeleton and thus increase overall buoyancy.

This chapter demonstrates that fish have adopted all of the above mechanisms to various degrees. Whereas some fish simply do not have any buoyancy device, others rely on a single strategy or structure as a buoyancy device, and still others use several strategies simultaneously. The situation becomes even more complex if we acknowledge the fact that low-density material such as lipids not only provides upthrust, but also may serve as an energy reserve. In terms of deep-sea fish, the constant low temperature, which cause a low-energy turnover in animals, must be considered. Furthermore, at high hydrostatic pressures, gas cavities lose their effectiveness as a buoyancy aid because gas density increases with increasing hydrostatic pressure. Physiological data on typical deep-sea fish are scarce, however, so mesopelagic species must be referred to as well.

II. THE PROBLEM OF BUOYANCY

A. Hydrostatic Pressure

Organisms living in water are exposed to hydrostatic pressure, which increases by about 1 atm for each 10 m of water depth. Fish living at the water surface experience hydrostatic pressure of 1 atm, whereas at a depth of 1000 m, hydrostatic pressure increases to 101 atm. Water is almost incompressible. With a small margin of error, the same appears to be true for lipids (Corner et al., 1969). Nevertheless, there may be small but significant changes in density induced by temperature-dependent changes

5. BUOYANCY AT DEPTH 197

in lipid fluidity encountered during vertical migrations. Thus, a fish without a swim bladder experiences only small changes in density with changes in depth.

Gas-filled cavities, with the exception of rigid-walled cavities such as the shell of *Nautilus* (Denton, 1960; Denton and Gilpin-Brown, 1966), will change volume in proportion to changes in hydrostatic pressure according to the gas law. Any change in the volume of a gas cavity in turn will change the buoyancy of the organism. Another important aspect relevant to deep-sea fish is the fact that the specific gravity of gas increases with gas pressure, so that the difference between gas density and water density decreases with increasing water depth. Based on a report of the National Research Council (1928), Alexander (1966a) calculated the specific gravity of oxygen as 0.6 kg liter^{-1} at a pressure of 500 atm, equivalent to a depth of about 5000 m.

The equilibrium pressure of gases dissolved in water also is modified by hydrostatic pressure, but compared with the changes observed in gas cavities, this influence can almost be neglected. Measurements of Enns *et al.* (1967) revealed an increase in physical CO_2 solubility in water of about 16% at a hydrostatic pressure of 100 atm. For oxygen, nitrogen, and argon, the increase was about 14%. If the data are extrapolated to a water depth of 10,000 m, the equilibrium partial pressure of dissolved gases in water would probably increase by a factor of 4, whereas the hydrostatic pressure would increase from 1 atm to 1001 atm. The increase in gas solubility is also enhanced by a decrease in temperature. A bulk of water moving a depth with a given content of dissolved gases thus will experience a decrease in gas partial pressure as a result of an increase in solubility associated with an increase in hydrostatic pressure and a decrease in temperature.

B. Energy Costs of Neutral Buoyancy

The density of body tissues is quite variable (Table I). Tissues with a high water content can be expected to have a density close to water, but accumulation of heavy ions (such as Ca^{2+}) or tissue mineralization significantly increases tissue densities. Thus, most tissues have a density greater than that of water, and bones usually have the highest density, about 1.3 to 1.5 kg liter^{-1}, followed by cartilage, skin, and muscle tissue.

Tissue density (ρ) and volume (V) determine the weight (m) of an animal in air

$$m = Vg\rho, \tag{1}$$

where g equals gravitational acceleration. When the animal is fully immersed in water, the water density (ρ_w) must be accounted for, and the animal weight is given as

Table I
Density of Various Fish Tissues

Tissue	Tissue density (kg liter^{-1})		
	Pleuronectes platessa[a]	*Myoxocephalus scorpius*[a]	*Scyliorhinus canicula*[b]
Skin	1.054	1.070	1.128
Fins	1.092	1.151	—
Muscle	1.048	1.062	1.071
Liver	1.040	1.062	1.072
Head	1.300	1.530	1.165
Axial skeleton	1.299	1.532	1.128

[a] Data from Webb (1990).
[b] Data from Bone and Roberts (1969).

$$m = Vg(\rho - \rho_w). \quad (2)$$

If the whole-body density of a specimen is equal to water density, the specimen has no weight in water (neutral buoyancy); if whole-body density is greater than water density, the specimen has some weight in water and will tend to sink. To achieve neutral buoyancy, weight must be balanced by lift. Thus, the lift (L) required is

$$L = Vg(\rho - \rho_w), \quad (3)$$

or, in terms of body mass (m),

$$L = mg(\rho - \rho_w)/\rho. \quad (4)$$

Very small animals sink slowly and may be kept afloat for some time by vertical eddies. Riley *et al.* (1949) calculated that a stable population (constant number of specimens at a given depth) is possible if the animals reproduce rapidly and are smaller than 150 μm with a mass of no more than about 2 μg (Alexander, 1990). This is especially true if specimens have special floating devices or parachutes, such as long antennae, which reduce sinking speed. These strategies increase the surface area to volume ratio of an organism, and because drag is proportional to the power of the surface area, the tendency to sink is reduced. Clearly, this strategy, which works well at deep sea, applies only to protists and small plankton, including larval fish. Larger animals need specialized buoyancy devices, or they need to generate hydrodynamic lift by muscular activity to occupy a pelagic niche in the water column.

5. BUOYANCY AT DEPTH

1. Hydrodynamic Lift

Fish produce hydrodynamic lift mainly by using their pectoral fins as hydrofoils. The metabolic power (E_m) needed to propel the hydrofoils through the water can be calculated from drag on the hydrofoils and speed,

$$E_m = 4L^2/\eta\pi\rho_w U\lambda^2. \tag{5}$$

where η is the efficiency coefficient for the conversion of metabolic energy to mechanical power, U is the speed, and λ is the span of the hydrofoil (Alexander, 1990). Thus, the additional power, needed to produce the lift necessary to remain at a certain water depth decreases with increasing swimming speed.

2. Volume of a Buoyancy Device

Another strategy to achieve neutral buoyancy is to build up and maintain a buoyancy device, that is, compensate for the high density of most tissues by including special structures or organs characterized by a very low density. To match the density of a fish of volume V_s and density ρ_s to water density, the buoyancy organ must have a density lower than that of water. The volume of this buoyancy organ (V_b) is dependent on its density (ρ_b), or the difference between water and tissue density. The compensation is complete if

$$(V_s\rho_s + V_b\rho_b)/(V_s + V_b) = \rho_w. \tag{6}$$

This gives the volume of the buoyancy organ as a fraction of the fish volume without this organ:

$$V_b/V_s = (\rho_s - \rho_w)/(\rho_w - \rho_b). \tag{7}$$

Much more informative, however, is the volume of the buoyancy organ as a fraction of the volume of the intact fish:

$$V_b/(V_s + V_b) = (\rho_s - \rho_w)/(\rho_s - \rho_b). \tag{8}$$

Thus, for a growing fish it is essential that the volume of the buoyancy organ increases in proportion to body volume to retain neutral buoyancy. After one distinguishes between fish volume without buoyancy organ and the volume of the buoyancy organ, it becomes apparent that the buoyancy structure may add considerable volume to the fish if the density difference between the buoyancy organ and the water is small.

Another important aspect is the location of the low-density material. It must be located in balanced portions around the center of gravity; otherwise it will hamper the trim of the animal, resulting in a head-up or head-down position.

3. ENERGY EXPENDITURE

Although one intuitively would expect that being neutrally buoyant is energetically much more efficient than continuous swimming, the quantitative analysis of this question has proved to be quite difficult (Alexander, 1972, 1990). The mechanical power necessary to propel a rigid body through the water is determined by the velocity of the body and the drag produced. The metabolic energy required can be estimated by including an appropriate efficiency coefficient for the conversion of metabolic energy in mechanical power. Fish, however, are not rigid bodies; they swim with undulatory movements, which modifies drag and therefore makes this analysis very difficult (Alexander, 1990).

A somewhat easier attempt to quantify the energy expenditure for neutral buoyancy is to use metabolic data such as oxygen consumption and swimming speed. In trout, oxygen consumption ($\dot{M}_{O_2}$) is correlated with speed (U) (Webb, 1971). Using these data, Alexander (1990) calculated the metabolic power (E) from total body volume (V) and speed (U):

$$E = 146(V)^{0.5} U^{2.5}. \tag{9}$$

If a buoyancy device is present in addition to body volume V_s, additional power is required,

$$E + E_b = 146(V_s + V_b)^{0.5} U^{2.5}, \tag{10}$$

with the additional power being

$$E_b \approx 2.3 m^{0.5} U^{2.5} (V_b/V_s). \tag{11}$$

The additional power necessary to swim with a larger volume due to a buoyancy organ increases with the ratio of V_b/V_s; thus, the larger the volume of the buoyancy organ, the more inefficient swimming will be. Given this trade-off, pelagic fish could limit their locomotory habits or use for buoyancy organs materials with lower densities and therefore lower volumes.

The preceding analysis does not include any energy requirement for organ growth and for maintenance of its volume. For fat deposition, an approximation can be calculated based on the heat of combustion of the fat body (H_b). The energy cost of fat deposition ($E_{b'}$) is then given as the product of H_b, the volume and the density of the fat body, and its relative growth rate (G):

$$E_{b'} = G H_b V_b \rho_b. \tag{12}$$

If we consider gas deposition, the situation is more complicated. The swim bladder has a flexible wall, and its volume and pressure change with changes in hydrostatic pressure. When a fish descends, the swim bladder volume and thus buoyancy can be kept constant by deposition of gas; when a fish ascends, gas must be resorbed to maintain buoyancy. The energy necessary to compress gas can be calculated according to the gas law as $RT \ln P_1/P_2$. Accordingly, the work (W) required to keep the volume constant by deposition of oxygen is given as

$$W = V_d RT \ln(Psb_{O_2}/PaO_2), \qquad (13)$$

where V_d is the volume of oxygen deposited, Psb_{O_2} is the partial pressure of oxygen (P_{O_2}) in the swim bladder, and Pa_{O_2} is arterial P_{O_2}; the efficiency of the process of gas deposition, however, is unknown (Alexander, 1971, 1972). Furthermore, the swim bladder wall has a gas permeability that is much lower than that of other tissues, but it is not completely impermeable to gases. For this reason the analysis must include the energy required for gas replacement due to the loss of gases to the surrounding tissue. The diffusional loss of gas from the swim bladder depends on the magnitude of the partial pressure gradient between the swim bladder lumen and the surrounding tissue and the surface area of the swim bladder. In the ocean, water P_{O_2} decreases with depth; at 1000 m it appears to be only a fraction of surface P_{O_2} (Alexander, 1972; Brooks and Saenger, 1991; Vetter et al., 1994). Swim bladder P_{O_2} in turn increases with depth, so that the diffusional loss of oxygen is dependent on water depth.

A quantitative energetic description comparing the various strategies of achieving neutral buoyancy thus remains incomplete or at least is based on a number of assumptions that are difficult to verify. Nevertheless, some important conclusions evolve from this analysis so far: to remain at a certain depth, it is generally desirable to use material of the lowest possible density as a buoyancy device. However, if a fish swims very fast, it is more economical to use hydrofoils instead of a buoyancy aid. Alexander (1990) calculated that for a fish with a body mass of 1 kg, it becomes more economical to use hydrodynamic lift instead of a swim bladder if the fish swims more than 0.75 m s^{-1}; if squalene is accumulated, it becomes more economical at a speed of 0.45 m s^{-1}. The speed, however, at which using fins as hydrofoils is more economical than a swim bladder or storage of squalene increases with body mass (Fig. 1). Given the low energy turnover in deep-sea fish due to the constant low temperature at depth, one might predict that deep-sea fish do not primarily rely on hydrodynamic lift to achieve neutral buoyancy, but use buoyancy devices or reduce their tissue density.

III. SWIM BLADDER FUNCTION

A. Morphology of the Swim Bladder

A very effective way to achieve neutral buoyancy is a gas-filled cavity. At least at moderate water depth, gas has an almost negligible density

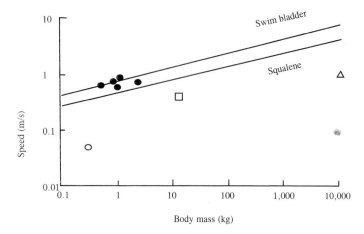

Fig. 1. Speeds at which use of fins to generate hydrodynamic lift becomes more energy efficient than use of a gas-filled swim bladder or squalene stores to achieve neutral buoyancy. ●, Swimming speeds and body mass of various scombroids that rely on hydrodynamic lift; ○, trout; □, wahoo, which have swim bladders; △, basking shark, which stores squalene. Redrawn from Alexander (1990).

compared to the density of water, and a gas cavity with a volume of about 5–6% of the body volume is sufficient to ensure neutral buoyancy in seawater [see Eq. (8)].

Embryonically the swim bladder originates as an unpaired dorsal outgrowth of the foregut. During development the connection to the gut may persist as the pneumatic duct (physostome fishes). In the great majority of teleosts, this duct is completely lost at an early stage during development and the swim bladder is a closed gas cavity (physoclist fishes). Deep-sea fish usually are physoclistic. In adult teleosts the structural diverity in general swim bladder morphology is remarkable and has been reviewed by Fänge (1953), Marshall (1960), and Steen (1970). In many species the swim bladder consists of two chambers: a thick-walled section in which gas can be deposited and a thin-walled chamber in which gas can be resorbed (e.g., Cyprinidae). In other fish the resorbing part of the swim bladder is reduced to a special section of the secretory bladder, called *oval,* which can be closed off by muscular activity (gadoid fishes). The eel is a physostome fish, but in the adult fish the pneumatic duct is transformed to a resorbing part of the swim bladder and is functionally closed.

The arrangement of blood vessels typically is characterized by the presence of a countercurrent system, a rete mirable. Figure 2A shows the swim bladder anatomy of the deep-sea teleost *Ichthyococcus ovatus.* The rete

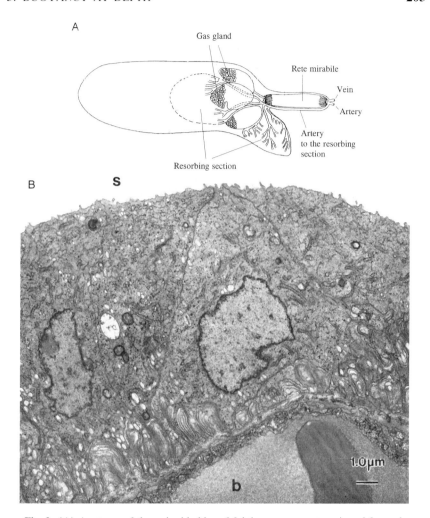

Fig. 2. (A) Anatomy of the swim bladder of *Ichthyococcus ovatus*, viewed from above. The three lobes of the gas gland are located on the bottom of the swim bladder. The blood supply to the resorbing section of the swim bladder bypasses the rete mirabile [adapted from Marshall (1960)] (B) Histology of swimbladder gas gland cells of the European eel *Anguilla anguilla;* b, blood vessel; n, nucleus; s, swim bladder lumen (Courtesy of J. Würtz, Zoology Department, Karlsruhe, Germany).

mirabile is made up of several 10,000 arterial and venous capillaries arranged so that each arterial capillary is surrounded by several venous capillaries and vice versa. The length of the capillaries in a rete can be several millimeters, and the diffusion distance between arterial and venous vessels is about 1 to 2 μm (Stray-Pedersen and Nicolaysen, 1975). In a bipolar rete the capillaries at the swim bladder pole reconvene to only a few larger arterial and venous vessels, which then again give rise to an additional capillary system supplying the swim bladder epithelium. In a unipolar rete the capillaries of the rete mirabile hardly reconvene to larger vessels and almost directly supply a special area of the swim bladder epithelium, in which so-called gas gland cells are located. The anatomical arrangement of a rete mirable supplying the gas gland typically is found in deep-sea fish, but it is not present in all swim bladders. Salmonids, for example, lack a rete mirabile.

The wall of the secretory bladder consists of a number of thin tissue layers, sometimes including thin layers of smooth muscle cells. The terminology of Fänge (1953) describes an inner epithelium, a muscularis mucosa, a submucosa, and a tunica externa. The tunica externa represents a dense connective tissue capsule. The submucosa usually is impregnated with guanine crystals (Lapennas and Schmidt-Nielsen, 1977; Kleckner, 1980) or may include layered lipid membranes (Brown and Copeland, 1978), rendering the swim bladder wall impermeable to gases and thus preventing diffusional loss of gases (Kutchai and Steen, 1971; Denton *et al.*, 1972; Lapennas and Schmidt-Nielsen, 1977; Kleckner, 1980). The muscularis mucosa, mainly consisting of smooth muscle cells, is present in physostome fishes and is found often in physoclist fishes.

The gas gland cells of the swim bladder are epithelial cells specialized for the production of acidic metabolites. Whereas in the eel (*Anguilla*) gas gland cells are spread over the whole internal epithelium of the secretory bladder, in many species (*Perca, Gadus*) gas gland cells are clustered together, forming a massive complex of several cell layers. In some species, a compact gas gland results from extensive secondary folding of a single layer of epithelium (*Gobius, Syngnathus*) (Woodland, 1911; Fänge, 1983). Gas gland cells are usually in intimate contact with an extensive capillary vascular system.

Gas gland cells are cubical or cylindrical, with a size ranging from 10 to 25 μm to giant cells of 50 to 100 μm or even more. The size of the individual cells appears not to be correlated to the water depth at which fish normally live (Fänge, 1953; Marshall, 1960), although the size of the gas gland tends to be larger in deep-sea fish. Gas gland cells are polarized with some small microvilli on the luminal side, whereas the basal side is often more densely vacuolated and shows a large number of infoldings (Fig. 2B) but lacks mitochondria. The significance of these foldings is not yet understood (Dorn, 1961; Copeland, 1969; Morris and Albright, 1975). The variable density of the granulated plasma of gas gland cells may represent variable functional states, and does not necessarily indicate the presence of different cell types (Dorn, 1961; Morris and Albright, 1975). Gas gland cells are characterized by the presence of only a few filamentous or elongated mitochondria with few tubular cristae (Dorn, 1961; Copeland, 1969; Jasinski and Kilarski, 1969; Morris and Albright, 1975).

B. Mechanisms of Gas Deposition

Gas is deposited into the swim bladder by passive diffusion from the blood. The high gas partial pressures necessary to establish the diffusion gradients between the blood and the swim bladder lumen are established by two mechanisms: the reduction of the effective gas-carrying capacity of swim bladder blood and the subsequent countercurrent concentration of gases in the rete mirabile.

Reduction of the effective gas-carrying capacity of swim bladder blood is brought about by the metabolic and secretory activity of the epithelial gas gland cells. Although the P_{O_2} usually is high in the swim bladder epithelium, gas gland cells are specialized for the anaerobic production of acidic metabolites (for review, see Pelster, 1995a). Glucose is the main fuel and is removed from the blood, whereas internal glycogen stores do not appear to be of major importance. Gas gland tissue of various species incubated *in vitro* or artificially perfused with saline solution has been shown to produce large amounts of lactate (Ball *et al.*, 1955; Deck, 1970; D'Aoust, 1970; Kutchai, 1971; Pelster *et al.*, 1989; Ewart and Driedzic, 1990; Pelster, 1995b). Even at hyperbaric oxygen pressures of about 50 atm, gas gland tissue of *Sebastodes miniatus* continued to produce lactic acid, indicating the absence of a Pasteur effect (D'Aoust, 1970). *In vivo* lactate formation has been demonstrated in only two species, namely, the barracuda *Sphyraena barracuda* and the European eel *Anguilla anguilla* (Steen, 1963a; Enns *et al.*, 1967; Kobayashi *et al.*, 1989a). A quantitative analysis of lactate and glucose metabolism of the active, gas-depositing swim bladder of the European eel revealed that about 75–80% of glucose taken up from the blood is converted into lactate (Pelster and Scheid, 1993).

The low mitochondrial density and activities of enzymes of the citric acid cycle or the respiratory chain suggest that aerobic glucose metabolism is of minor importance and perhaps is almost negligible (Dorn, 1961; Copeland, 1969; Jasinski and Kilarski, 1969; Boström *et al.*, 1972; Morris and Albright, 1975; Ewart and Driedzic, 1990; Pelster and Scheid, 1991; Walsh and Milligan, 1993). Indeed, evaluation of glucose metabolism as well as O_2 and CO_2 exchange in the swim bladder of the European eel suggests that only 1% of the glucose removed from the blood is oxidized (Pelster and Scheid, 1992, 1993). Gas gland cells produce significant amounts of CO_2 by the decarboxylation reaction of the enzyme 6-phosphogluconate dehydrogenase in the pentose phosphate shunt (Walsh and Milligan, 1993; Pelster *et al.*, 1994).

Figure 3 shows the metabolic pathways involved in glucose metabolism, based on results obtained from European and American eels. End products of glucose metabolism are lactic acid and CO_2, which are released into the blood. CO_2 readily diffuses along its partial pressure gradient into the blood to lower blood pH. Carbonic anhydrase activity, typically found in gas gland cells (Fänge, 1953; Skinazi, 1953; Maetz, 1956; Dorn, 1961; D'Aoust, 1970; Kutchai, 1971), is responsible for a rapid equilibrium of the reaction:

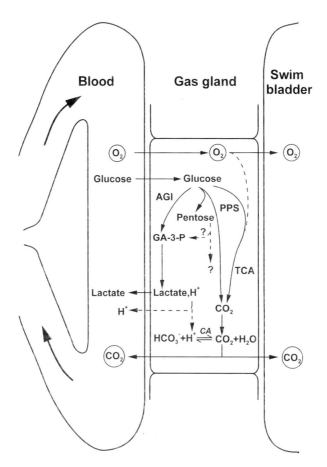

Fig. 3. Present concept for pathways of glucose metabolism in swim bladder gas gland cells. AGl, Anaerobic glycolysis; CA, carbonic anhydrase; PPS, pentose phosphate shunt; TCA, tricarboxylic acid cycle. Modified from Pelster (1995a). Reprinted from *Biochem. Mol. Biol. Fishes* **4**, B. Pelster, Metabolism of the swimbladder tissue, 101–118. Copyright 1995 with kind permission of Elsevier Science-NL, Sara Burgerhartstraat 25, 1055 KV Amsterdam, The Netherlands.

5. BUOYANCY AT DEPTH

$$H_2O + CO_2 \rightleftharpoons H^+ + HCO_3^-.$$

Inhibition of carbonic anhydrase has been shown to reduce the rate of acid release from cultured gas gland cells, and there appears to be cytoplasmatic as well as membrane-bound carbonic anhydrase activity (Pelster, 1995c). Further pathways for the release of protons from gas gland cells include Na^+-dependent carriers, such as Na^+/H^+ exchange and Na^+-dependent anion exchange, and a proton ATPase (adenosine triphosphatase) (Klenk and Pelster, 1995; Pelster, 1995c). Gas gland cells release acid over a wide pH range; pH values between 6.6 and 7.8 have been measured in swim bladder blood after passage through the gas gland. It is tempting, therefore, to speculate that the various mechanisms for proton secretion are determined by their pH dependence.

In the blood, CO_2 and lactic acid reduce the effective gas-carrying capacity of the blood, resulting in an increase in gas partial pressure in the blood (the single concentrating effect) (Kuhn et al., 1963) (Fig. 4A). An increase in blood lactate concentration causes a decrease in the physical

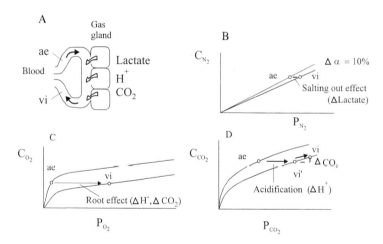

Fig. 4. Mechanisms that reduce the effective gas-carrying capacity in swim bladder blood (single concentrating effect). (A) Metabolic end products of glucose metabolism, mainly lactic acid and CO_2. Both metabolites are released into the blood, initiating the single concentrating effect. ae, Arterial efflux; vi, venous influx. (B) The increase in blood lactate concentration reduces the physical solubility of gas according to the salting-out effect. Based on our present knowledge, a salting-out effect of about 1% can be expected, resulting in an increase in gas partial pressure of 1%. The graph shows a 10% decrease for clarification of the principle. (C) Acidification induces a severe increase in P_{O_2} via the Root effect. (D) Acidification shifts the equilibrium of the CO_2/HCO_3^- reaction toward formation of CO_2, and CO_2 is produced in the metabolism. Both contribute to a marked increase P_{CO_2}. Redrawn from Pelster et al. (1990).

solubility of any gas due to the salting-out effect (Fig. 4B). Whereas typically molar concentrations of solutes are used to demonstrate the magnitude of the salting-out effect (Enns *et al.*, 1967; Gerth and Hemmingsen, 1982), the increase in blood lactate concentration measured during passage of the gas gland cells ranges from 5 to 10 mmol liter^{-1} or even lower. This, however, allows only for a decrease in physical gas solubility of no more than 1 to 2%, with a concomitant increase in gas partial pressures of the same magnitude (Pelster *et al.*, 1988). According to our present knowledge, this is the only way to induce an increase in gas partial pressures for inert gases such as nitrogen and argon.

For CO_2, the situation is different. The acidification of the blood during passage through the gas gland induces an increase in P_{CO_2} (Fig. 4C). Carbonic anhydrase activity within the red blood cells and probably also in the gas gland cell membrane (Pelster, 1995c) ensures a rapid equilibrium of the CO_2/HCO_3^- system in the swim bladder. In addition, CO_2 is produced by metabolism and released into the swim bladder as well as into the blood. Therefore a large increase in gas partial pressure for CO_2 can be expected (Fig. 4D); Kobayashi *et al.* (1990) indeed observed an increase in blood P_{CO_2} from 31 ± 7 torr to 62 ± 18 torr during passage of the gas gland cells in the European eel.

The largest increase in gas partial pressure, however, has to be expected for oxygen. The hemoglobin of many fish is characterized by the presence of a Root effect (Root, 1931; Brittain, 1987; Riggs, 1988; Pelster and Weber, 1991), that is, a decrease in hemoglobin oxygen-carrying capacity with decreasing pH (Fig. 4C). Although allosteric effectors typically modify the stability of the deoxygenated state of the hemoglobin, recent studies on the Root-effect hemoglobin of the spot *Leiostomus xanthurus* suggest that substitution of a number of amino acids destabilizes the oxygenated R state of the Root-effect hemoglobin at low pH, promoting the transition to the deoxygenated T state (Mylvaganam *et al.*, 1996). Fishes equipped with a swim bladder typically possess Root-effect hemoglobins, and the acidification of the blood during passage through the gas gland induces the transition in these hemoglobins from the oxygenated R state to the deoxygenated T state. Based on *in vitro* hemoglobin oxygen-binding curves and *in vivo* measurements of blood pH in the swim bladder of the European eel, Pelster and Weber (1991) proposed that 40% of the hemoglobin can easily be deoxygenated. A hemoglobin content of several millimoles per liter allows for a large increase in P_{O_2} in swim bladder blood.

The release of lactic acid and CO_2 from gas gland cells thus causes an increase in the gas partial pressure of all gases in the blood, and following partial pressure gradients, gases will enter the swim bladder by diffusion. In addition, gas partial pressures in venous blood returning to the countercurrent system are higher than those in the arterial blood supplying the

5. BUOYANCY AT DEPTH

swim bladder epithelium. Thus, back-diffusion of gas from the venous to the arterial capillaries of the rete mirabile results in a countercurrent concentration of gases in the swim bladder.

The basic principle of countercurrent concentration as outlined by Kuhn et al. (1963) has been accepted for the rete mirabile of the swim bladder. The experimental and theoretical studies of Kobayashi et al. (1989a,b) extended this basic model by proving that the rete capillaries are permeable not only to gases but also to metabolites such as lactate, and that the countercurrent concentration of lactate enhances the salting-out effect. Figure 5 shows a theoretical plot of the concentrating ability of a countercurrent system for an inert gas (i.e., without chemical binding of the gas). The concentrating ability of a countercurrent system appears to depend on the conductance ratio $D/\dot{Q} \cdot \alpha$ (where D is the diffusing capacity of the barrier between venous and arterial capillaries in the rete mirabile, $\dot{Q}$ is blood perfusion, and α is physical gas solubility), the magnitude of the salting-out effect [i.e., the solubility ratio in venous and arterial blood (α_v/α_a)], and the permeability ratio of the rete, $F/(D/\alpha_a)$ (where F is the rate of solute transfer). The enhancement in arterial inert gas partial pressure in the rete clearly increases with the magnitude of the salting-out effect: the larger the decrease in solubility and thus the initial increase in gas partial

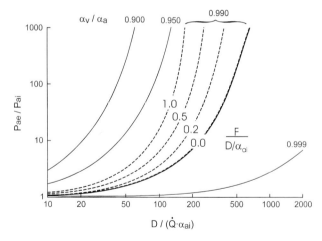

Fig. 5. Efficiency of the rete in enhancing inert gas partial pressure, calculated as the ratio of partial pressure in the arterial efflux and influx (P_{ae}/P_{ai}). The efficiency is given by the conductance ratio $(D/\dot{Q} \cdot \alpha)$, the magnitude of the salting-out effect (α_v/α_a), and the permeability ratio of the rete $[F/(D/\alpha_a)]$; see text for further explanations. Modified from *Respir. Physiol.* **78**; H. Kobayashi, B. Pelster, and P. Scheid. Solute back-diffusion raises the gas concentrating efficiency in counter-current flow, 45–57. Copyright 1989 with kind permission of Elsevier Science-NL, Sara Burgerhartstraat 25, 1055 KV Amsterdam, The Netherlands.

pressure (the single concentrating effect), the higher the maximum gas partial pressure achieved in the rete. Accordingly, very high gas partial pressures for CO_2 and O_2 can be generated in the rete, and during periods of gas deposition, these two gases are deposited mainly into the swim bladder (Fänge, 1983; Kobayashi et al., 1990).

The enhancing influence of the permeability ratio $[F/(D/\alpha_a)]$ is due to the fact that solute back-diffusion in the rete initiates the salting-out effect in arterial capillaries and thereby causes an increase in gas partial pressure in addition to the increase induced by back-diffusion of gas molecules. Similarly, arterial P_{O_2} in the rete mirabile is enhanced not only by back-diffusion of O_2 but also by back-diffusion of CO_2. Back-diffusion of CO_2 in the rete acidifies the arterial blood and increases arterial P_{O_2} by initiating the Root effect (Kobayashi et al., 1990). In an active, gas-depositing swim bladder, the P_{CO_2} gradient for back-diffusion of CO_2 inevitably exists, because CO_2 is produced in the metabolism of gas gland cells and released into the blood. Although CO_2 is not the main gas deposited into the swim bladder, CO_2 production in the pentose phosphate shunt and CO_2 back-diffusion in the countercurrent system appear to play a pivotal role in the functioning of the swim bladder.

Figure 5 also reveals that even with a small single concentrating effect, high gas partial pressures can be achieved in a countercurrent system if the diffusing capacity is high. The diffusing capacity largely depends on the geometry of the barrier between the capillaries. Any increase in surface area of the capillaries enhances the diffusing capacity. As shown in Fig. 6,

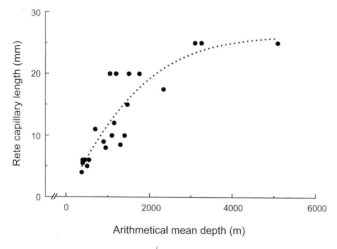

Fig. 6. Length of rete capillaries in relation to the arithmetical mean depth of species occurrence. Data from Marshall (1972).

prepared based on a study by Marshall (1972), the length of rete capillaries increases with medium depth of occurrence of a species.

Model calculations (Kuhn *et al.*, 1963; Enns *et al.*, 1967; Sund, 1977; Kobayashi *et al.*, 1989b), typically based on parameters taken from the eel with a capillary length of about 4 to 6 mm, clearly show that in the swim bladder, gas partial pressures can be generated sufficiently high to explain the occurrence of fish with a gas-filled swim bladder at a depth of several thousand meters. In deep-sea fish, the concentrating ability of the countercurrent system should be even higher as a result of the increased length of the rete capillaries (Fig. 6), which increases the diffusing capacity of the rete mirabile.

C. Resorption of Gas

In the resorbing section of the swim bladder (the oval), gases are resorbed into the blood by diffusion along the partial pressure gradient. Total gas partial pressure in arterial blood after passage through the gills is close to ambient (i.e., close to 1 atm or less), and therefore gas partial pressures in the swim bladder are higher than those in blood. If the oval is open or a bolus of gas is released into the resorbing section of the swim bladder, gases diffuse into the richly vascularized epithelia lining this section of the bladder (Denton, 1961; Steen, 1963b). However, different gas solubilities result in different rates of removal. As a result, the more soluble gases (i.e., CO_2 and O_2) are preferentially resorbed, leaving behind the less soluble inert gases (Piiper, 1965). Consequently, CO_2 makes up some 20–30% of newly deposited gas (Meesters and Nagel, 1935; Wittenberg *et al.*, 1964; Kobayashi *et al.*, 1990), but, under steady-state conditions, contributes no more than 1–2% to swim bladder gas.

D. Depth Limitations on the Utility of a Gas-Filled Swim Bladder

The specific gravity of gas increases with gas pressure, so the difference between swim bladder gas density and water density decreases with increasing water depth. This relationship considerably decreases the effectiveness of a swim bladder at depth. The maximum depth at which fish with gas-filled swim bladders have been found is about 5000–7000 m (Nybelin, 1957; Marshall, 1960; Nielsen and Munk, 1964). According to the calculation of Alexander (1966a), the density of O_2, representing the main swim bladder gas at this depth, is about 0.6–0.65 kg liter^{-1}. Although this value is much higher than O_2 density at moderate water depth, it still is below lipid density. Therefore, even at a depth of 5000–6000 m, a swim bladder appears to be more effective than lipid accumulation in terms of achieving neutral buoyancy.

Not only does the specific gravity of gas increase with depth, but the partial pressure gradient between the gas cavity and the surrounding water increases as well, which enhances diffusional loss of gas from the swim bladder. This may be balanced in part by an increased gas impermeability of the swim bladder wall at great depth (Denton et al., 1972; Kleckner, 1980). Nevertheless, the advantage of a gas-filled cavity as a buoyancy organ is clearly diminished at great depth because the energy expenditure to retain a gas-filled bladder increases with increasing water depth.

Apart from this energetical aspect O_2 uptake may even become limiting in terms of gas deposition. A fish descending 100 m h^{-1} encounters a hydrostatic pressure increase of 10 atm. To keep the swim bladder volume constant, it needs to increase by 10-fold the swim bladder volume, measured at a pressure of 1 atm. If the swim bladder volume is about 5% of the body volume, which is typical for marine fish, and O_2 is the main gas deposited, the descent requires about 500 ml kg^{-1} h^{-1} of O_2, which is far above the normal $\dot{M}_{O_2}$ of a fish (Johansen, 1982) and probably will exceed the gas exchange capacity, at least in hypoxic water layers present at depths of about 1000 m (Brooks and Saenger, 1991).

These considerations raise another interesting and unresolved question: Do fish with a swim bladder perform vertical migrations in a fully compensated state with neutral buoyancy? Energetically it appears to be most efficient if fish performing extensive vertical migrations use hydrodynamic lift to compensate for the change in hydrostatic pressure and keep the swim bladder volume constant, adjusted to neutral buoyancy at the upper water level (Alexander, 1972; Gee, 1983). Myctophids are well known for their migratory behavior, and acoustical analysis of sound-scattering layers appears to support this idea, suggesting that the migrations occur with either constant swim bladder volume or constant masses of gas (Vent and Pickwell, 1977; Kalish et al., 1986). Due to the flexibility of the swim bladder wall, it is hardly possible that swim bladder volume can be kept constant in the face of changing hydrostatic pressure. On the other hand, some myctophids have been found with gas-filled swim bladders only in surface water during the night, whereas in deeper water the occurrence of inflated swim bladders decreased significantly (Neighbors, 1992).

An important aspect is the physiology and the mechanism of the Root effect, which is not yet completely understood. If one assumes that with increasing hyperbaric oxygen pressure the hemoglobin oxygen-binding capacity asymptotically reaches the point at which the respiratory pigment is completely saturated with O_2, then there must be an oxygen partial pressure at which the Root effect is no longer functional (Noble et al., 1975). In the

European eel, for example, Sund (1977) predicted that the Root effect is abolished at a P_{O_2} of 10 atm. In this case, at a P_{O_2} above 10 atm, acidification of eel blood would no longer release oxygen from the hemoglobin, and the mechanisms of gas deposition into the swim bladder would essentially be reduced to the deposition of inert gases and of CO_2. Because at great depth O_2 is the main gas deposited into the swim bladder, this basically would be the maximum depth for the deposition of gas. Any further compression of the swim bladder caused by an increase in hydrostatic pressure could hardly be balanced by gas deposition.

Experimental evidence of the existence of an upper limit for the functioning of the Root effect, however, is not decisive. Even with an oxygen partial pressure of 140 atm, Scholander and Van Dam (1954) were not able to saturate hemoglobin of black grouper *Epinephelus mystacinus* or *Alphestes* sp. completely. Furthermore, the presence of fish with gas-filled swim bladders at a depth of several thousand meters demonstrates that it is possible to deposit gas against a gas pressure of several hundred atmospheres (Nybelin, 1957; Nielsen and Munk, 1964).

E. Lipid-Filled Swim Bladders

In midwater and deep-sea fishes, large amounts of lipid may be present in the swim bladder. Basically, there are two strategies for the accumulation of lipid in the swim bladder: fat investment of regressed swim bladders and fat-filled swim bladders that are fully functional in terms of gas deposition (Phleger, 1991).

In fat-invested swim bladders, fat, mainly consisting of wax esters, is accumulated between the peritoneum and the tunica externa. The lipid found in regressed swim bladders of myctophids consists of more than 90% wax esters (Butler and Pearcy, 1972), and in the coelacanth *Latimeria chalumnae,* wax esters make up 97% of swim bladder fat (Nevenzel *et al.,* 1966). Another example of a fat-invested swim bladder is the orange roughy *Hoplostethus atlanticus* (Phleger and Grigor, 1990). More than 90% of the swim bladder lipid is wax ester, and it appears to be exctracellular, contained within a three-dimensional network of collagen fibers. This suggests that the lipid of fat-invested swim bladders is not available for intermediary metabolism but is deposited purely to reduce the overall density of the fish. Regressed, fat-invested swim bladders appear to be common among midwater fishes that undertake extended vertical migrations, which are difficult to perform in a state of neutral buoyancy if a gas-filled swim bladder is present. Interestingly, histological analysis of gas gland cells and the rete mirabile in *Myctophum punctatum* did not reveal any degeneration of these tissues, although the swim bladder was much smaller and the swim bladder wall was thickened (Kleckner and Gibbs, 1972; Kleckner, 1974; cited in Neighbors, 1992).

Many deep-sea fishes have a fat-filled swim bladder (Morris and Culkin, 1989; Phleger, 1991). These swim bladders are fully functional in terms of O_2 deposition. Because the O_2 content of swim bladders increases with increasing water depth, these swim bladders mainly contain O_2. The lipid accumulated in these swim bladders consists mostly of cholesterol and phospholipid; the majority of the phospholipids and fatty acids are unsaturated (Phleger and Benson, 1971; Phleger et al., 1978). In *Antimora rostrata* as well as in *Bassozetus* species and two species of the genus Barathrodemus, the lipid appears to have a bilayer membrane configuration. Ultrastructural analysis of the membranous lipids of *Coryphaenoides* and *Parabassogigas* revealed that they exist as sheets of typical bilayered membranes (Phleger and Holtz, 1973). Accordingly, large quantities of membranes can be isolated from these swim bladders (Josephson et al., 1975; Phleger et al., 1978). These membrane lipids apparently are synthesized in the swim bladder tissue, and high oxygen tensions enhance cholesterol biosynthesis in fish (Kayama et al., 1971; Phleger, 1971, 1975b; Phleger et al., 1973; Phleger et al., 1977).

In terms of buoyancy, the presence of large quantities of cholesterol is somewhat difficult to explain, because the density of cholesterol, at 1.067 kg liter^{-1}, is significantly higher than seawater density. Accumulation of cholesterol therefore does not provide any lift. In phospholipid membranes, however, the addition of cholesterol reduces the diffusion constant for gases approximately 10-fold (Finkelstein, 1976; Wittenberg et al., 1980). These considerations suggest that the presence of lipid with the incorporation of cholesterol in the swim bladder of deep-sea fishes would render the swim bladder wall more impermeable to gases and thus reduce diffusional loss of gas. Alternatively, O_2 is dissolved in the fat fraction, which would reduce the back-pressure on the Root effect (Phleger, 1972; cited in Phleger, 1991). Dissolving O_2 in lipid would indeed be advantageous if it were to reduce the density of the lipid. It will not, however, reduce the pressure in the gas phase, which is determined by the hydrostatic pressure.

IV. LIPID ACCUMULATION

A. Density of Lipids

Lipids are accumulated not only in the swim bladder but also in other tissues. Figure 7 shows the various lipids implicated in the buoyancy of fishes. The most widespread lipids are triacylglycerol, alkyl diacylglycerol,

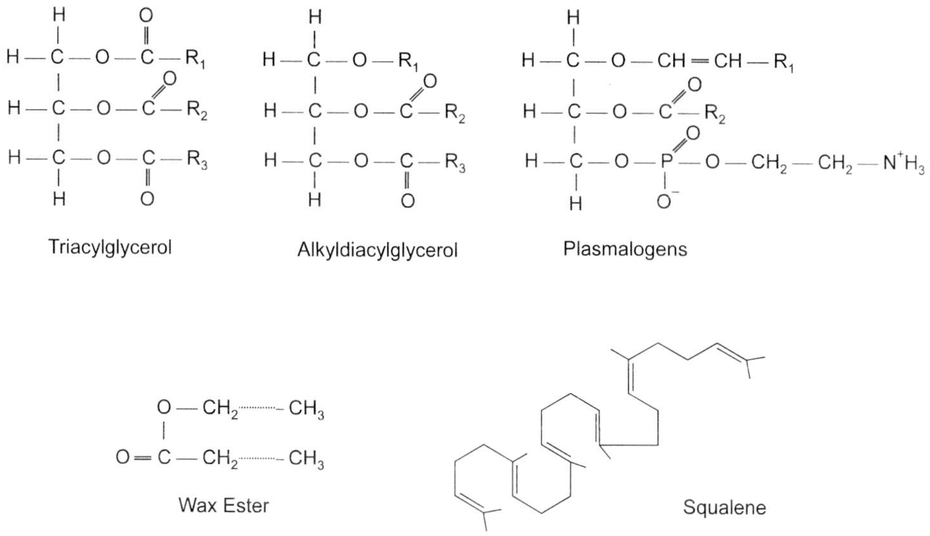

5. BUOYANCY AT DEPTH

Fig. 7. Molecular structure of lipids implicated to buoyancy in fishes.

wax ester, and squalene. Plasmalogens are far less common, and very few species contain pristane, a very low-density lipid derived from squalene. Table II presents density values for these lipids, although it should be kept in mind that density varies slightly with composition. The increase in lipid density with increasing hydrostatic pressure is in the range of 1–2% for a pressure change from 1 to 200 atm (Clarke, 1978a) and thus often is ne-

Table II
Specific Gravities of Various Lipids Accumulated in Fishes

Lipid	Density (kg liter^{-1})
Triacylglycerol[a]	0.93
Alkyl diacylglycerol[a]	0.91
Wax ester[a]	0.86
Squalene	0.86
Cholesterol	1.065
Pristane	0.78

[a] Gravity varies slightly with the chain length of the fatty acids and alcohol and with the degree of unsaturation (Sargent, 1989; Phleger, 1991).

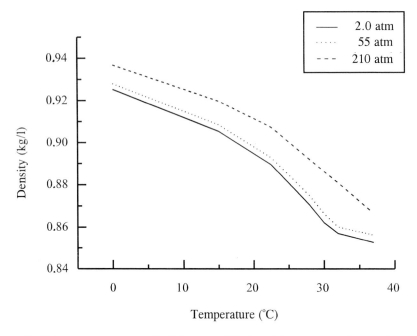

Fig. 8. Increase in spermaceti oil density, which is mainly composed of wax ester and triglyceride, with decreasing temperature. Data from Clarke (1978a).

glected. Lipid density also increases with decreasing temperature (Fig. 8), and wax esters may even solidify at the low temperatures encountered at depth. The density of sperm whale spermaceti oil, for example, which is composed mainly of wax ester and triglycerides, increases from 0.853 kg liter^{-1} at 37°C to 0.925 kg liter^{-1} at 0°C (Clarke, 1978a).

Metabolically these lipids can be synthesized in various tissues (Nevenzel, 1970, 1989; Grigor et al., 1990), but they can also be derived from the diet. It should be noted that depending on the constitution and density of the food, gut contents can change whole-body density. Dietary administration of labeled palmitic acid or of labeled acetate resulted in the appearance of the label in lipid stores within a few hours (Kayama and Nevenzel, 1974; Phleger et al., 1976; Phleger, 1988a; Grigor et al., 1990). On the other hand, starvation also has been shown to cause a reduction in lipid stores (Benson and Lee, 1975; Phleger, 1987, 1988a; Phleger and Laub, 1989). Triacylglycerols have an especially high turnover rate, whereas the metabolic turnover rate of alkyl diacylglycerol and wax ester, synthesized via the reaction of a fatty acid with a long-chain alcohol, is much slower. Squalene is an intermediate of the cholesterol synthesis pathway. It is ubiquitous in fish but accumulates in especially high levels in elasmobranchs and in various deep-sea fish. Metabolically, squalene is inert compared to the other lipids and can be converted only to cholesterol. Accumulation of squalene therefore is connected primarily to buoyancy adjustment. Storage of other lipids for buoyancy may provide an energy reserve, but if the fish in periods of food storage uses up the lipid reserves, it will become less buoyant. In this case, to retain neutral buoyancy, the fish must increase hydrodynamic lift by expending more energy for muscular activity.

Lipids are stored in various tissues, ranging from subcutaneous stores to stores in liver and bone tissue; occasionally lipids even appear to be in extracellular lipid sacs (see following discussion). The fatty acids and long-chain alcohols used for synthesis have been analyzed in great detail in some species. The main components of triacylglycerol are usually hexadecanoic acid (16:0), oleic acid (18:1), and octadecanoic acid (18:0). In wax esters, the long-chain alcohol usually consists of 30–42 carbon atoms, whereas again hexadecanoic acid, oleic acid, and octadecanoic acid make up more than 90% of the fatty acids (Nevenzel *et al.*, 1965, 1969; Nevenzel, 1970; Patton, 1975; Phleger and Grimes, 1976; Hayashi, 1987; Hayashi and Kashiki, 1988; Phleger and Laub, 1989).

According to the principle of constant fluidity, the degree of unsaturation of the fatty acids varies with water temperature, and thus deeper living species tend to have a higher degree of unsaturated fatty acids in their lipids (Phleger, 1975a; Patton, 1975; Van Vleet *et al.*, 1984; Cossins and MacDonald, 1986).

B. Lipid Droplets in Eggs and Larvae

Compared to adult fish, eggs and larvae have the advantage that skeletal elements, typically the most dense tissues of all, are not yet developed or are only starting to develop. Nevertheless, a planktonic lifestyle demands neutral buoyancy to retain a certain water depth, and eggs and larvae of deep-sea fish typically are pelagic. Oil droplets or oil globules are present in the plasma of many eggs, such as the eggs of ling, turbot, and grenadier. Although Tocher and Sargent (1984) did not find differences in the lipid content of pelagic and demersal marine eggs, the water content in pelagic eggs appears to be higher than that in demersal eggs (Yin and Blaxter, 1987). In planktonic larvae of the Antarctic fish *Pleuragramma antarcticum*, which live in water layers down to 50 to 100 m, lipid accumulation starts only with the onset of skeleton ossification (Hubold and Tomo, 1989). This finding supports the notion that the eggs and larvae of marine fish do not primarily use lipid accumulation to achieve neutral buoyancy. In many species, dead or dying pelagic eggs tend to sink, indicating that osmotic and ionic regulation plays a crucial role in achieving neutral buoyancy in these eggs. Craik and Harvey (1987) calculated that about 90% of the buoyancy in marine pelagic eggs is obtained by high water content. This strategy appears to be especially important for eggs of deep-sea fish, which cannot afford to float or sink.

Nevertheless, oil droplet lipids in eggs probably have a density of about 0.86 to 0.93 kg liter^{-1} and thus provide lift. In eggs of the eel, for example, oil globules cause the eggs to ascend (Balon, 1975). The situation is different in fresh water. Due to the higher osmolarity of the body fluids, eggs are denser than the environmental water and the importance of oil droplets to achieve neutral buoyancy increases. For example, eggs of the Amur snakehead *Ophiocephalus argus warpachowskii*, the macropod *Marcropodus opercularis*, or the gourami *Colisia lalia* achieve neutral buoyancy by means of an enormous oil droplet (Craik and Harvey, 1987).

C. Lipid Accumulation in the Liver

Although the liver usually makes up about 2–4% of the body weight, in sharks it may contribute up to 20–25%. In a number of Florida sharks, Baldridge (1970) measured liver-free body densities ranging from 1.051 to 1.089 kg liter^{-1} Those species with a large liver (up to 16.9% of total body weight) were very close to neutral buoyancy. The larger the relative size of the liver, the more lipid is stored and the lower the density of the liver tissue (Bone and Roberts, 1969; Baldridge, 1970). If the liver makes up more than 10% of the total body weight, the species usually is close to neutral buoyancy (Bone and Roberts, 1969). In this situation the liver clearly represents the main lipid store of the species and may comprise up to 95% of total lipids (Van Vleet *et al.*, 1984; Phleger, 1988b).

High squalene contents in elasmobranch liver lipid stores occur in five families of sharks (Nevenzel, 1989). Several members of the deep-sea squaloids (e.g., *Centrophorus squamosus*, *Centrophorus granulosus*, *Centroscymus coelolepis*, *Dalatias cacea*, *Dalatias licha*, and *Etmopterus princeps*) have bulky livers that store large amounts of squalene (Corner *et al.*, 1969; Sargent *et al.*, 1973; Hayashi and Takagi, 1981; Van Vleet *et al.*, 1984). In *Centrophorus uyato* squalene accounts for up 90% of the liver mass. In other shark families, in rays, and in chimaeras, liver lipid mainly consists of diacylglyceryl ether, with only traces of squalene. Triglycerides and wax esters may also be accumulated in liver tissue.

The presence of significant liver lipid stores among the elasmobranchs is not related to the systematical position of a species (Bone and Roberts, 1969; Baldridge, 1970; Nevenzel, 1989). The actual composition of liver lipids varies among species and may also vary within a species depending on the season and the location (Springer, 1967; cited in Bone and Roberts, 1969; Corner et al., 1969; Hayashi and Takagi, 1981). This variability indicates that lipid stores may be used as energy reserves and that their composition in turn may depend on diet.

The density of lipids is only about 10–15% lower than sea-water density. A typical shark would have to accumulate about 250 g of squalene per 1000 g of body weight to retain neutral buoyancy [see Eq. (8)]. Accordingly, to retain neutral buoyancy, any increase in body weight must be accompanied by an appropriate increase in lipid stores, which in turn requires a delicate control system. Malins and Barone (1969) addressed this question by artificially disturbing the equilibrium between weight and buoyancy plus hydrodynamic lift in the dogfish *Squalus acanthias*. The liver of *S. acanthias* contains 62–76% lipid, mostly triglycerides (TGs) and diacylglyceryl ethers (DAGEs). Whereas in control animals the ratio DAGE/TG was 0.73 ± 0.20, in a group of dogfish in which the body weight was artificially increased for 2 days with lead weights the ratio significantly increased to 1.29 ± 0.23. The authors postulate a regulatory mechanism involving the selective metabolism of DAGE and TG that allows for buoyancy control in dogfish.

Lipid storage in the liver is also found within other groups of cartilaginous fishes. In the electric ray *Torpedo nobiliana*, the liver makes up about 20% of the animal's volume and contains 70% oil with a density of 0.91 kg liter^{-1}, which significantly contributes to the low density of the animal (Roberts, 1969). In water its weight is only 0.4% of its weight in air. In *Torpedo marmorata*, a member of the same genus, the liver is much smaller and contains only 2% oil, giving a much higher overall density. In deep-sea ratfish *Hydrolagus novaezealandiae*, the liver yields a lipid content of 64%, with 65.8% diacylglyceryl ethers, 10.4% triglycerides, and 10.5% fatty acids, hydrocarbons, and sterols (Hayashi and Takagi, 1980).

A large liver used for lipid storage is not only found in elasmobranchs. In the coelacanth *Latimeria chalumnae*, the liver contains 67.7% lipid with 8.2% wax ester and significantly contributes to the buoyancy status of this species (Nevenzel et al., 1966). Even a few teleosts, such as redlip blenny *Ophioblennius atlanticus* larvae and *Laemonema longipes*, accumulate lipid in their livers and rely on this strategy to reduce whole-body density (Nursall, 1989; Hayashi and Kashiki, 1988).

D. Bone Lipids

The skeleton is usually the tissue with the highest density. Several teleosts, however, use their bones to store lipids and thus significantly reduce their density. Occasionally, the lipid content even reduces the skeleton density below seawater density (Phleger, 1975a). In general, triacylglycerol is the main bone lipid, with minor contributions by cholesterol and phospholipid. Bone lipids are present in many families (Phleger and Grimes, 1976; Phleger, 1987, 1988b). The major fatty acids of the triglycerides are palmitate (16:0), palmitoleate (16:1), stearate (18:0), and oleate (18:1) (Phleger, 1975a; Lee et al., 1975; Phleger and Grimes, 1976; Phleger, 1991).

Typical sites for the storage of bone lipids are the spine and skull, but lipids may also be found in other locations. In the hawkfish *Cirrhitus pinnulatus*, the skull contains 90% lipid (percentage of dry weight) and floats in seawater (Phleger, 1975a); the giant hawkfish *Cirrhitus rivulatus* also has an oil-filled skull, with 23.9% lipid (percentage of dry weight) (Phleger, 1987). In *Peprilus simillimus* and *Anoplopoma fimbria*, the skull contains 68 and 60% lipid (percentage of dry weight), respectively (Lee et al., 1975).

Because lipids may be the major constituent of the skeleton, or at least of some skeletal bones, bone lipids may be the main lipid store of the organism. In sheepshead wrasse *Pimelometopon pulchrum* and in sablefish *A. fimbria*, bone lipid comprises 79–93% and 52–82% of total body lipid, respectively (Phleger et al., 1976). *Acanthurus chirurgus* stores 81% of the total body lipid in bones, whereas bone lipid is usually less than 1% of dry weight in land mammals (Phleger, 1988b).

Bone lipids appear to be available for intermediary metabolism. *Anoplopoma fimbria*, for example, may use bone lipids as an energy reserve (Phleger, 1987), and dietary palmitic acid was incorporated into bone lipids in less than 12 h following administration (Phleger et al., 1976). The composition and magnitude of bone lipid storage may be related to food availability and food composition (Phleger, 1975a, 1987, 1988b).

A further example of a species that uses lipid storage in the skull and bones is the castor-oil fish *Ruvettus pretiosus* (Bone, 1972). Interestingly, in this species the lipid consists mainly of wax esters of cetyl and oleyl alcohols. The frontal bone and the vertebral certrum contain 30 and 21% lipid, respectively. Some of the dermal roofing bones and the skull are little more than girder systems enclosing oil sacs. Bones of the orange roughy *Hoplostethus atlanticus* also appear to contain wax esters (Grigor et al., 1983; Phleger and Laub, 1989).

Lipid storage in bones can be found in conjunction with the storage of lipids in other organs and with the presence of a swim bladder. An extraordinary example of lipid storage in a whole variety of tissues is the castor-oil fish *R. pretiosus* (Nevenzel et al., 1965) (see following discussion). In *Peprilus simillimus* and in *Schedophilus medusophagus*, 32 and 20% of the total lipid is stored in bones, respectively; in both species the liver also contains appreciable lipid stores. The wrasse *Cheilinus rhodochrous* and the rockfish *Sebastes ruberrimus* have gas-filled swim bladders and oil-filled bones (Lee et al., 1975). The swordfish *Xiphias gladius* has a swim bladder and a high lipid content with porous fatty bones (Carey and Robison, 1981).

E. Lipid Accumulation in Other Tissues

Apart from liver and bones, lipids are stored in several other body tissues (e.g., muscle, intestine, subcutaneously). Typically the lipid is stored in adipocytes, but examples of the extracellular storage of lipids in oil sacs are also found. Lipid stored in adipocytes is readily available for metabolism; the availability of extracellular lipid stores, however, is questionable. The lipid stored in these various tissues consists mainly of triacylglycerols and wax esters, but the composition, as well as the degree of saturation of the fatty acids, appears to be much more variable than the lipid composition in bone.

The castor-oil fish *R. pretiosus* stores extensive amounts of lipid in various tissues. The ctenoid scales contain oil-filled cells. The lipid content of the integument amounts to 32.3% (Bone, 1972). On a wet-weight basis, the muscle tissue of the castor-oil fish has a lipid content of about 14.7%, which consists predominantly of wax esters of 34–36 carbon atoms (Nevenzel et al., 1965). In the eulachon *Thaleichthys pacificus*, both the whole body and liver contain about 20% wet-weight lipid (Ackman et al., 1968). The lipid consists mainly of triglyceride, with a small amount of squalene (12% of whole-body lipid, 18% of liver lipid). Because of its high lipid content, the fish after drying is suitable for burning as a "candle fish" (Ackman et al., 1968).

The Antarctic notothenioid fishes are mostly bottom-dwellers and lack a swim bladder *Pleuragramma antarcticum*, *Dissostichus mawsoni*, and *Aethotaxis mitopteryx*, however, have achieved neutral buoyancy by reducing the mineralization of the skeleton and by accumulation of lipid (Eastman, 1985). Whereas *P. antarcticum* accumulates lipids in special lipid sacs, *D. mawsoni* and *A. mitopteryx* possess a subcutaneous layer of adipose tissue. In *D. mawsoni*, the giant Antarctic cod, the subcutaneous lipid layer has a thickness of 2 to 8 mm, comprising 4.75% of the body wet weight and 23% of the dry weight. In addition, white muscle tissue contains 23% lipid, mainly triglycerides (Eastman and DeVries, 1981; Eastman, 1985, 1988; Clarke et al., 1984).

Extensive subcutaneous lipid depots are found in the pelagic teleost *Maurolicus muelleri* and in the mesopelagic teleost *Benthosema glaciale* (Falk-Petersen et al., 1986a). *Maurolicus muelleri* also is a good example of the storage of lipid intramuscularly and in the digestive tract. About 60% of the dry weight is lipid. The lipid is mainly triglycerides, stored in conventional adipocytes. In *B. glaciale,* the situation is similar, but here wax ester comprises 77% of the lipid fraction.

It has been reported that the fluidity of the lipid stored in the head of the sperm whale varies with depth (Clarke, 1978b), and this condition also appears to apply to teleost fish. The orange roughy *Hoplostethus atlanticus* accumulates about 70% of the total lipid stores in the muscle tissue and skin (Phleger and Grigor, 1990). The lipid is mainly wax ester (about 95%). At the surface (14°C) the fish is positively buoyant, but at the depth of occurrence—typically it is caught at 1000 m, where the temperature is 6°C—the lipid is expected to be partly (17%) solid. This change is fluidity would change lipid density and give neutral buoyancy to the fish at depth.

A few species appear to rely on extracellular lipid stores. For example, juvenile *Lumpenus maculatus* have large oil sacs, mainly consisting of triacylglycerol, situated on the ventral part of the fish from the pectoral fins to the anus (Falk-Petersen *et al.,* 1986b). Muscle lipid (40% of dry weight) consists of 50% triacylglycerols and 35% wax esters. These lipid stores appear to be extracellular, but apparently each polygonal unit is enclosed by a kind of envelope containing numerous nuclei, resembling a cell syncytium. The authors speculate that these units might be special adipocytes in an arrangement that allows for the mobilization of these lipids. Juveniles of this species are pelagic, whereas adults are strictly demersal, indicating that these lipid stores are metabolized during metamorphosis.

Further examples of extracellular storage of lipids are the coelacanth *Latimeria chalumnae* (Nevenzel *et al.,* 1966) and the Antarctic fish *Pleuragramma antarcticum* (Eastman, 1985). Muscle tissue of the coelacanth contains 30–71% dry-weight lipid, deposited extracellularly; 90% of the lipid is wax ester (Nevenzel *et al.,* 1966). In *P. antarcticum,* there are about 100 to 200 subcutaneous lipid sacs (0.2–1.2 mm diameter) along the sides of the body and especially in the pectoral region. Lipid sacs (0.5–3.0 mm) are also found proximal to the bases of the dorsal and anal fins and adjacent to the dorsal and ventral median septa (Fig. 9). Lipid accumulation starts at the onset of skeleton ossification, and the lipid is composed mainly of triglycerides made up of oleic acid, myristic acid, palmitoleic acid, and palmitic acid; there are no wax esters (DeVries and Eastman, 1978; Eastman, 1985, 1988). In species caught at the Antarctic Peninsula, however, Reinhardt and Van Vleet (1986) found wax esters that were not found in McMurdo species (Eastman, 1988). Electron microscopical analysis revealed that the lipid sacs of *P. antarcticum* consist of several white adipocytes arranged circumferentially around large lipid droplets (Eastman and DeVries, 1989), which could be the clue for metabolization of the lipids. These observations suggest that lipid stores that appear to be extracellular are still available as energy reserve.

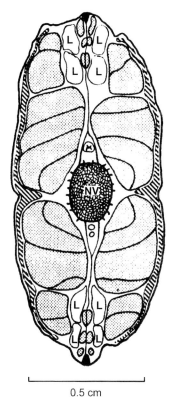

0.5 cm

Fig. 9. Cross section of *Pleuragramma antarcticum* showing the location of lipid sacs (L), which are especially concentrated at the basis of the dorsal and anal fins. Reprinted with permission from *Nature*, A. L. DeVries and J. T. Eastman, Lipid sacs as a buoyancy adaptation in an Antarctic fish, **271**, 352–353. Copyright 1978 Macmillan Magazines Limited.

V. WATERY TISSUES

A. Basic Principle

Plasma osmolarity of most vertebrates, including freshwater and marine teleosts, is about 300 mOsm. In elasmobranchs the situation is different. Plasma osmolarity of rays and sharks is adjusted to values close to seawater osmolarity by accumulation of urea. Water density increases with increasing salinity, and fluids of salinity lower than that of the surrounding water therefore usually provide lift. Therefore the plasma of freshwater fish is denser than water density and body fluids of marine elasmobranchs have a density close to seawater density, but the plasma of marine teleosts and thus of deep-sea fish is less dense than seawater and provides lift. The same is true for other body fluids of marine teleosts that have an ionic composition similar to plasma and are hypoosmolar compared to seawater. The difference in seawater and teleost plasma density is very small, and, according to Eq. (8), a very large volume of fluid would be necessary to achieve near-neutral buoyancy, as demonstrated impressively by deep-sea squids (Denton et al., 1969).

B. Eggs and Larvae

Pelagic marine eggs usually are near neutrally buoyant for most of their development but have a tendency to become denser toward hatching time (Coombs et al., 1985). The water content in pelagic eggs ranges from 90 to 92% and the lipid content usually varies between 10 and 15% of dry weight (Craik and Harvey, 1987); in demersal eggs the water content tends to be lower (Yin and Blaxter, 1987).

During vitellogenesis, most of the yolk material is synthesized in the liver in the form of yolk precursor protein, vitellogenin, and is transported in plasma and absorbed by oocytes (Wallace, 1985). During postvitellogenic meiotic maturation, termed *ripening*, little or no further yolk is accumulated, but the oocyte undergoes characteristic changes in appearance and structure. A massive water influx takes place, leading to the characteristic high water content of eggs (Wallace and Selman, 1981). Increases in K^+ content and sometimes in Na^+ content as well as proteolysis are the driving forces for the osmotic water uptake (Craik and Harvey, 1987).

The perivitelline space is formed just after spawning. Depending on the permeability of the chorion, water is absorbed into the perivitelline space and into the egg membrane; the membrane then swells into a gelatinous substance. The egg membrane is isosmotic to the environment and represents a neutrally buoyant outer shell to the ovoplasm. The vitelline membrane surrounding the yolk largely restricts the exchange between yolk and perivitelline fluid (Coombs et al., 1985), although it appears to remain slightly permeable (May, 1974). Thus the ovoplasm has an osmolarity similar to that of adult cells and is hypoosmotic compared to the environmental water. The degree of change in density is mainly dependent on the volume change of the "swimming belt," the perivitelline fluid. In many eggs (e.g., eggs of *Ophidium barbatum*, *Carapus* sp., *Scorpaena* sp., *Histrio histrio*) the egg membrane forms a compact veil in which the eggs are embedded, and the whole structure floats like a raft on the surface of the sea (Balon, 1975). Special temporary appendages occasionally optimize the surface-to-volume ratio. In addition to (or instead of) an oil globule, several marine species have a special sinus in the anterior part of the enlarged dorsal fin fold that takes up water as the yolk is absorbed (Balon, 1975).

5. BUOYANCY AT DEPTH

Because the buoyancy status of marine eggs is mainly determined by water content, water content should vary with the salinity of the external medium. Eggs of a number of Baltic fishes adjust their buoyancy (water intake into the egg capsule) according to the external salinity (Kändler and Tan, 1965a,b; cited in Rosenthal and Alderdice, 1976), and at low salinity (15%), eggs of the sciaenid fish *Bairdiella icistia* are larger and have a higher water content than at full-strength salinity (May, 1974). In the pilchard *Sardina pilchardus,* adjustments in density were observed during salinity changes from 35 to 33% (Coombs *et al.,* 1985).

Changes in the buoyancy status of developing embryos and larvae have been reported with the degree of yolk depletion and after the onset of feeding with the availability of food (Yin and Blaxter, 1987). In the pilchard *S. pilchardus,* there appears to be a seasonal effect; autumn eggs, compared to summar eggs, showed a slightly higher density (Coombs *et al.,* 1985). This observation may also be related to food availability and the amount of lipid reserves in the eggs. Changes in the buoyancy status during starvation could also reflect a lack of energy that is necessary to retain osmotic gradients between egg plasma and the surrounding water.

C. Reduction of Skeletal Density

Water can be accumulated not only in fluids but also in tissues. Water content of tissues averages between 60 and 80%. Fish that have a much higher water content in the tissues have so-called watery tissues. Watery muscle tissue is much softer than muscle tissue of normal water content and has a lower density. Water accumulation in the skeleton is achieved by reduced mineralization, that is, by a reduction in the ash content of the skeleton. Typically, the density of watery tissues is still higher than water density, but it is significantly lower than the density of normal tissue. Thus, although water accumulation often does not result in neutral buoyancy, it significantly reduces the weight of the fish in water.

A reduction in skeleton weight can be achieved by reducing the size and thickness of the bones and by reducing the mineral content of the bones. The high density of the skeleton is related to the high content of heavy ions such as Ca^{2+} and phosphate or sulfate; reducing their content in bones significantly decreases bone density.

In deep-sea fish, both strategies have been adopted. In the antarctic fish *Pleuragramma antarcticum,* the vertebrae are not amphicelous but are merely a thin collar of bone surrounding and barely constricting the persisting notochord. Neural and haemal arches and the spines are reduced, and the ribs are very small (DeVries and Eastman, 1978). In *Dissostichus mawsoni,* the vertebrae are unconstricted and the size of the vertebral processes is reduced (Eastman and DeVries, 1982). Parts of the skeleton, such as dorsal aspects of the neurocranium, the pectoral girdle, and sections of the caudal skeleton, are replaced by cartilage.

The three neutrally buoyant notothenioids—*Pleuragramma antarcticum, Aethotaxis mitopteryx*, and *Dissostichus mawsoni*—are characterized by reduced mineralization of the skeleton. The ash content of the skeleton is less than 0.6% of the body weight, whereas in most other members of the family it is 0.7–3.8% (DeVries and Eastman, 1978; Eastman and DeVries, 1982; Eastman, 1985). Usually the ash content of a teleost skeleton is about 2%.

Most nothotenioids have ctenoid scales, which again are less mineralized in the neutrally buoyant species. In *D. mawsoni*, the posterior margin is unmineralized, and ctenii are lacking (Eastman and DeVries, 1982).

Another well-known example of a reduced skeleton is the lumpsucker. The skeleton of the lumpsucker *Cyclopterus lumpus* is cartilaginous and almost uncalcified (Davenport and Kjorsvik, 1986). The density of the vertebral column is about 1.05 kg liter^{-1}, compared to 1.229 kg liter^{-1} in plaice, for example. Even the cartilage of the lumpsucker is less dense than usual.

D. Watery Muscle

Whereas the density of muscle tissue generally is about 1.06–1.08 kg liter^{-1} in female lumpsuckers the density of the muscle tissue is as low as 1.024 kg liter^{-1}. The large dorsal muscle is especially loose fiberd, watery, and low in osmolarity, with a density of only 1.019 kg liter^{-1} (Davenport and Kjorsvik, 1986). The authors suggest that these watery muscles have a reduced locomotory ability. In males the water accumulation in muscle tissue is less pronounced and the muscles have a firmer appearance, although they have an increased lipid content.

Watery muscles are common among deep-sea fish. The deep-sea ophidiid *Acanthonus armatus* has reduced tissue components and is only a little denser than the water. The muscle is loosely packed and gelatinous, and there is no significant storage of lipid (Horn *et al.*, 1978). *Acanthonus armatus* also accumulates hypoosmolar fluids. The head is very large, and the cranial activity (10% of head volume) contains a fluid (7–8.5 ml) of low osmolarity (294 mOsm). Na^+ and K^+ concentrations of this fluid are lower than those in plasma, giving a density of only 1.008 kg liter^{-1}. The cranial fluid thus does contribute to buoyancy of the fish and to the typical horizontal to slightly head-up position of this species (Horn *et al.*, 1978).

E. Gelatinous Masses

Gelatinous material forms the bulk of the core of the leptocephalus larvae (Pfeiler, 1986). This jellylike material mainly consists of glycosaminoglycans and has a very low density. It contributes significantly to the low density of the pelagic larvae. Yancey *et al.* (1989) analyzed several species of deep-sea fish for the presence of gelatinous layers and found four species of nonmigrating fish without a swim bladder (*Bathylagus pacificus, Bathylagus milleri, Tactostoma macropus,* and *Chauliodus macouni*) with large deposits of gelatinous material that stained for glycosaminoglycans. Glycosaminoglycans are hygroscopic and thus responsible in part for the high water content of these gelatinous masses. The material is located between the muscle cells, along the dorsal midline, and surrounding the spine. *Bathylagus pacificus* and *Bathylagus milleri* have a subcutaneous layer. In *B. pacificus* the water content is 96%, with a low ion content, resulting in a density less than seawater density.

The dorsal hump of the lumpsucker *Cyclopterus lumpus* is made up mainly of subcutaneous jelly, which contributes significantly to the low density of the species (Davenport and Kjorsvik, 1986).

Gelatinous masses are also found in elasmobranchs. A gelatinous layer of watery, jellylike tissue that floats in seawater is present in the nose of the sharks *Cetorhinus* and *Prionace;* it is also found underneath the skin of the skate *Torpedo nobiliana* (Bone and Roberts, 1969).

VI. HYDRODYNAMIC LIFT

Not only low-density structures are suitable to achieve neutral buoyancy, a high tissue density can also be compensated for by hydrodynamic lift. Small plankton can be kept in suspension by eddies, and this is particularly facilitated by parachute devices such as long antennae. Ciliary activity also allows small organisms to remain suspended. Larvae of the lancelet *Branchiostoma lanceolatum* (Cephalochordata) hover almost motionless in midwater by means of beating epidermal cilia in metachronal waves that pass from anterior to posterior at about 0.3 m sec^{-1}. If the ciliary movements are interrupted by brief exposure to 0.1% glutaraldehyde, the larvae start sinking (Stokes and Holland, 1995).

Larger organisms probably are not able to hover by ciliary action but may occasionally be able to hover using their pectoral fins. The mandarin fish *Synchropus picturatus* (weight, 5–10 g) is negatively buoyant and demersal, but it is often seen hovering close to the bottom or to coral while feeding. It generates lift with its pectoral fins and when it is close to the ground, the "ground effect" causes a 30–60% reduction in power required for hovering, which is comparable to the effect used by helicopters and insects (Blake, 1979). The coelacanth *Latimeria chalumnae* is a nocturnal drift-hunter, moving slowly in upwelling and downwelling currents. Being neutrally buoyant or at least close to neutrally buoyant by means of lipid accumulation, it uses the paired fins as hydrofoils to stabilize and to correct the drift motion (Fricke *et al.*, 1987).

During swimming, lift is primarily produced by the pectoral fins (Harris 1936, 1937; Magnuson, 1970, 1978; Alexander, 1990). The fins are used as hydrofoils, acting the same way as an airplane that is supported by hydrodynamic lift on its wings. Analogous to wings on an airplane, water flows faster over the upper surface of the pectorals than over the lower surface, creating a higher pressure on the lower surface. The pressure difference produces a net lift, directly proportional to the area of the hydrofoils, and proportional to the swimming speed [Eq. (5)]. Long fins are especially economical because in relation to the vortices induced at the tips of the fins (which equal induced drag) they produce more lift. The aerodynamic design of the pectoral fins, however, is certainly compromised by their structural strength and the ability of the fish body to carry them. The lift produced from the pectorals acts perpendicularly and, depending on the position of the pectorals, usually acts anteriorly to the center of gravity (Magnuson, 1970). It thus lifts up the anterior part of the fish.

Sharks, sturgeons, and scombrid fishes are commonly known to swim more or less continuously at high speed. A comparison of pelagic and bottom-dwelling sharks reveals that pelagic sharks have a lower density, without achieving neutral buoyancy (Bone and Roberts, 1969). Also, sharks close to neutral buoyancy have smaller pectorals (Corner *et al.*, 1969). To achieve neutral buoyancy, the denser sharks need more hydrodynamic lift, which is obtained by increasing the size of the hydrofoils, or the pectoral fins. The fins of selachians and sturgeons cannot be folded and project permanently from the body. Pectoral fins of scombridae are not fixed, and the lift produced by these fins varies with their extension. Thus, at high speed, the pectoral fins are extended less, because less hydrodynamic lift is required for hydrostatic equilibrium (Magnuson, 1970).

Additional lift is produced by the peduncular keel and heterocercal tails, as demonstrated in *Acipenser sturio* (Alexander, 1966b), *Scyliorhinus canicula, Galeorhinus galeus* (Alexander, 1965), and also *Heterodontus portusjacksoni* and *Squalus megalops* (Simons, 1970). Water passing the fish flows diagonally across the keel. The keel, with its sinusoid movement, travels faster than the fish. Both contribute to the generation of hydrodynamic lift by the peduncular keel (Magnuson, 1970). Hydrodynamic lift produced by the tail acts behind the center of gravity. This action appears to be necessary for longitudinal stability, to keep the swimming fish from continuously rising. In fish with heterocercal tails, the larger dorsal section of the tail generates lift, whereas the smaller ventral section reduces lift. In sharks the tail is equipped with radial muscles in the hypochordal regions. Thus, by virtue of these muscles the ventral lobe of the tail could become a "horizontal trim" of the fish (Simons, 1970).

A round, symmetrical body swimming exactly with a zero angle cannot provide any hydrodynamic lift. If the fish swims at a positive angle—like mackerel do—the body can act as hydrofoil, although the effect is probably small or even negligible compared to that achieved with the pectoral fins, which may provide 70–80% of the hydrodynamic lift (Alexander, 1965; Magnuson, 1970, 1978).

Scombroids (e.g., tuna, bonito, and mackerel) swim continuously; they do not stop swimming. The minimum speed observed in adult *Acanthocybium solanderi* was "only" 0.33 body length per sec (bl sec^{-1}); many others usually do not swim less than 1 bl sec^{-1}. The minimum speed required to prevent the fish from sinking decreases with increasing fork length. Small fish with a 10-cm fork length need to swim 3 bl sec^{-1} (30 cm sec^{-1}) to prevent sinking. Larger fish (70-cm fork length) must swim 1.2 bl sec^{-1} (84 cm sec^{-1} (Magnuson, 1970). This minimum swimming speed is similar to that of several other fish species and even that for dolphins (see Magnuson, 1970, for references).

A few tuna species make use of additional buoyancy devices. For example, yellowfin tuna *Thunnus albacores,* albacore *Thunnus alalunga,* and Pacific mackerel *Scomber japonicus* have a swim bladder, although typically it is too small to provide neutral buoyancy (Magnuson, 1978).

VII. CONCLUSIONS

The high density of most body tissues can be compensated by various strategies, ranging from muscular activity to accumulation of low-density material like water, lipid, or gas. The energetical advantage of neutral buoyancy is clearly demonstrated by the extent, at which fishes have adopted all these strategies, suitable for the individual way of life or adjusted to the constraints of the biotop. For a bottom-dwelling fish it is not "useful" to be neutrally buoyant, for a fish invading the open water column above the bottom it might be useful. Similarly, for a pelagic fish performing extended vertical migrations or hunting its prey at high swimming speed, a swim bladder might not be the best choice to achieve neutral buoyancy; for a fish travelling at slow speed or hovering at a reef it might be an energetical advantage to have a swim bladder. Accordingly, the various strategies to achieve neutral buoyancy have been adopted irrespective of the systematical allocation of a species. Most deep-sea fish apparently achieve neutral buoyancy by reducing tissue density even at the expense of their locomotory ability (watery muscles, gelatinous masses). Accumulation of low density material like lipid or gases is also widespread, but with increasing depth, species accumulating lipid appear to outnumber species with a gas-filled swim bladder. A gas-filled swim bladder is rarely observed in species living below 1000 m, although it has been found in species caught at a depth of 5000–7000 m. Some of the fast swimming teleosts or elasmobranchs referring to hydrodynamic lift in order to achieve neutral buoyancy can be found at considerable water depth, but this strategy does not appear to be typical for deep-sea fish.

REFERENCES

Ackman, R. G., Addison, R. F., and Eaton, C. A. (1968). Unusual occurrence of squalene in a fish, the eulachon *Thaleichthys pacificus*. *Nature (London)* **220**, 1033–1034.

Alexander, R. M. (1965). The lift produced by the heterocercal tails of selachii. *J. Exp. Biol.* **43**, 131–138.

Alexander, R. M. (1966a). Physical aspects of swimbladder function. *Biol. Rev.* **41**, 141–176.

Alexander, R. M. (1966b). Lift produced by the heterocercal tail of Acipenser. *Nature* **210**, 1049–1050.

Alexander, R. M. (1971). Swimbladder gas secretion and energy expenditure in vertically migrating fishes. *In* "Proceedings of the International Symposium on Biological Sound Scattering in the Ocean" (G. B. Farquhar, ed.), pp. 75–86. Government Printing Office, Washington, D.C.

Alexander, R. M. (1972). The energetics of vertical migration by fishes. *In* The effect of pressure. *Symp. Soc. Exp. Biol.* **25**, 273–294.

Alexander, R. M. (1990). Size, speed and buoyancy adaptations in aquatic animals. *Am. Zool.* **30**, 189–196.

Baldridge, H. D. (1970). Sinking factors and average densities of Florida sharks as functions of liver buoyancy. *Copeia*, 744–754.

Ball, E. G., Strittmatter, C. F., and Cooper, O. (1955). Metabolic studies on the gas gland of the swim bladder. *Biol. Bull.* **108**, 1–17.

Balon, E. K. (1975). Reproductive guilds of fishes: A proposal and definition. *J. Fish. Res. Bd. Can.* **32**, 821–864.

Benson, A. A., and Lee, R. F. (1975). The role of wax in oceanic food chains. *Sci. Am.* **232**, 76–86.

Blake, R. W. (1979). The energetics of hovering in the mandarin fish (*Synchropus picturatus*). *J. Exp. Biol.* **82**, 25–33.

Bone, Q. (1972). Buoyancy and hydrodynamic functions of integument in the castor oil fish, *Ruvettus pretiosus* (Pisces: Gempylidae). *Copeia*, 78–87.

Bone, Q., and Roberts, B. L. (1969). The density of elasmobranchs. *J. Mar. Biol. Assoc. U.K.* **49**, 913–937.

Boström, S. L., Fänge, R., and Johansson, R. G. (1972). Enzyme activity patterns in gas gland tissue of the swimbladder of the cod (*Gadus morrhua*). *Comp. Biochem. Physiol.* **43B**, 473–478.

Brittain, T. (1987). The Root effect. *Comp. Biochem. Physiol.* **86B**, 473–481.

Brooks, A. L., and Saenger, R. A. (1991). Vertical size-depth distribution properties of midwater fish off Bermuda, with comparative reviews for other open ocean areas. *Can. J. Fish. Aquat. Sci.* **48**, 694–721.

Brown, S. D., and Copeland, D. E. (1978). Layered membranes: A diffusion barrier to gases in teleostean swimbladders. *Tissue Cell* **10**, 785–796.

Butler, J. L., and Pearcy, W. G. (1972). Swimbladder morphology and specific gravity of myctophids off Oregon. *J. Fish. Res. Bd. Can.* **29**, 1145–1150.

Carey, F. G., and Robison, B. H. (1981). Daily patterns in the activities of swordfish. *Xiphias gladius*, observed by acoustic telemetry. *Fish. Bull.* **79**, 277–292.

Clarke, A., Doherty, N., DeVries, A. L., and Eastman, J. T. (1984). Lipid content and composition of three species of Antarctic fish in relation to buoyancy. *Polar Biol.* **3**, 77–83.

Clarke, M. R. (1978a). Physical properties of spermaceti oil in the sperm whale. *J. Mar. Biol. Assoc. U.K.* **58**, 19–26.

Clarke, M. R. (1978b). Buoyancy control as a function of the spermaceti organ in the sperm whale. *J. Mar. Biol. Assoc. U.K.* **58**, 27–71.

Coombs, S. H., Fosh, C. A., and Keen, M. A. (1985). The buoyancy and vertical distribution of eggs of sprat (*Sprattus sprattus*) and pilchard (*Sardina pilchardus*). *J. Mar. Biol. Assoc. U.K.* **65**, 461–474.

Copeland, D. E. (1969). Fine structural study of gas secretion in the physoclistous swim bladder of *Fundulus heteroclitus* and *Gadus callarias* and in the euphysoclistous swim bladder of *Opsanus tau*. *Zeitschrift für Zellforschung* **93**, 305–331.

Corner, E. D. S., Denton, E. J., Forster, F. R. S., and Forster, G. R. (1969). On the buoyancy of some deep-sea sharks. *Proc. R. Soc. London B* **171**, 415–429.

Cossins, A. R., and MacDonald, A. G. (1986). Homeoviscous adaptation under pressure. III. The fatty acid composition of liver mitochondrial phospholipids of deep-sea fish. *Biochem. Biophys. Acta* **860**, 325–335.

Craik, J. C. A., and Harvey, S. M. (1987). The causes of buoyancy in eggs of marine teleosts. *J. Mar. Biol. Assoc. U.K.* **67**, 169–182.

D'Aoust, B. G. (1970). The role of lactic acid in gas secretion in the teleost swimbladder. *Comp. Biochem. Physiol.* **32**, 637–668.

Davenport, J., and Kjorsvik, E. (1986). Buoyancy in the lumpsucker *Cyclopterus lumpus*. *J. Mar. Biol. Assoc. U.K.* **66**, 159–174.

Deck, J. E. (1970). Lactic acid production by the swimbladder gas gland *in vitro* as influenced by glucagon and epinephrine. *Comp. Biochem. Physiol.* **34**, 317–324.

Denton, E. (1960). The buoyancy of marine animals. *Sci. Am.* **203**, 119–129.

Denton, E. J. (1961). The buoyancy of fish and cephalopods. *Prog. Biophys. Biophys. Chem.* **1**, 178–234.

Denton, E. J., and Gilpin-Brown, J. B. (1966). On the buoyancy of the pearly nautilus. *J. Mar. Biol. Assoc. U.K.* **46**, 723–759.

Denton, E. J., Gilpin-Brown, J. B., and Shaw, T. I. (1969). A buoyancy mechanism found in cranchid squid. *Proc. R. Soc. London B* **174**, 271–279.

Denton, E. J., Liddicoat, J. D., and Taylor, D. W. (1972). The permeability to gases of the swimbladder of the conger eel (*Conger conger*). *J. Mar. Biol. Assoc. U.K.* **52**, 727–746.

DeVries, A. L., and Eastman, J. T. (1978). Lipid sacs as a buoyancy adaptation in an Antarctic fish. *Nature* (*London*) **271**, 352–353.

Dorn, E. (1961). Über den Feinbau der Schwimmblase von *Anguilla vulgaris* L. Licht- und elektronenmikroskopische Untersuchungen. *Zeitschrift für Zellforschung* **55**, 849–912.

Eastman, J. T. (1985). The evolution of neutrally buoyant Notothenioid fishes: Their specializations and potential interactions in the Antarctic marine food web. *In* "Antarctic Nutrient Cycles and Food Webs" (W. R. Siegfried, P. R. Condy, and R. M. Laws, eds.), pp. 430–436. Springer-Verlag, Berlin, Heidelberg.

Eastman, J. T. (1988). Lipid storage systems and the biology of two neutrally buoyant Antarctic Notothenioid fishes. *Comp. Biochem. Physiol.* **90B**, 529–537.

Eastman, J. T., and DeVries, A. L. (1981). Buoyancy adaptations in a swim-bladderless Antarctic fish. *J. Morphol.* **167**, 91–102.

Eastman, J. T., and DeVries, A. L. (1982). Buoyancy studies of Notothenioid fishes in McMurdo Sound, Antarctica. *Copeia* **1982**, 385–393.

Eastman, J. T., and DeVries, A. L. (1989). Ultrastructure of the lipid sac wall in the Antarctic Notothenioid fish *Pleuragramma antarcticum*. *Polar Biology* **9**, 333–335.

Enns, T., Douglas, E., and Scholander, P. F. (1967). Role of the swimbladder rete of fish in secretion of inert gas and oxygen. *Adv. Biol. Med. Phys.* **11**, 231–244.

Ewart, H. S., and Driedzic, W. R. (1990). Enzyme activity levels underestimate lactate production rates in cod (*Gadus morhua*) gas gland. *Can. J. Zool.* **68**, 193–197.

Falk-Petersen, I.-B., Falk-Petersen, S., and Sargent, J. R. (1986a). Nature, origin and possible roles of lipid deposits in *Maurolicus muelleri* (Gmelin) and *Benthosema glaciale* (Reinhart) from Ullsfjorden, Northern Norway. *Polar Biol.* **5**, 235–240.

Falk-Petersen, S., Falk-Petersen, I.-B., and Sargent, J. R. (1986b). Structure and function of an unusal lipid storage organ in the Arctic fish *Lumpenus maculatus* Fries. *Sarsia* **71**, 1–6.

Fänge, R. (1953). The mechanisms of gas transport in the euphysoclist swimbladder. *Acta Physiol. Scand.* **30**, 1–133.

Fänge, R. (1983). Gas exchange in fish swim bladder. *Rev. Physiol. Biochem. Pharmacol.* **97**, 111–158.

Finkelstein, A. (1976). Water and nonelectrolyte permeability of lipid bilayer membranes. *J. Gen. Physiol.* **68**, 127–165.

Fricke, H., Reinicke, O., Hofer, H., and Nachtigall, W. (1987). Locomotion of the coelacanth *Latimeria chalumnae* in its natural environment. *Nature (London)* **329**, 331–333.

Gee, J. H. (1983). Ecologic implications of buoyancy control in fish. *In* "Fish Biomechanics" (P. W. Webb, ed.), pp. 140–176. Praeger Scientific, New York.

Gerth, W. A., and Hemmingsen, E. A. (1982). Limits of gas secretion by the salting-out effect in the fish swimbladder rete. *J. Comp. Physiol.* **146**, 129–136.

Grigor, M. R., Sutherland, W. H., and Phleger, C. F. (1990). Wax ester metabolism in the orange roughy (*Hoplostethus atlanticus*) (Bericiformes: Trachichthydae). *Mar. Biol.* **105**, 223–227.

Grigor, M. R., Thomas, C. R., Jones, P. D., and Buisson, D. H. (1983). Occurrence of wax esters in the tissues of the orange roughy (*Hoplostethus atlanticus*). *Lipids* **18**, 585–588.

Harris, J. E. (1936). The role of the fins in the equilibrium of the swimming fish. I. Wind-tunnel tests on a model of *Mustelus canis* (Mitchill). *J. Exp. Biol.* **13**, 476–493.

Harris, J. E. (1937). The role of the fins in the equilibrium of the swimming fish. II. The role of the pelvic fins. *J. Exp. Biol.* **15**, 32–47.

Hayashi, K. (1987). Liquid wax esters in liver oils of the deep-sea teleost fish *Laemonema longipes*. *Bull. Jpn. Soc. Sci. Fish.* **53**, 2263–2267.

Hayashi, K., and Kashiki, I. (1988). Level and composition of wax esters in the different tissues of deep-sea teleost fish *Laemonema longipes*. *Bull. Jpn. Soc. Sci. Fish.* **54**, 135–140.

Hayashi, K., and Takagi, T. (1980). Composition of diacyl glyceryl ethers in the liver lipids of ratfish. *Hydrolagus novaezealandiae*. *Bull. Jpn. Soc. Sci. Fish.* **46**, 855–861.

Hayashi, K., and Takagi, T. (1981). Distribution of squalene and diacyl glyceryl ethers in the different tissues of deep-sea shark, *Dalatias licha. Bull. Jpn. Soc. Sci. Fish.* **47,** 281–288.

Horn, M. H., Grimes, P. W., Phleger, C. F., and McClanahan, L. L. (1978). Buoyancy function of the enlarged fluid-filled cranium in the deep-sea ophidiid fish *Acanthonus armatus. Mar. Biol.* **46,** 335–339.

Hubold, G., and Tomo, A. P. (1989). Age and growth of Antarctic silverfish. (*Pleuragramma antarcticum* Boulenger 1902), from the southern Weddell Sea and Antarctic Peninsula. *Polar Biol.* **9,** 205–212.

Jasinski, A., and Kilarski, W. (1969). On the fine structure of the gas gland in some fishes. *Zeitschrift für Zellforschung* **102,** 333–356.

Johansen, K. (1982). Respiratory gas exchange of vertebrate gills. *In* "Gills" (D. F. Houlihan, J. C. Rankin, and T. J. Shuttleworth, eds.), Society for Experimental Biology Seminar Series, 16, pp. 99–128. Cambridge Univ. Press, Cambridge.

Josephson, R. V., Holtz, R. B., Misock, J. P., and Phleger, C. F. (1975). Composition and partial protein characterization of swimbaldder foam from deep-sea fish *Coryphaenoides acrolepis* and *Antimora rostrata. Comp. Biochem. Physiol.* **52B,** 91–95.

Kalish, J. M., Greenlaw, C. F., Pearcy, W. G., and Holliday, D. V. (1986). The biological and acoustical structure of sound scattering layers off Oregon. *Deep-Sea Res.* **33,** 631–653.

Kändler, R., and Tan, E. O. (1965a). Investigations of the osmoregulation in pelagic eggs of gadoid and flatfishes in the Baltic. I. Changes in volume and specific gravity at different salinities. *Int. Counc. Explor. Sea Counc. Meet.* **43.**

Kändler, R., and Tan, E. O. (1965b). Investigations of the osmoregulation in pelagic eggs of gadoid and flatfishes in the Baltic. II. Changes in chemical composition at different salinities. *Int. Counc. Explor. Sea Counc. Meet.* **44.**

Kayama, M., and Nevenzel, J. C. (1974). Wax ester biosynthesis by midwater marine animals. *Marine Biol.* **24,** 279–285.

Kayama, M., Zafar, S., Rizvi, W., and Asakawa, S. (1971). Biosynthesis of squalene and cholesterol in the fish. I. *In vitro* studies in acetate incorporation. *J. Fac. Fish. Anim. Husb. Hiroshima Univ.* **10,** 1.

Kleckner, R. C. (1974). "Swimbladder Morphology of Mediterranean Sea Mesopelagic Fishes. M.S. thesis, Univ. of RI, Kingston, RI.

Kleckner, R. C. (1980). Swimbladder wall guanine enhancement related to migratory depth in silver phase *Anguilla rostrata. Comp. Biochem. Physiol.* **65A,** 351–354.

Kleckner, R. C., and Gibbs, R. H., Jr. (1972). Swimbladder structure of Mediterranean midwater fishes and a method of comparing swimbladder data with acoustic profiles. *In* "Mediterranean Biological Studies, Final Report," pp. 230–281. Smithsonian Institution, Washington, D.C.

Klenk, M., and Pelster, B. (1995). ATPase activities and energy charge in swimbladder tissue of the European eel, *Anguilla anguilla. Verhandlungen der Deutschen Zoologischen Gesellschaft* **88,** 112.

Kobayashi, H., Pelster, B., and Scheid, P. (1989a). Water and lactate movement in the swimbladder of the eel, *Anguilla anguilla. Respir. Physiol.* **78,** 45–57.

Kobayashi, H., Pelster, B., and Scheid, P. (1989b). Solute back-diffusion raises the gas concentrating efficiency in counter-current flow. *Respir. Physiol.* **78,** 59–71.

Kobayashi, H., Pelster, B., and Scheid, R. (1990). CO_2 back-diffusion in the rete aids O_2 secretion in the swimbladder of the eel. *Respir. Physiol.* **79,** 231–242.

Kuhn, W., Ramel, A., Kuhn, H. J., and Marti, E. (1963). The filling mechanism of the swimbladder. Generation of high gas pressures through hairpin countercurrent multiplication. *Experientia* **19,** 497–511.

Kutchai, H. (1971). Role of carbonic anhydrase in lactate secretion by the swimbladder. *Comp. Biochem. Physiol.* **39A,** 357–359.

Kutchai, H., and Steen, J. B. (1971). The permeability of the swimbladder. *Comp. Biochem. Physiol.* **39A,** 119–123.

Lapennas, G. N., and Schmidt-Nielsen, K. (1977). Swimbladder permeability to oxygen. *J. Exp. Biol.* **67,** 175–196.

Lee, R. F., Phleger, C. F., and Horn, M. H. (1975). Composition of oil in fish bones: Possible function in neutral buoyancy. *Comp. Biochem. Physiol.* **50B,** 13–16.

Maetz, J. (1956). Le role biologique de l'anhydrase carbonique chex quelques téléostéens. *Supplement au Bulletin Biologique de France et de Belgique Les Presses Universitaieres de France,* 1–129.

Magnuson, J. J. (1970). Hydrostatic equilibrium of *Euthynnus affinis,* a pelagic teleost without a gas bladder. *Copeia,* 56–85.

Magnuson, J. J. (1978). Locomotion by scombrid fishes: Hydromechanics, morphology, and behaviour. *In* "Fish Physiology" (W. S. Hoar and D. J. Randall, eds.), Vol. 7, pp. 239–313. Academic Press, New York.

Malins, D. C., and Barone, A. (1969). Glyceryl ether metabolism: Regulation of buoyancy in dogfish *Squalus acantias. Science* **167,** 79–80.

Marshall, N. B. (1960). Swimbladder structure of deep-sea fishes in relation to their systematics and biology. *Discovery Reports* **31,** 1–122.

Marshall, N. B. (1972). Swimbladder organization and depth ranges of deep-sea teleosts. *Symposium of the Society of Experimental Biology* **26,** 261–272.

May, R. C. (1974). Factors affecting buoyancy in the eggs of *Bairdiella icistia* (Pisces: Sciaenidae). *Marine Biology* **28,** 55–59.

Meesters, A., and Nagel, F. G. P. (1935). Über Sekretion und Resorption in der Schwimmblase des Flußbarsches. *Zeitschrift für vergleichende Physiologie* **21,** 646–657.

Morris, R. J., and Culkin, F. (1989). Fish. *In* "Marine Biogenic Lipids, Fats and Oils" (R. G. Ackman, ed.), pp. 145–178. CRC Press, Boca Raton, Florida.

Morris, S. M., and Albright, J. T. (1975). The ultrastructure of the swimbladder of the toadfish, *Opsanus tau* L. *Cell Tissue Res.* **164,** 85–104.

Mylvaganam, S. E., Bonaventura, C., Bonaventura, J., and Getzoff, E. D. (1996). Structural basis for the Root effect in haemoglobin. *Nature Struct. Biol.* **3,** 275–283.

National Research Council of the USA (1928). "International Critical Tables of Numerical Data, Physics, Chemistry and Technology," vol. 3. McGraw-Hill, New York.

Neighbors, M. A. (1992). Occurrence of inflated swimbladders in five species of lanternfishes (family Myctophidae) from waters off southern California. *Mar. Biol.* **114,** 355–363.

Nevenzel, J. C. (1970). Occurrence, function and biosynthesis of wax esters in marine organisms. *Lipids* **5,** 308–319.

Nevenzel, J. C. (1989). Biogenic hydrocarbons of marine organisms. *In* "Marine Biogenic Lipids, Fats, and Oils" (R. G. Ackman, ed.), Vol. 1, pp. 3–197. CRC Press, Boca Raton, Florida.

Nevenzel, J. C., Rodegker, W., and Mead, J. F. (1965). The lipids of *Ruvettus pretiosus* muscle and liver. *Biochemistry* **4,** 1589–1594.

Nevenzel, J. C., Rodegker, W., and Mead, J. F. (1966). Lipids of the living Coelacanth, *Latimeria chalumnae. Science* **152,** 1753–1755.

Nevenzel, J. C., Rodegker, W., Robinson, J. S., and Kayama, M. (1969). The lipids of some lantern fishes (Family Myctophidae). *Comp. Biochem. Physiol.* **31,** 25–36.

Nielsen, J. G., and Munk, O. (1964). A hadal fish (*Bassogigas profundissimus*) with a functional swimbladder. *Nature (London)* **204,** 594–595.

Noble, W. R., Pennelly, R. R., and Riggs, A. (1975). Studies of the functional properties of the hemoglobin from the benthic fish, *Antimora rostrata. Comp. Biochem. Physiol.* **52B,** 75–81.
Nursall, J. R. (1989). Buoyancy is provided by lipids of larval redlip Blennies, *Ophioblennius atlanticus* (Teleostei: Blenniidae). *Copeia,* 614–621.
Nybelin, O. (1957). Deep-sea bottom fishes. *Report of the Swedish Deep Sea Expendition 2, Zoology* **20,** 247–345.
Patton, J. S. (1975). The effect of pressure and temperature on phospholipid and triglyceride fatty acids of fish white muscle: A comparison of deepwater and surface marine species. *Comp. Biochm. Physiol.* **52B,** 105–110.
Pelster, B. (1995a). Metabolism of the swimbladder tissue. *Biochem. Mol. Biol. Fishes* **4,** 101–118.
Pelster, B. (1995b). Lactate production in isolated swim bladder tissue of the European eel *Anguilla anguilla. Physiol. Zool.* **68,** 634–646.
Pelster, B. (1995c). Mechanisms of acid release in isolated gas gland cells of the European eel *Anguilla anguilla. Am. J. Physiol.* **269,** R793–R799.
Pelster, B., and Scheid, P. (1991). Activities of enzymes for glucose catabolism in the swimbladder of the European eel *Anguilla anguilla. J. Exp. Biol.* **156,** 207–213.
Pelster, B., and Scheid, P. (1992). The influence of gas gland metabolism and blood flow on gas deposition into the swimbladder of the European eel *Anguilla anguilla. J. Exp. Biol.* **173,** 205–216.
Pelster, B., and Scheid, P. (1993). Glucose metabolism of the swimbladder tissue of the European eel *Anguilla anguilla. J. Exp. Biol.* **185,** 169–178.
Pelster, B., and Weber, R. E. (1991). The physiology of the Root effect. *Adv. Comp. Environ. Physiol.* **8,** 51–77.
Pelster, B., Hicks, J., and Driedzic, W. R. (1994). Contribution of the pentose phosphate shunt to the formation of CO_2 in swimbladder tissue of the eel. *J. Exp. Biol.* **197,** 119–128.
Pelster, B., Kobayashi, H., and Scheid, P. (1988). Solubility of nitrogen and argon in eel whole blood and its relationship to pH. *J. Exp. Biol.* **135,** 243–252.
Pelster, B., Kobayashi, H., and Scheid, P. (1989). Metabolism of the perfused swimbladder of European eel: Oxygen, carbon dioxide, glucose and lactate balance. *J. Exp. Biol.* **144,** 495–506.
Pelster, B., Kobayashi, H., and Scheid, P. (1990). Reduction of gas solubility in the fish swimbladder. *In* "Oxygen Transport to Tissue XII" (J. Piper, T. K. Goldstick, and M. Meyer, eds.), pp. 725–733. Plenum, New York.
Pfeiler, E. (1986). Towards an explanation of the developmental strategy in leptocephalous larvae of marine teleost fishes. *Environ. Biol. Fishes* **15,** 3–13.
Phleger, C. F. (1971). Pressure effects on cholesterol and lipid synthesis by the swimbladder of an abyssal Coryphaenoides species. *Am. Zool.* **11,** 559–570.
Phleger, C. F. (1975a). Bone lipids of Kona Coast reef fish: Skull buoyancy in the hawkfish, *Cirrhites pinnulatus. Comp. Bioch. Physiol.* **52B,** 101–104.
Phleger, C. F. (1975b). Lipids synthesis by *Antimora rostrata,* and abyssal codling from the Kona coast. *Comp. Biochem. Physiol.* **52B,** 97–99.
Phleger, C. F. (1972). Cholesterol and Hyperbaric Oxygen in Swimbladders of Deep Sea Fishes. Ph.D. Thesis, Univ. of CA, San Diego.
Phleger, C. F. (1987). Bone lipids of tropical reef fishes. *Comp. Biochem. Physiol.* **86B,** 509–512.
Phleger, C. G. (1988a). The importance of skull lipid as an energy reserve during starvation in the ocean sturgeon, *Acanthurus bahianus: Comp. Biochem. Physiol.* **91A,** 97–100.
Phleger, C. F. (1988b). Bone lipids of Jamaican reef fishes. *Comp. Biochem. Physiol.* **90B,** 279–283.

Phleger, C. F. (1991). Biochemical aspects of buoyancy in fishes. *In* "Biochemistry and Molecular Biology of Fishes" (P. W. Hochachka and T. P. Mommsen, eds.), pp. 209–247. Elsevier, Amsterdam.

Phleger, C. F., and Benson, A. A. (1971). Cholesterol and hyperbaric oxygen in swimbladder of deep-sea fishes. *Nature (London)* **230,** 122.

Phleger, C. F., and Grigor, M. R. (1990). Role of wax esters in determining buoyancy in *Hoplostethus atlanticus* (Beryciformes: Trachichthyidae). *Mar. Biol.* **105,** 229–233.

Phleger, C. F., and Grimes, P. W. (1976). Bone lipids of marine fishes. *Physiol. Chem. Phys.* **8,** 447–456.

Phleger, C. F., and Holtz, R. B. (1973). The membranous lining of the swimbladder in deep sea fishes.–I. Morphology and chemical composition. *Comp. Biochem. Physiol.* **45B,** 867–873.

Phleger, C. F., and Laub, R. J. (1989). Skeletal fatty acids in fish from different depths off Jamaica. *Comp. Biochem. Physiol.* **94B,** 329–334.

Phleger, C. F., Benson, A. A., and Yayanos, A. A. (1973). Pressure effect of squalene-2,3-oxide cyclization in fish. *Comp. Biochem. Physiol.* **45B,** 241–247.

Phleger, C. F., Patton, J., Grimes, P., and Lee, R. F. (1976). Fish-bone oil: Percent total body lipid and carbon-14 uptake following feeding of 1-[14] C-palmitic acid. *Mar. Biol.* **35,** 85–89.

Phleger, C. F., Holtz, R., and Grimes, P. W. (1977). Membrane biosynthesis in swimbladders of deep sea fishes *Coryphaenoides acrolepis* and *Antimora rostrata Comp. Biochem. Physiol.* **56B,** 25–30.

Phleger, C. F., Grimes, P. W., Pesely, A., and Horn, M. H. (1978). Swimbladder lipids of five species of deep benthopelagic. Atlantic ocean fishes. *Bull. Mar. Sci.* **28,** 198–202.

Piiper, J. (1965). Physiological equilibria of gas cavities in the body. *In* "Handbook of Physiology, Respiration." (W. O. Fenn and H. Rahn, eds.), pp. 1205–1218. American Physiological Society, Bethesda, Maryland.

Reinhardt, S. B., and Van Vleet, E. S. (1986). Lipid composition of twenty-two species of Antarctic midwater zooplankton and fish. *Mar. Biol.* **91,** 149–159.

Riggs, A. F. (1988). The Bohr effect. *Annu. Rev. Physiol.* **50,** 181–204.

Riley, G. A., Stommel, H., and Bumpus, D. F. (1949). Quantitative ecology of the plankton of the western North Atlantic. *Bulletin of the Bingham Oceanographic Collection. Yale University* **12,** 1–169.

Roberts, B. L. (1969). The buoyancy and locomotory movements of electric rays. *J. Mar. Biol. Assoc. U.K.* **49,** 621–640.

Root, R. W. (1931). The respiratory function of the blood of marine fishes. *Biol. Bull.* **61,** 427–456.

Rosenthal, H., and Alderdice, D. F. (1976). Sublethal effects of environmental stressors, natural and pollutional, on marine fish eggs and larvae. *J. Fish. Res. Bd. Can.* **33,** 2047–2065.

Sargent, J. R. (1989). Ether-linked glycerides in marine animals. *In* "Marine Biogenic Lipids, Fats, and Oils" (R. G. Ackman, ed.), pp. 175–197. CRC Press, Boca Raton, Florida.

Sargent, J. R., Gatten, R. R., and McIntosh, R. (1973). The distribution of neutral lipids in shark tissues. *J. Mar. Biol. Assoc. U.K.* **53,** 649–656.

Scholander, P. F., and Van Dam, L. (1954). Secretion of gases against high pressures in the swimbladder of deep sea fishes. I. Oxygen dissociation in blood. *Biol. Bull.* **107,** 247–259.

Simons, J. R. (1970). The direction of the thrust produced by the heterocercal tails of two dissimilar elasmobranchs: The port Jackson shark, *Heterodontus portusjacksoni* (Meyer), and the piked dogfish, *Squalus megalops* (MacLeay). *J. Exp. Biol.* **52,** 95–107.

Skinazi, L. (1953). L'anhydrase carbonique dans deux Téléostéens voisins. Inhibition de la sécrétion des gaz de la vessie natatoire chez la perche par les sulfamides. *Comptes Rendue des Seances de la Societe de Biologie et de ses Filiales* **147,** 295–299.

Springer, S. (1967). Social organization of shark populations. *In* "Sharks, Skates and Rays" (P. W. Gilbert, R. F. Mathewson, and D. P. Rall, eds.), pp. 149–174. Johns Hopkins Press, Baltimore.

Steen, J. B. (1963a). The physiology of the swimbladder in the eel *Anguilla vulgaris*. III. The mechanism of gas secretion. *Acta Physiol. Scand.* **59,** 221–241.

Steen, J. B. (1963b). The physiology of the swimbladder in the eel *Anguilla vulgaris*. II. The reabsorption of gases. *Acta Physiol. Scand.* **58,** 138–149.

Steen, J. B. (1970). The swim bladder as a hydrostatic organ. *In* "Fish Physiology" (W. S. Hoar and D. J. Randall, eds.), pp. 413–443. Academic Press, New York.

Stokes, M. D., and Holland, N. D. (1995). Ciliary hovering in larval lancelets (=Amphioxus). *Biol. Bull.* **188,** 231–233.

Stray-Pedersen, S., and Nicolaysen, A. (1975). Qualitative and quantitative studies of the capillary structure in the rete mirabile of the eel, *Anguilla vulgaris* L. *Acta Physiol. Scand.* **94,** 339–357.

Sund, T. (1977). A mathematical model for counter-current multiplication in the swim-bladder. *J. Physiol. (London)* **267,** 679–696.

Tocher, D. R., and Sargent, J. R. (1984). Analyses of lipids and fatty acids in ripe roes of some northwest European marine fish. *Lipids* **19,** 492–499.

Van Vleet, E. S., Candileri, S., McNeillie, J., Reinhardt, S. B., Conkright, M. E., and Zwissler, A. (1984). Neutral lipid components of eleven species of Caribbean sharks. *Comp. Biochem. Physiol.* **79B,** 549–554.

Vent, R. J., and Pickwell, G. V. (1977). Acoustic volume scattering measurements with related biological and chemical observations in the northeastern tropical Pacific. *In* "Oceanic Sound Scattering Prediction" (N. R. Andersen and B. J. Zahuranec, eds.), pp. 697–716. Plenum, New York.

Vetter, R. D., Lynn, E. A., Garza, M., and Costa, A. S. (1994). Depth zonation and metabolic adaptation in Dover sole, *Microstomus pacificus*, and other deep-living flatfishes: Factors that affect the sole. *Mar. Biol.* **120,** 145–159.

Wallace, R. A. (1985). Vitellogenesis and oocyte growth in nonmammalian vertebrates. *Dev. Biol.* **1,** 127–177.

Wallace, R. A., and Selman, K. (1981). Cellular and dynamic aspects of oocyte growth in teleosts. *Am. Zool.* **21,** 325–343.

Walsh, P. J., and Milligan, C. L. (1993). Roles of buffering capacity and pentose phosphate pathway activity in the gas gland of the gulf toadfish. *Opsanus beta*. *J. Exp. Biol.* **176,** 311–316.

Webb, P. W. (1971). The swimming energetics of trout. II: Oxygen consumption and swimming efficiency. *J. Exp. Biol.* **55,** 521–540.

Webb, P. W. (1990). How does benthic living affect body volume, tissue composition, and density of fishes? *Can. J. Zool.* **68,** 1250–1255.

Wittenberg, J. B., Schwend, M. J., and Wittenberg, B. A. (1964). The secretion of oxygen into the swim-bladder of fish. III. The role of carbon dioxide. *J. Gen. Physiol.* **48,** 337–355.

Wittenberg, J. B., Copeland, D. E., Haedrich, R. L., and Child, J. S. (1980). The swimbladder of deep-sea fish: The swimbladder wall is a lipid-rich barrier to oxygen diffusion. *J. Mar. Biol. Assoc. U.K.* **60,** 263–276.

Woodland, W. N. F. (1911). On the structure and function of the gas glands and retia mirabilia associated with the gas bladder of some teleostean fishes. *Proc. Zool. Soc. London* **1,** 183–248.

Yancey, P. H., Lawrence-Berrey, R., and Douglas, M. D. (1989). Adaptations in mesopelagic fishes. I. Buoyant glycosaminoglycan layers in species without diel vertical migrations. *Mar. Biol.* **103,** 453–459.

Yin, M. C., and Blaxter, J. H. S. (1987). Temperature, salinity tolerance, and buoyancy during early development and starvation of Clyde and North Sea herring, cod, and flounder larvae. *J. Exp. Mar. Biol. Ecol.* **107,** 279–290.

6

BIOCHEMISTRY AT DEPTH

ALLEN G. GIBBS

I. Introduction
II. Effects of Pressure on Biochemical Systems: Protein Interactions and Enzyme Kinetics
   A. Thermodynamics of Pressure Effects
   B. Summary
III. Tolerance Adaptations: Maintenance of Biochemical Function in the Deep Sea
   A. Protein–Protein Interactions
   B. Enzyme–Substrate Interactions
   C. Membrane Proteins and Lipid–Protein Interactions
   D. Integrating Cellular Processes: G-Protein-Mediated Signal Transduction
   E. Summary
IV. Capacity Adaptation: Biochemical Correlates of Organismal Metabolism
   A. Depth-Related Patterns in Metabolic Rates
   B. Biochemical Consequences of Reduced Metabolism
   C. Summary
V. Future Directions: Phylogenetic and Molecular Approaches
   References

I. INTRODUCTION

Fishes are the charismatic megafauna of the deep sea and, as such, have received considerable attention from deep-sea biologists. Their environment is characterized by low temperatures (2–4°C), lack of light, and high pressures. Hydrostatic pressure increases by approximately 1 MPa (10^6 Pa) for every 100 m increase in depth. [Several different measurement units for pressure have been used in the literature. In recent years, most journals have begun to require the Système International (SI) unit for pressure, the Pascal (Pa). The following conversion factors can be used: 1 atm = 1 bar = 14.7 psi = 101,325 Pa.] Thus, in the deepest trenches, pressures can be over 1000 times greater than at sea level. Invertebrates and bacteria have been collected near the bottom of the Marianas Trench (depth

11,043 m, pressure ~110 MPa) (Yayanos, 1995), demonstrating that life can exist at these pressures.

Pressures of only a few megapascals are sufficient to greatly perturb the behavior of shallow-living animals. Exposure to high pressures results in hyperactivity, convulsions, torpor, and eventual death (Jannasch et al., 1987). Few data are available on the effects of reduced pressure on deep-sea fishes, but those individuals who are brought to the surface alive and apparently undamaged generally do not remain so for long. Physiological parameters such as cardiovascular and nervous system function of deep-sea fishes are greatly perturbed at atmospheric pressure (Harper et al., 1987; Macdonald et al., 1987; Pennec et al., 1988). These observations indicate that pressure is an important factor in determining the distributions of marine organisms. Thus, understanding mechanisms of pressure tolerance is critical to understanding adaptation of organisms to the deep sea.

Physiological studies of deep-sea life are severely restricted by the requirement for high pressure. Special materials (e.g., titanium) are needed for hyperbaric chambers, and the thickness of the walls required to prevent rupture increases more rapidly than the internal diameter of the chamber (A. A. Yayanos, personal communication). Thus, for an organism as large as a fish, the bulk and expense of equipment needed simply for animal maintenance can be prohibitive. Although a few such studies have been conducted (Macdonald and Gilchrist, 1980; Yayanos, 1981), bacteria are the only organisms routinely maintained in the laboratory under high pressure.

Partly out of necessity, physiologists interested in the deep sea have used biochemical analyses to investigate mechanisms of adaptation (Siebenaller and Somero, 1989; Somero, 1992). By examining the effects of high pressure and low temperature on isolated macromolecules from species inhabiting different depths, it has been possible to gain insight into the mechanisms of adaptation to the deep sea. This review will concentrate on pressure adaptation of biological molecules, especially proteins and membrane lipids. Temperature adaptation will be addressed to a lesser extent, but several excellent reviews have been published (Hazel and Williams, 1990; Cossins, 1994; Hazel, 1995; Somero, 1995; Johnston and Bennett, 1996). An additional topic of importance in the deep sea is metabolic rates, and biochemical approaches to this question will also be discussed.

An important issue arising in comparative biology in recent years is the choice of study organisms, especially the role of phylogeny in the evolution of physiological systems (Garland and Adolph, 1994; Garland and Carter, 1994). The development of phylogenetically based analytical techniques has had almost no impact upon deep-sea biochemists and physiologists. This is due in large part to the fact that researchers are constrained by the availability of their organisms. Deep-sea fishes and other animals are diffi-

6. BIOCHEMISTRY AT DEPTH

cult and expensive to obtain. In many cases, phylogenetically appropriate groups of organisms may not be available. The result is that there has not been a single study performed in a rigorous phylogenetic context. As we will see, many "good" examples of pressure adaptation may suffer from phylogenetic artifacts. Rather than fault researchers for doing the best they can under the circumstances, I will point out a few significant studies that may suffer from phylogenetic problems in the course of this review, in order to illustrate the limitations of our understanding.

This review will take a bottom-up approach. Following a brief description of the effects of pressure on biochemical reactions, I will discuss increasingly complex processes: protein–protein interactions, enzyme–substrate binding, lipid–protein interactions, signal transduction, and biochemical correlates of organismal metabolism. Examples from taxa other than fish will be included in those cases where little or nothing is known about fish, or where other organisms provide clearer understanding of the mechanisms of biochemical adaptation. Several general issues will arise repeatedly in regard to a given biochemical process: Do physiologically relevant pressures have significant effects? Do proteins from deep-sea fish respond to pressure differently than do homologous proteins in shallow-living species? Which functional characteristics of proteins exhibit pressure adaptation, and how are these differences achieved? How and why are enzyme activities regulated at certain levels?

It is clear that our understanding of biochemical adaptation to the deep sea is fragmentary. The deep sea is the largest habitat on earth (in terms of volume), yet most studies have used only a limited subset of the species occurring off the coasts of North America. From these species, only a handful of proteins have been studied. Both species and enzymes have often been chosen primarily on the basis of availability of material. The results to date demonstrate that biochemical adaptation to the deep sea has occurred, but certainly do not encompass the entire range of mechanisms. At the end of this review, I will suggest a few areas in which recently developed experimental and analytical techniques can provide greater understanding of both the mechanisms and the evolution of biochemical adaptation in the deep sea.

II. EFFECTS OF PRESSURE ON BIOCHEMICAL SYSTEMS: PROTEIN INTERACTIONS AND ENZYME KINETICS

By comparison with temperature, pressure has been the forgotten thermodynamic variable in biology. Advances in instrumentation have now made it possible to apply almost any biochemical or biophysical technique at high pressure, among them nuclear magnetic resonance (NMR), X-ray crystallography, gel electrophoresis, fluorescence polarization, and infrared spectroscopy (see references in Mozhaev et al., 1996). Most of these studies have been aimed at understanding the gross effects of high pressure on the properties of proteins, or have used pressure as a means of probing enzyme mechanisms (Heremans, 1982; Jaenicke, 1983; Weber, 1992; Silva and Weber, 1993; Mozhaev et al., 1996). Pressures of over 100 MPa (corresponding to depths of >10,000 m) are typically used. However, we shall see that much lower, environmentally relevant pressures can also have important, albeit more subtle, effects on enzyme function.

A. Thermodynamics of Pressure Effects

An understanding of the biochemical effects of high pressure requires an understanding of the effects of pressure on macromolecular structure. Pressure exerts its effects through volume changes. The Gibbs free energy change (ΔG) associated with a chemical reaction is given by (Morild, 1981)

$$\Delta G = \Delta U - T\Delta S + P\Delta V$$

where ΔU is the change in internal energy of the system, T is the absolute temperature, ΔS is the entropy change, P is the pressure, and ΔV is the difference in volume between products and reactants. By Le Chatelier's principle, application of pressure will tend to shift equilibria toward the lower volume state. The equations governing this phenomenon are

$$K_{eq} = RT \exp(-\Delta G),$$
$$(\delta \ln K_{eq}/\delta P)_T = -\Delta V/RT,$$

where K_{eq} is the equilibrium constant and R is the gas constant. Thus, the volume change associated with a reaction can be calculated from the slope of a plot of ln K_{eq} versus pressure, the high pressure equivalent of an Arrhenius plot. It must be stressed that these equations include the effects of pressure on all parts of the system, including interactions of proteins with other proteins, membrane lipids, and small molecules such as water and solutes. Each of these will make its own incremental contribution to the overall volume change. Thus, any consideration of pressure effects on enzymes must take into account the microscopic milieu of the protein as well.

The functional properties of proteins are determined by their three-dimensional structure, which depends on hundreds of weak bonds, including ionic interactions and salt bridges, hydrophobic interactions, van der Waals interactions, and hydrogen bonds. Formation of ionic bonds or hydrophobic

6. BIOCHEMISTRY AT DEPTH

interactions usually involves an increase in system volume; thus, pressure disrupts these bonds. The overall contribution of hydrogen bonds to protein volume is unclear; hydrogen bonds form with a decrease in volume (Low and Somero, 1975a), but breaking an internal hydrogen bond in a protein leaves its components free to hydrogen bond with water. Thus, the net volume change associated with differences in hydrogen bonding between the native and unfolded states is uncertain. Native proteins also contain empty spaces, which will tend to be filled by water molecules when the protein is unfolded (Rashin et al., 1986). These cavities may be the major contributor to the fact that the difference in volume between the folded, native conformation of a typical protein and the unfolded, random-coil state is usually greater than 100 ml/mol, so that high pressures will tend to denature proteins (Weber and Drickamer, 1983; Silva and Weber, 1993). However, the pressures required to unfold proteins typically exceed 100 MPa (corresponding to depths of over 10,000 m). Thus, protein denaturation is not expected to be an important locus of pressure sensitivity under physiological conditions.

Most biological processes are out of equilibrium, so that kinetic rate theory applies. For reaction rates, the relevant thermodynamic equations are (Morild, 1981)

$$A \xrightarrow{k} B; \quad k = RT \exp(-\Delta G^{\ddagger})$$
$$\delta \ln k/\delta P)_T = -\Delta V^{\ddagger}/RT,$$

where k is the rate constant for the reaction, and the double dagger ($\ddagger$) indicates the activation energy ($\Delta G^{\ddagger}$) or activation volume ($\Delta V^{\ddagger}$), i.e., the difference in free energy or volume, respectively, between the transition state and ground state. Thus, for a simple chemical reaction with a single transition state, the reaction will proceed more slowly at high pressure if the transition state has a larger volume than the reactants (i.e., $\Delta V^{\ddagger}$ is positive). Enzymatic reactions generally involve more complicated, multistep reactions, so the effects of pressure can be used only to calculate an *apparent* activation volume for the reaction. Apparent activation volumes may be positive or negative, but most enzymes are inhibited by high pressure (Morild, 1981).

The volume change associated with a reaction can change with pressure, leading to nonlinear plots of $\ln K_{eq}$ or $\ln k$ versus pressure. For an equilibrium reaction, this may result from a difference in compressibility between the products and reactants. Both will have a smaller volume at high pressure, but the volume of one may change more rapidly. This will result in a different volume change of the reaction as pressure increases. The relevant thermodynamic parameter is the absolute compressibility, given by the change in volume with pressure (Morild, 1981):

$$\kappa = -(\delta V/\delta P)_T.$$

The difference in compressibility between reactants and products can be calculated from the second derivative of a plot of ln K_{eq} versus pressure (Morild, 1981):

$$(\delta^2 \ln K_{eq}/\delta P^2)_T = \Delta\kappa/RT.$$

If a ln K_{eq} versus pressure plot is concave down ($\Delta\kappa > 0$), then the reactants are more compressible than the products. The opposite is true for a concave-up plot. Similar comments apply to the effects of pressure on reaction rates and the transition and ground states of reactions, with the complication that other factors can lead to nonlinear pressure effects on reaction rates. For example, a change in the rate-limiting step of a multistep reaction mechanism will result in nonlinearity, even if each step individually is linearly related to pressure.

B. Summary

The effects of pressure on biochemical processes are determined by the effects of pressure on all of the hundreds of weak bonds that contribute to protein structure, as well as the interactions of proteins with water, solutes, and other components of their cellular environment. At atmospheric pressure, volume changes associated with most biochemical reactions are negligible compared with the overall Gibbs free energy change; thus, volume effects can usually be ignored. Moderate pressures will affect equilibrium processes such as protein polymerization, and kinetic processes such as enzyme reactions. Very high pressures (>100 MPa) cause protein denaturation.

Deep-sea fishes generally have much larger ranges in depth of occurrence than do shallow-living species; thus, they are more likely to experience large changes in pressure over their life-span. A general theme appearing in biochemical studies of deep-sea fishes is that adaptation to high and variable pressures has entailed the evolution of pressure-insensitive forms of enzymes, rather than enzymes adapted for function at a specific range of high pressures.

III. TOLERANCE ADAPTATIONS: MAINTENANCE OF BIOCHEMICAL FUNCTION IN THE DEEP SEA

Organismal adaptations to the environment can be categorized as "tolerance" adaptations, which enable an organism to survive in a given environment, and "capacity" adaptations, involving regulation of rates of physiological processes at appropriate levels. In the context of deep-sea biochemistry, tolerance adaptations encompass those associated with changes in the primary structure of proteins enabling function at high pressure, as well as changes in small molecules (e.g., lipids) that affect protein stability and activity.

A. Protein–Protein Interactions

Interactions between subunits of multimeric proteins provide a relatively straightforward example of the biochemical effects of pressure, because they can be studied as simple equilibria between monomers and polymers (Weber, 1992). A study by Swezey and Somero (1985) illustrates well several aspects of the effects of pressure on protein aggregation and biochemical adaptation to pressure. The authors compared skeletal muscle actins purified from fishes living at different depths. Actin is a major structural element in muscles and in the cytoskeleton in general, for which interconversions between monomers (G-actin) and the polymerized form (F-actin) play important functional roles.

Swezey and Somero (1985) found that the association constant for polymerization of actin from *Coryphaenoides armatus*, a macrourid occurring at depths of 1900–4800 m (pressure 19–48 MPa), is relatively unaffected by pressure, with a volume change associated with polymerization of less than 10 ml/mol. By contrast, actins from shallower living species, including a congener, *Coryphaenoides acrolepis*, are much more pressure sensitive (Fig. 1). At atmospheric pressure, Swezey and Somero (1985) calculated that actin polymerization in these species results in an increase in volume of ~60 ml/mol. Even greater volume changes are associated with actins from terrestrial vertebrates: 63–139 ml/mol at atmospheric pressure. Most of the volume decrease associated with depolymerization of actin probably reflects changes in hydration and the filling-in of void volumes at the interfacial surfaces (Kornblatt *et al.*, 1993; Silva and Weber, 1993).

The responses of actin to pressure are nonlinear. Actins from *C. acrolepis* and chicken are very pressure sensitive at moderate pressures (<20 MPa), but at higher pressures they are as pressure insensitive as the *C. armatus* homolog (Swezey and Somero, 1985). Nonlinear pressure responses can have several bases, but the most straightforward explanation in the case of actin is that the monomer is more compressible than the filamentous form. In actin, we have our first example of a common theme running through studies of biochemical adaptation in the deep sea. Pressure adaptation of proteins involves evolution of pressure-insensitive homologs, not proteins optimized for function at some intermediate depth range.

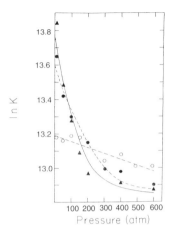

Fig. 1. Effects of pressure on actin polymerization at 4°C. The association constant, K, equals $1/C_c$, where C_c is the critical monomer concentration. Habitat pressures are plotted for different species: ○, *Coryphaenoides armatus* (19–48 MPa); ●, *Coryphaenoides acrolepis* (7–20 MPa); and ▲, chicken (0.1 MPa). Reprinted with permission from Swezey and Somero (1985). Copyright 1985 American Chemical Society.

This makes intuitive sense; deep-sea fishes may undergo extensive vertical migrations on diurnal and longer time scales (Stein and Pearcy, 1982; Stein, 1985; Wakefield and Smith, 1990). Pressure insensitivity of biochemical processes allows organisms to survive under a wider absolute pressure range than can shallow-living species. Alternatively, one might hypothesize that deep-sea fishes could synthesize depth-specific isoforms of enzymes, but the only studies addressing this question found no supporting evidence (Siebenaller, 1978, 1984a).

Hennessey and Siebenaller (1985) compared the effects of pressure on the aggregation state of the tetrameric enzyme lactate dehydrogenase (LDH) in six macrourid fishes, including five members of the genus *Coryphaenoides*. They found a correlation between depth of occurrence and the pressure at which LDH is 50% inactivated. However, their assumption that inactivation was due solely to subunit dissociation is not supported by the fact that pressure release did not result in complete recovery of activity. One possibility is that the monomers adopted a new, inactive conformation at high pressure, which would have remained inactive after reaggregation at atmospheric pressure. Weber and colleagues (Ruan and Weber, 1989, 1993; Silva and Weber, 1993) have proposed that "conformational drift" of monomers may be a general cause of loss of enzyme activity in multimeric enzymes. The fundamental idea behind this proposal is that monomeric proteins, which exist in equilibrium between native and unfolded states, can adopt inactive conformations that resemble the native conformation closely enough that polymerization can occur.

If conformational drift is an important factor in loss of the functional properties of proteins, then one would expect that proteins from deep-sea species might exhibit less tendency to unfold. In accordance with this idea, Swezey and Somero (1982a, 1985) found a close correlation between body temperature and the thermal stability of actin, across a wide range of terrestrial and shallow-water vertebrates. However, actins from two deep-sea species did not fit this pattern; they were just as resistant to thermal denaturation as homologs from mammals and the desert iguana, *Dipsosaurus dorsalis*. Swezey and Somero speculated that deep-sea proteins may generally have a greater number of weak bonds stabilizing their structure, so that adaptation to maintain integrity under high pressures has had the concomitant effect of increasing thermal stability. This has turned out not to be a general finding; eye lens proteins of the same deep-sea species are no more temperature-resistant than those of shallow-living fishes living at similar temperatures (McFall-Ngai and Horowitz, 1990). Differences in polymerization thermodynamics of actin described by Swezey and Somero (1982a) may reflect pressure adaptation, but it is clear that temperature and pressure exert differing proximate effects and selective forces on protein structure.

B. Enzyme–Substrate Interactions

Substrate binding and catalysis can involve large conformational changes in enzymes, with the simultaneous breaking and making of many weak bonds. Given the large number of bonds affected, it is not surprising that substrate binding can be extremely sensitive to pressure. This is well demonstrated in the case of dehydrogenases isolated from fish skeletal muscle. Much of this work has used the congeneric scorpaenids, *Sebastolobus alascanus* and *Sebastolobus altivelis*, which occur off the western coast of North America. These benthic species are morphologically and ecologically similar and experience similar thermal regimes. Although there is considerable overlap in their overall depth ranges, *S. alascanus* is most abundant above 500 m, whereas *S. altivelis* occurs primarily below this depth. Thus, this species pair provides an opportunity to explore pressure adaptation with minimal concerns about potential confounding factors (e.g., temperature, phylogeny, ecology).

No enzyme has been more intensively studied by environmental biochemists than LDH. Lactate dehydrogenase in fish skeletal muscle is responsible for regeneration of NAD^+ during anaerobic metabolism (e.g., burst movements) by the reduction of pyruvate to lactate. At atmospheric pressure, LDH homologs from the *Sebastolobus* species have similar affinities (apparent K_m values) for pyruvate and NADH. However, K_m values for the *S. alascanus* homolog increase rapidly with pressure up to 6.9 MPa, corresponding to a depth of 680 m (Siebenaller and Somero, 1978) (Fig. 2). The K_m for NADH approximately doubles in this range and continues to rise up to 20 MPa. The LDH homolog from *S. altivelis* is much less sensitive to pressure; binding of pyruvate is unaffected, and the K_m for NADH increases by approximately one-third at 6.8 MPa and remains constant at higher pressures. Maximal activities (V_{max}) are relatively unaffected by pressure, but LDH from *S. altivelis* is slightly less inhibited (11% loss

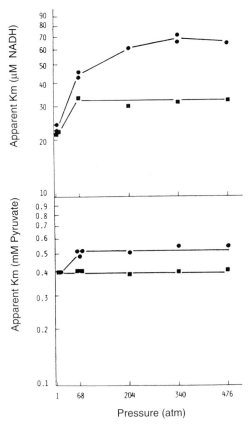

Fig. 2. Effects of pressure on the apparent K_m values for pyruvate and NADH of the *Sebastolobus* LDH homologs, measured at 5°C. ●, *S. alascanus*; ■, *S. altivelis*. From Siebenaller and Somero (1978).

6. BIOCHEMISTRY AT DEPTH

of activity at 34 MPa vs. 17% activity loss for *S. alascanus* LDH) (Siebenaller and Somero, 1979).

Under physiological conditions, the effects of pressure on V_{max} of LDH are minor. Changes in substrate-binding properties are more likely to have significant organismal effects, because *in vivo* substrate levels are below saturating. The comparison of LDH kinetic properties in *Sebastolobus* suggests that environmental pressures as low as 5 MPa have been sufficient to select for homologs for which substrate binding is relatively unaffected by pressure. This is only one enzyme from two species, however. Garland and Adolph (1994) have pointed out the dangers of such limited comparisons. To assess the generality of pressure adaptation of enzyme catalytic properties, what is needed is information about additional enzymes from a greater variety of species.

Other dehydrogenases from the *Sebastolobus* congeners exhibit differences in their responses to pressure similar to those of LDH (Siebenaller, 1984b). These include glyceraldehyde 3-phosphate dehydrogenase (GAPDH) and two isozymes of cytoplasmic malate dehydrogenase (MDH). Each of these enzymes, like LDH, is a Rossmann fold dehydrogenase, containing a structurally similar cofactor binding site. In each case, the K_m for coenzyme increases with pressure for the homolog from *S. alascanus*, whereas that of the deeper-living congener is not affected (Fig. 3). Thus, there appears to have been convergent evolution of pressure responses in multiple Rossmann fold dehydrogenases in the *Sebastolobus* congeners.

Do other species exhibit similar patterns of pressure adaptation? Unfortunately, convenient species pairs (or better still, species groups) such as the *Sebastolobus* spp. are rare. However, a broad comparison of LDH homologs isolated from fishes in several families, from shallow water and the deep sea, suggests that high pressure has repeatedly selected for pressure-insensitive forms (Siebenaller and Somero, 1979). In an even broader phylogenetic context, Dahlhoff and Somero (1991) studied the effects of pressure on MDH in 15 species from four invertebrate phyla. They found that, for 10 species occurring at habitat pressures greater than 5–10 MPa (depths of 500–1000 m), the K_m for NADH was unaffected by pressure, whereas MDHs from all five shallow-living species exhibited higher K_m values above atmospheric pressure. Thus, broad phylogenetic patterns mirror those found in *Sebastolobus*, consistent with an adaptive explanation for differences in pressure responses.

Unlike most deep-sea fishes, species living near the hydrothermal vents may sometimes be exposed to warm water. High temperatures increase K_m values for LDH (Somero, 1995). Dahlhoff *et al.* (1990) compared the effects of temperature and pressure on the kinetic properties of LDH from two vent fishes (a bythitid, *Bythites hollisi*, and a zoarcid, *Thermarces andersoni*)

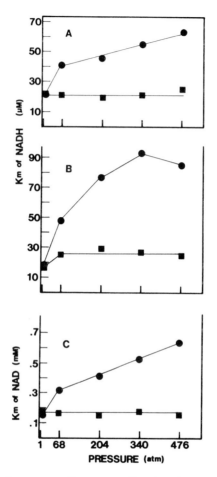

Fig. 3. The effects of pressure on the apparent K_m values for coenzyme binding of NAD-dependent dehydrogenases from *S. alascanus* (●) and *S. altivelis* (■). The enzymes studied were (A) MDH-1, (B) MDH-2, and (C) GAPDH. All assays at 5°C. From *J. Comp. Physiol. B.*, Pressure-adaptive differences in NAD-dependent dehydrogenases of congeneric marine fishes living at different depths. J. F. Siebenaller, **154**, 443–448, Fig. 1, 1984. Copyright Springer-Verlag.

and a rattail, *C. armatus,* which is common at similar depths away from the vents. At 5°C, the apparent K_m for NADH for all three LDH homologs was unaffected by pressures beyond the physiological range (25 MPa). At higher temperatures, the K_m for the rattail and bythitid enzymes increased significantly with pressure. In contrast, the kinetic properties of the zoarcid enzyme were relatively unaffected by temperature–pressure combinations

6. BIOCHEMISTRY AT DEPTH   251

up to 20°C and 34 MPa. Based on reports of observations from submersibles, Dahlhoff *et al.* (1990) hypothesized that *T. andersoni* experiences higher temperatures than do the other vent species, and has evolved an LDH homolog adapted to both high temperature and high pressure. Similar conclusions were reached in a study of invertebrate malate dehydrogenases (Dahlhoff and Somero, 1991). Homologous MDHs from species found in warm hydrothermal vent waters had temperature- and pressure-insensitive kinetic properties.

Another example of the potential importance of protein–ligand interactions in the deep sea comes from studies of hemoglobin and myoglobin. Although they are not enzymes, the thermodynamic principles governing the properties of oxygen-binding proteins are identical to those of enzymes. Deep-sea fishes face three challenges with regard to hemoglobin function. First, in many areas of the ocean, biological and abiotic factors contribute to the establishment of zones of extremely low oxygen (Childress, 1995). At the organismal level, fishes and invertebrates living in oxygen-minimum layers are able to regulate O_2 consumption at very low environmental oxygen levels, partly by having hemoglobin with relatively high affinities for O_2 (Sanders and Childress, 1990; Yang *et al.*, 1992). Second, fishes with swim bladders fill them with gas mixtures highly enriched with O_2, and the percentage of oxygen in the swim bladder increases with depth (Pelster and Scheid, 1992). Hemoglobins from some deep-sea fishes containing swim bladders exhibit extremely large Root effects (Noble *et al.*, 1986), which would be of value for secretion of O_2 into the swim bladder at high pressure (Pelster and Scheid, 1992). Third, pressure will affect oxygen binding. The large conformational changes associated with O_2 binding and release suggest that hemoglobin may be an inherently pressure-sensitive molecule.

The only studies of pressure effects on hemoglobin or myoglobin have used mammalian homologs. Early research indicated that the net change in volume associated with O_2 binding is negligible (Johnson and Schlegel, 1948), so that pressure will have little effect on the affinity for oxygen. On the other hand, Ogunmola *et al.* (1976) found that azide and other charged molecules bind the heme group of myoglobin with a net volume change of −10 ml/mol (i.e., binding is favored at high pressures). One might expect binding of a hydrophobic molecule such as oxygen to have an even more negative volume change. Charged molecules such as azide are surrounded by a compact, structured shell of water molecules. Binding of azide should result in the dispersal of this shell into a higher volume bulk phase, tending to make the net volume change less negative. Because dissolved oxygen does not have this compact layer, the net volume change on binding should be more negative than for azide, and high pressure should increase the affinity of myoglobin for oxygen. The discrepancy between this prediction and the available literature data is likely due to differences in techniques used, and more direct comparisons need to be done.

Kinetic analyses have shown that oxygen and carbon monoxide binding to the heme moiety of hemoglobin is faster at high pressures (i.e. the apparent activation volume is negative) (Unno et al., 1990, 1991). Similar results have been obtained with myoglobin (Adachi and Morishima, 1989). It is interesting to note that myoglobin from sperm whale is less affected by pressure than are the dog or human homologs (Adachi and Morishima, 1989). Sperm whales are known to dive to depths of at least 1100 m (pressure 11 MPa) (Heezen, 1957), and so must sometimes experience high hydrostatic pressures, but the data are too limited to say whether sperm whales have evolved a pressure-insensitive myoglobin. Noble et al. (1986) found that hemoglobins from deep-sea, swim-bladder-containing fishes exhibited biphasic CO binding kinetics and low cooperativity, and suggested that these properties were associated with the enhanced Root effect seen in these species. Unfortunately, no high-pressure studies of O_2 binding proteins from deep-sea fishes have been performed, but the detailed knowledge available on the structure and function of vertebrate oxygen-binding proteins, including homologs from fish (Mylvaganam et al., 1996), would seem to make these proteins ideal subjects for future mechanistic studies of pressure adaptation.

C. Membrane Proteins and Lipid–Protein Interactions

The structural and functional properties of proteins depend on the microscopic environment of the protein. For example, differences in ionic strength or ion composition can significantly affect the pressure responses of enzymes through changes in protein hydration (Low and Somero, 1975b). Deep-sea fishes osmoregulate in a manner similar to shallow-water species (Blaxter et al., 1971; Shelton et al., 1985), but potential interacting effects of pressure and solute composition on enzyme function have received little attention. Instead, the importance of the protein microenvironment at high pressures is best demonstrated by studies of membrane processes in deep-sea fishes.

Cell membranes are strongly perturbed by pressure, and a variety of behavioral evidence suggests that the pressure tolerance limits of organisms are determined by the effects of pressure on membrane function. These conclusions are largely based on the similar counteracting effects of temperature and pressure on behavioral, cellular, and membrane phenomena. For example, the cellular effects of high pressure can often be reversed by increasing temperature (Wann and Macdonald, 1980). The apparent importance of pressure effects on membrane processes have made lipid–protein interactions the subject of numerous biochemical and biophysical studies.

Pressure affects membranes by compressing the bilayer laterally, so that the membrane actually becomes thicker (Stamatoff et al., 1978; Braganza and Worcester, 1986). The surface area per phospholipid molecule decreases, and the closer packing of acyl chains results in reduced molecular mobility and greater van der Waals interactions. Membrane fluidity, as measured by numerous techniques, is thereby reduced. These effects are very similar to those of reduced temperature. For many membrane biophysical properties, the effects of a 100-MPa (1000-atm) pressure increase are equivalent to those exerted by a 15–30°C decrease in temperature (Macdonald, 1984). Thus, as far as membrane properties are concerned, conditions at the bottom of the Marianas Trench (11,000 m depth, 110 MPa, 4°C) are equivalent to about −20°C at atmospheric pressure.

One might predict that organisms would respond to the membrane-ordering effects of high pressure with changes in membrane lipid composition and fluidity mirroring those at low temperatures (Macdonald and Cossins, 1985; Cossins and Macdonald, 1989). This appears to be the case; species from greater depths have mitochondrial membranes containing a greater proportion of unsaturated fatty acids (Phleger and Laub, 1975; Avrova, 1984; Cossins and Macdonald, 1986) (Fig. 4), and brain myelin

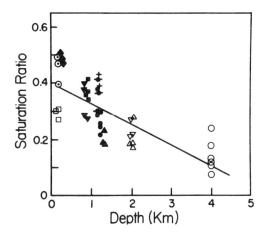

Fig. 4. Relationship between depth of capture and the saturation ratio for phosphatidylethanolamines prepared from liver mitochondrial membranes. The saturation ratio is the ratio of the weight percentages of saturated fatty acids to unsaturated fatty acids. Note that each symbol represents a different species of North Atlantic fish. Reprinted from *Biochim. Biophys. Acta* **860**, A. R. Cossins and A. G. Macdonald. Homeoviscous adaptation under pressure: III. The fatty acid composition of liver mitochondrial phospholipids of deep-sea fish, 325–335. Copyright 1986, with kind permission of Elsevier Science-NL, Sara Burgerhartstraat 25, 1055 KV Amsterdam, The Netherlands.

membranes are more fluid in deep-sea fishes than in shallow-living species (Cossins and Macdonald, 1984; Behan et al., 1992).

Many fishes undergo large vertical migrations, from diurnal movements of hundreds of meters to longer term migrations of thousands of meters (Stein and Pearcy, 1982; Stein, 1985; Wakefield and Smith, 1990). Every 1000 m increase in depth will affect membrane physical properties to the same extent as a temperature decrease of ~2°C. A common organismal response to reduced temperature is an increase in lipid unsaturation and membrane fluidity (Hazel and Williams, 1990), due to desaturase activation and biosynthesis (Tiku et al., 1996). Membrane acclimation to pressure (e.g., comparisons of conspecifics collected at different depths) has not been examined in fishes. Bacteria have been shown to increase the proportion of unsaturated fatty acids at higher growth pressures (DeLong and Yayanos, 1985, 1986; Wirsen et al., 1987; Kamimura et al., 1993), but the effects of these changes on membrane physical properties have not been assessed. Kaneshiro and Clark (1995) found that the deep-sea thermophile *Methanococcus jannaschii* exhibits pressure-dependent changes in the proportions of three isopranoid ether lipids. Despite the unusual composition of these archaebacterial membranes, temperature and pressure had opposing effects on lipid physical properties, of a magnitude similar to phospholipid membrane systems: 20°C per 100 MPa.

An important question regarding pressure adaptation of membrane processes is whether differences in membrane composition and fluidity really have an effect on membrane functional properties. Two types of evidence have been taken as supporting this hypothesis. Unfortunately, both are correlative in their approach and do not provide strong evidence for the role of homeoviscous adaptation.

The first approach relies on the observation that many membrane enzymes exhibit nonlinear pressure dependence. Plots of the natural log of the activity versus pressure can frequently be fitted to two lines, and sometimes the breakpoint is reasonably close to the breakpoint for some measured physical property of the membrane (e.g., fluidity, fluid–gel phase transition). This has been taken as evidence that membrane lipid properties determine the activities of certain membrane enzymes (Ceuterick et al., 1978; Heremans and Wuytack, 1980). Similar approaches have been applied to nonlinear Arrhenius plots, and objections raised there apply equally in the case of pressure. A biphasic plot is often statistically unjustified (Silvius and McElhaney, 1981); a continuous curve is an equally good fit. In addition, nonlinear relationships can arise from several causes besides changes in membrane properties (Klein, 1982): a change in the rate-limiting step of a multistep reaction, a nonzero heat capacity of activation (e.g., a difference in heat capacity between the transition and ground states), phase separation

6. BIOCHEMISTRY AT DEPTH

of the membrane, etc. These concerns are borne out by the fact that even monomeric soluble enzymes can exhibit nonlinear responses to pressure (Gross et al., 1993).

The second line of reasoning relies on the fact that similar counteracting effects of temperature and pressure (20°C vs. 100 MPa) are exhibited by a variety of membrane enzymes (Macdonald, 1984). This idea has even been extended to organismal levels (Airriess and Childress, 1994). However, there is no reason why other biochemical processes could not exhibit similar temperature and pressure effects. These correlations are merely suggestive, and additional experimental evidence is required to demonstrate membrane adaptation.

Given these difficulties, is there other evidence for homeoviscous adaptation to pressure? The strongest evidence comes from the sodium pump, Na^+,K^+-adenosine triphosphatase (Na^+,K^+-ATPase) (Gibbs, 1995), which plays an important role in osmoregulation in marine fishes. Comparison of pressure responses of gill Na^+,K^+-ATPase activities in fish living under different temperature and pressure regimes reveals a correlation between presumed membrane fluidity and the degree of inhibition by pressure of the enzyme (Gibbs and Somero, 1989) (Fig. 5). The order of increasing sensitivity to pressure matches the order of decreasing expected membrane fluidity: deep sea, cold < deep sea, warm (hydrothermal vent) < shallow,

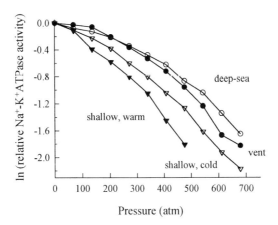

Fig. 5. Effects of pressure on gill Na^+,K^+-ATPase activities in fishes from different habitats. Deep-sea species include species living at 2–4°C at depths exceeding 2000 m; vent fishes include two species occurring near warm hydrothermal vents; shallow, cold indicates eastern Pacific fishes found at depths of less than 2000 m; and shallow, warm species were from surface waters near Hawaii. All assays at 10°C. Data from Gibbs and Somero (1989), with permission of the Company of Biologists Ltd.

cold < shallow, warm. In this case, "deep sea" indicates species occurring at depths greater than 2000 m. No differences are observed among species found at shallower depths. The effects of pressure are nonlinear, becoming greater at higher pressures, but there is no evidence of a breakpoint. Thus, the apparent activation volume increases with pressure, although $\Delta V^{\ddagger}$ values are similar (30–60 ml/mol) in all species at their respective habitat pressures. Note that the kinetic parameter of interest here is the maximal activity. Unlike the dehydrogenases, the affinity of Na^+,K^+-ATPase for two of its substrates, ATP and sodium ion, is unaffected by pressure up to 48 MPa, and does not differ among fishes from different depths (Gibbs and Somero, 1989).

A potential phylogenetic problem arises in this case. The "deep-sea" species comprise two congeners, *C. armatus* and *C. leptolepis*. The explanation that members of the genus *Coryphaenoides* generally have pressure-insensitive Na^+,K^+-ATPases is contradicted by the fact that the homolog from *C. acrolepis*, a shallower living species, exhibits pressure dependence similar to that of other fishes from its depth range. The phylogenetic argument is still not answered; Wilson *et al.* (1991), using peptide mapping of LDH, found that *C. armatus* and *C. leptolepis* were more closely related to each other than to other members of the genus. Thus, one can not rigorously distinguish pressure adaptation from phylogenetic relatedness in this example. Additional evidence for pressure adaptation comes from the observation that Na^+,K^+-ATPase also exhibits reduced pressure sensitivity in two deep-sea hydrothermal vent fishes, including a zoarcid and a bythitid (Gibbs and Somero, 1989). However, no shallow-living members of these families were studied, and the scarcity of specimens resulted in only a single individual being assayed for each of the vent species.

Both the correlation between pressure dependence of Na^+,K^+-ATPase and presumed membrane fluidity and the conservation of $\Delta V^{\ddagger}$ values at physiological pressures are consistent with the hypothesis that membrane lipid properties are responsible for both interspecific differences and nonlinear pressure dependence (Gibbs and Somero, 1989). Additional support comes from measurements of both membrane fluidity and Na^+,K^+-ATPase activity. For both parameters, the effects of a pressure increase of 50 MPa can be offset by increasing the temperature by 10°C. Thus, temperature–pressure combinations giving the same membrane fluidity also result in the same Na^+,K^+-ATPase activity (Chong *et al.*, 1985; Gibbs and Somero, 1990a). The concerns expressed previously apply here as well; in the absence of additional experimental evidence, the correlation between membrane fluidity and Na^+,K^+-ATPase activity is merely suggestive of a causal relationship.

In a more direct test of the role of membrane adaptation, Gibbs and Somero (1990a) used a lipid replacement procedure to change the membrane environment of Na^+,K^+-ATPases from shallow-living and deep-sea fish. Membrane fluidity was not measured after treatment, but Na^+,K^+-ATPase was less inhibited by pressure when placed in a membrane environment containing phospholipids from deep-sea fish, and more pressure-sensitive in less fluid membranes (Fig. 6). Thus, it can be concluded that the pressure dependence of Na^+,K^+-ATPase is partly determined by its membrane milieu. However, when homologs from different species were placed in the same membrane environment, Na^+,K^+-ATPase from *C. armatus* remained less pressure inhibited than did Na^+,K^+-ATPase from sablefish

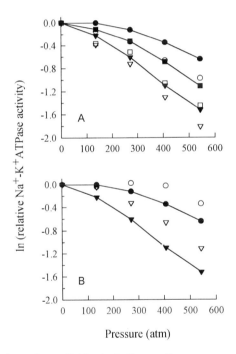

Fig. 6. Effects of membrane lipid substitution on the pressure responses of Na^+,K^+-ATPase. Native lipids were removed by gentle detergent treatment and replaced with (A) chicken egg phosphatidylcholine or (B) phospholipids prepared from gills of *Coryphaenoides armatus*. Filled symbols indicate pressure responses before lipid substitution; open symbols were assays after substitution. ●, *Coryphaenoides armatus* (deep sea, cold); ■, *Anoplopoma fimbria* (shallow, cold); ▲, *Sphyraena barracuda* (shallow, warm). All assays performed at 17.5°C. Data from Gibbs and Somero (1990a). From *J. Comp. Physiol. B,* Pressure adaptation of teleost gill Na^+,K^+-adenosine triphosphatase: Role of the lipid and protein moieties, A. Gibbs and G. N. Somero, **160,** 431–439, Figs. 6 and 7, 1990. Copyright Springer-Verlag.

(*Anoplopoma fimbria*) and barracuda (*Sphyraena barracuda*) (Fig. 6). This demonstrates that membrane lipid changes alone are not responsible for pressure adaptation of the sodium pump, but that primary structure differences are also involved (Gibbs and Somero, 1990a; Gibbs, 1995).

D. Integrating Cellular Processes: G-Protein-Mediated Signal Transduction

The examples of biochemical adaptation to pressure described thus far involve individual proteins studied in isolation. Cellular processes require the integrated function of many biochemical steps (e.g., 10 enzymes are involved in glycolysis). Pressure perturbation of any of these can disrupt the entire pathway; alternatively, relatively minor effects on each step may lead to cumulative effects on the overall process. Only one such biochemical process has received significant experimental attention: Siebenaller and Murray (1995) performed numerous studies of the effects of pressure on G-protein-mediated signal transduction, using the *Sebastolobus* congeners as their primary study system. This system involves each of the biochemical interactions outlined above: protein–protein interactions, substrate binding, and lipid–protein interactions, and thus serves as a useful model for each of these separately and for integration of cellular processes at high pressure.

Siebenaller and Murray (1995) have concentrated on the A_1 adenosine receptor–G protein–adenylyl cyclase pathway, which has important roles in nervous system function. The first step in the process is the binding of adenosine to the A_1 receptor in the plasma membrane. This causes a conformational change in the receptor to stabilize binding of a heterotrimeric inhibitory guanine nucleotide-binding protein (G_i). The receptor–G_i interaction increases the affinity of the α subunit of G_i for guanosine triphosphate (GTP), causing it to dissociate from the complex. The α subunit can then bind and inhibit adenylyl cyclase. The net effect is that adenosine binding to the A_1 receptor results in decreased levels of cyclic adenosine monophosphate (cAMP). Although G proteins are involved in many cellular processes in addition to adenylyl cyclase inhibition, the effects of pressure have been studied only in this context. The complexity of the system has made it difficult to distinguish exactly which steps are most affected by pressure or are pressure adapted, but the evidence to date indicates that environmental pressures of 5 MPa have significant effects on signal transduction pathways and may have been sufficient to select for pressure-adapted components of the pathways.

The role of this signal transduction pathway is to modulate intracellular cAMP levels by changing the activity of adenylyl cyclase, thus the pressure responses of adenylyl cyclase are of great importance. The direct effects

6. BIOCHEMISTRY AT DEPTH

of pressure on this enzyme are similar in the *Sebastolobus* species (Siebenaller *et al.*, 1991) (Fig. 7). Activity is significantly decreased by application of 13.7 MPa, and at 41 MPa activity is reduced by more than one-third. Adenylyl cyclase activity from a deeper living morid cod, *Antimora rostrata*, is unaffected by 27.5 MPa, suggesting that higher habitat pressures may select for pressure-insensitive homologs (Siebenaller and Murray, 1990).

Although maximal adenylyl cyclase activity in the *Sebatolobus* congeners exhibits similar pressure responses, the effects of pressure on affinity for substrate may differ. For technical reasons, the ATP analog 2-deoxy-ATP has been used in these assays. The apparent K_m for 2-deoxy-ATP increases with pressure in both species, but is more pressure sensitive in the shallower living *S. alascanus* (Siebenaller *et al.*, 1991). To the extent that this ATP analog serves as a valid substitute for ATP, this finding is

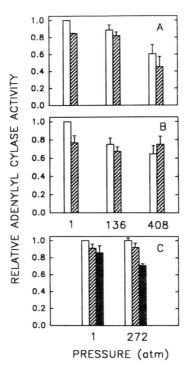

Fig. 7. The effects of pressure on adenylyl cyclase activity in brain membranes from (A) *Sebastolobus alascanus*, (B) *Sebastolobus altivelis*, and (C) *Antimora rostrata*. Open bars, no added agonist; hatched bars, 100 μM CPA; filled bars, 100 μM NECA. Data are normalized to basal activity at atmospheric pressure and 5°C. CPA and NECA are A_1 adenosine receptor agonists. Data from Siebenaller and Murray (1990) and Siebenaller *et al.* (1991).

similar to the pattern observed for substrate binding in the dehydrogenases. The enzyme from the deeper living species, *S. altivelis,* is relatively pressure-insensitive.

The G-protein-mediated coupling of the A_1 adenosine receptor to adenylyl cyclase has been studied using an A_1 receptor-specific agonist, N^6-cyclopentyladenosine (CPA) (Siebenaller *et al.,* 1991). Addition of CPA to brain membranes from *Sebastolobus* spp. inhibits adenylyl cyclase, indicating that an intact G-protein-mediated pathway exists in these fishes. However, the extent of inhibition is reduced at 13.7 MPa, suggesting that pressure disrupts one or more steps: agonist binding to the A_1 receptor or G protein interactions with either the receptor or adenylyl cyclase. It is impossible to distinguish which (or all) interactions are affected by pressure. In membranes prepared from *A. rostrata,* 27.5 MPa did not affect the efficacy of CPA (Siebenaller and Murray, 1990). Thus, it appears that this deeper living species has evolved a pressure-insensitive signal transduction pathway, although the precise mechanism by which this has been achieved is still unknown.

The effects of pressure on G protein polymerization state and interactions with the receptor have been studied using pertussis toxin. Pertussis toxin catalyzes the transfer of an ADP-ribosyl moiety to a cysteine residue of the α subunit of both inhibitory and stimulatory G proteins, but only when they are in the heterotrimeric form (Neer *et al.,* 1984). In the presence of guanosine diphosphate (GDP), which stabilizes the trimer, labeling of α subunits from *S. alascanus* is reduced by half at 21 MPa, whereas no differences are observed for the homolog from *S. altivelis* up to 35 MPa (Siebenaller and Murray, 1994a) (Fig. 8). These results are consistent with the idea that G protein subunit interactions are more disrupted by pressure in the shallower living species. Alternatively, receptor–G protein interactions may be stabilized, so that the α subunit is inaccessible to pertussis toxin.

An important regulatory role in this pathway is played by the high-affinity (low K_m) GTPase activity of the G protein α subunit; hydrolysis of bound GTP causes the deinhibition of adenylyl cyclase. Pressure has direct stimulatory effects on GTPase activity, by increasing V_{max} and reducing K_m (Siebenaller and Murray, 1994b). Pressure also has indirect effects; high pressures reduce the stimulation of GTPase activity by A^1 receptor agonists. These effects are similar in the *Sebastolobus* congeners, and no apparent pressure-adaptive differences are evident.

Because both the A_1 adenosine receptor and adenylyl cyclase are integral membrane proteins, one might expect changes in membrane lipids to play a role in pressure adaptation. However, no significant differences in composition have been found between brain membranes in the *Sebastolobus* species (Siebenaller *et al.,* 1991). The maximal pressures experienced

6. BIOCHEMISTRY AT DEPTH

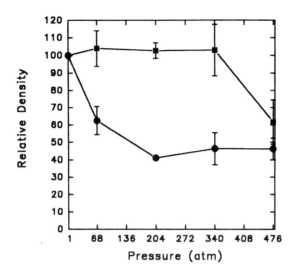

Fig. 8. Effects of pressure on pertussis toxin-catalyzed ADP-ribosylation of G protein α subunits, in the presence of 100 μM GDP, measured at 5°C. ●, *Sebastolobus alascanus;* ■, *Sebastolobus altivelis.* Normalized to atmospheric values. Reprinted by permission of the publisher from The effects of hydrostatic pressure on pertussis toxin-catalyzed ribosylation of guanine nucleotide-binding proteins from two congeneric marine fish. J. F. Siebenaller and T. F. Murray. *Comp. Biochem. Physiol. B* **108,** 423–430. Copyright 1994 by Elsevier Science Inc.

by these fish (up to 16 MPa) would be expected to affect membrane properties about as much as a temperature decrease of only 3–4°C, so a role for homeoviscous responses in pressure adaptation of signal transduction may not be detectable if it exists.

Interactions between the α subunit and pertussis toxin have also been studied in the presence of GTPγS, a nonhydrolyzable GTP analog that promotes dissociation of the α subunit from the trimer. Whereas GTPγS reduces ADP-ribosylation in membranes from *S. alascanus,* it has little effect on the *S. altivelis* homolog (Murray and Siebenaller, 1993). However, pressure inhibits ADP-ribosylation to a similar extent in both species. The explanation for this behavior remains unclear. Receptor–G protein interactions, binding of guanine nucleotides, and G protein subunit interactions could all have an effect.

Studies of the A_1 receptor–G protein–adenylyl cyclase system are revealing in a number of ways. Physiologically relevant pressures affect several steps, and some of these exhibit interspecific differences consistent with adaptation to pressure (K_m of adenylyl cyclase for 2-deoxy-ATP; ADP-ribosylation by pertussis toxin in the presence of GDP). Other parts of the signal transduction pathway do not appear to differ between the *Sebastolobus* congeners, in spite of the fact that pressure may disrupt them (GTPase activity of G proteins; membrane lipids). For some parameters, apparent pressure adaptation is only seen in a deeper living species, *A. rostrata* (adenylyl cyclase activity; G-protein-mediated coupling between the A_1 receptor and adenylyl cyclase). One interpretation is that hydrostatic pressure exerts a greater selective force on certain components of the system than others.

An alternative explanation concerns the species studied. The *Sebastolobus* spp. are very closely related; in an electrophoretic survey, 10 of 20 enzyme loci appeared identical (Siebenaller, 1978). Also, average heterozygosities were less than 5%, a low value for fishes. Selection for pressure-adapted proteins can not occur in the absence of genetically based variation. Lack of interspecific differences does not imply that pressure effects are not important, if hydrostatic pressure has no variation on which to select. Additionally, any observed differences are subject to criticism on the grounds that they may reflect a chance difference between founding populations, linkage effects, or genetic correlations among characters (Garland and Carter, 1994). This concern is especially important when there is no clear adaptive explanation for a phenomenon (e.g., differences in ADP-ribosylation in the presence of GTPγS). Thus, although the *Sebastolobus* species pair has proved useful for understanding pressure adaptation of some proteins (especially the Rossmann fold dehydrogenases), they may be too closely related to reveal biochemical adaptation in some cases, and some differences may be spurious.

E. Summary

Pressure has significant effects on enzyme and protein function. The properties subject to selection differ from one protein to another. Apparent K_m values appear to be pressure adapted for adenylyl cyclase and the Rossmann fold dehydrogenases, whereas Na^+,K^+-ATPase exhibits pressure adaptation of maximal activity. A general pattern is that homologous proteins from deep-sea fishes exhibit insensitivity to pressure, not optimization for a particular habitat pressure. This makes intuitive sense; deep-sea species encounter much larger absolute pressure ranges than do shallow-living fishes, and there is no evidence for production of different isoforms at different pressures (Siebenaller, 1978, 1984a). The environmental pressure apparently necessary to select for pressure-adapted homologs differs among proteins. Pressures as low as 5 MPa appear sufficient to select for pressure insensitivity of dehydrogenases and some aspects of signal transduction, whereas Na^+,K^+-ATPase and actin exhibit no differences among species living at pressures less than 20 MPa. Also important for protein function is the intracellular environment. Membrane lipids exert significant effects on the pressure responses of integral membrane proteins, and protein function requires maintenance of a proper membrane lipid environment. Finally, signal transduction pathways, which depend on the coordinated function of several proteins, exhibit apparent pressure adaptation of some, but not all, steps. This may reflect either the lack of importance of pressure effects on a particular process, or the limited number of species examined so far.

IV. CAPACITY ADAPTATION: BIOCHEMICAL CORRELATES OF ORGANISMAL METABOLISM

The mechanisms of biochemical adaptation previously discussed are important in allowing deep-sea fishes to survive at high and variable pressures. Fishes must also regulate their overall metabolism in response to environmental factors other than pressure, for instance temperature. Several researchers have observed a depth-related decline in metabolic rates in fishes (Torres *et al.*, 1979; Donnelly and Torres, 1988; Torres and Somero, 1988a,b) and in midwater crustaceans (Childress, 1975; Childress *et al.*, 1990a; Cowles *et al.*, 1991). Metabolic rates may be 15- to 20-fold lower in individuals collected 1 km below the surface, with a continued but less pronounced decrease at greater depths. This has fostered the idea that deep-sea fishes are generally sluggish, sit-and-wait predators, although a few species may forage actively (Priede *et al.*, 1990, 1991). The causes of reduced metabolism have been difficult to distinguish, for several reasons. Important environmental variables exhibit concurrent changes with depth: pressure, temperature, light, biomass (food availability), oxygen concentration, etc. Metabolic measurements have been performed under a wide variety of conditions, including *in situ* (Smith and Hessler, 1974; Smith, 1978; Smith and Laver, 1981; Smith and Brown, 1983), on shipboard at 1 atm (Torres *et al.*, 1979; Donnelly and Torres, 1988; Torres and Somero, 1988a), and in hyperbaric chambers (Belman and Gordon, 1979). Moreover, additional work suggests that any patterns may be taxon specific, and many studies have had significant phylogenetic biases. Childress (1995) and Childress and Thuesen (1995) have reviewed this field and evaluated competing hypotheses. Their reviews provide a much more complete analysis than is possible here; I will briefly outline their main conclusions and discuss the biochemical evidence as it relates to metabolic rates in deep-sea fishes.

A. Depth-Related Patterns in Metabolic Rates

Research on fishes has been biased toward midwater and benthopelagic species. These and midwater crustaceans exhibit reduced metabolic rates relative to shallow-water species, whereas benthic organisms may not (Smith, 1983). Some pelagic invertebrate groups also do not exhibit any apparent trends in metabolic rates (Thuesen and Childress, 1993, 1994). Several factors have been proposed to be responsible for reduced metabolic rates in midwater fishes: high pressure, low oxygen, low temperature, low food availability, and reduced ambient light levels.

Direct effects of pressure can be ruled out as the primary cause of reduced metabolic rates and enzyme activities. The largest gradient in physiological parameters is observed in the upper few hundred meters (Childress, 1995), where habitat pressures are only a few megapascals. These moderate pressures generally have no measurable effects on metabolic rates of marine species (Belman and Gordon, 1979; Childress and Thuesen, 1993; but see Bailey et al., 1994). Maximal activities of most enzymes are only slightly affected, although substrate binding may be disturbed (see above). Oxygen levels can also be discounted, because the decline in metabolic rates continues steadily through oxygen minimum zones to greater depths (Childress and Thuesen, 1995).

Temperature differences of only a few degrees Celcius significantly affect both metabolic rates and enzyme activities. In most places in the ocean, temperature also decreases rapidly in the upper few hundred meters. Torres et al. (1979), working in the eastern temperate Pacific, concluded that higher temperatures in shallow water explained little of the depth-related variation in metabolism of midwater fishes. This study, however, suffered from a potential phylogenetic artifact. Species were distinguished on the basis of their minimal depth of occurrence; thus, species undergoing diurnal vertical migrations near the surface were treated as being shallow-living species. Five of the seven shallow-occurring species studied were vertically migrating myctophids, and all had relatively high metabolic rates. A later study by Donnelly and Torres (1988) noted that myctophids had higher metabolic rates than did vertical migrators from other fish taxa (Fig. 9). They concluded that temperature effects alone could account for the depth-related decrease in metabolism of midwater fishes in the Gulf of Mexico, where the thermal gradient between the surface and deeper water is more pronounced than in the eastern Pacific. In order to minimize the effects of temperature, Torres and Somero (1988a) measured metabolic rates in fishes from a nearly isothermal water column in the Antarctic. They found a depth-related decline in metabolism similar to that found in other regions, suggesting that temperature was not driving depth-related differences in metabolism (Torres and Somero, 1988b). In summary, the effects of temperature are pervasive, important, and poorly understood. Temperature surely contributes to the depth-related decline in metabolic rates, but is not the sole cause.

Another hypothesis for the depth-related decrease in metabolic rates concerns the (lack of) availability of food. For any organism to survive, grow, and reproduce, it must acquire more energy than it consumes in metabolism. The deep sea is a food-limited (i.e., energy-limited) environment, thus one might expect that deep-living fishes might minimize energetic requirements by reducing the rates of energy-consuming processes. However, one would then expect that other phyla would exhibit similar patterns. This is true for crustaceans, but not for other invertebrate taxa (Thuesen and Childress, 1993, 1994). In addition, one would expect to find the lowest metabolic rates in areas with very low primary productivity. A comparison of metabolic rates in midwater crustaceans (whose depth-related pattern is similar to that of fishes) from near California and near Hawaii found the reverse situation. Crustaceans from the lower productivity waters near Hawaii actually had higher metabolic rates than did those near California (Childress, 1975; Cowles *et al.*, 1991). A confounding factor in testing the food limitation hypothesis is that metabolic rates are dependent on the recent feeding history (Sullivan and Somero, 1983; Yang and Somero, 1993).

Childress and Thuesen have argued that predator–prey relationships, particularly visual interactions, are a major cause of reduced metabolism in deep-living fishes (Childress, 1995, Childress and Thuesen, 1995). In this view, reduced light levels have resulted in relaxed selection for locomotory capabilities, since potential prey or predators will be detected only at short distances. Evidence in support of this idea comes from the fact that midwater crustaceans, which, like fish, appear to be primarily visual predators, also have reduced metabolic rates. Taxa with less well-developed eyes (e.g. chaetognaths, medusae) do not exhibit depth-related declines in metabolism (Thuesen and Childress, 1993, 1994). Within benthic crustacea, only the visually-oriented caridian decapods have reduced metabolic rates in deep-living species (Childress *et al.*, 1990a). One is left to conclude that depth-related reductions in metabolic rates of fishes probably result from a combination of the effects of temperature, food availability, and light regime, whose relative importance may differ from one region to another.

B. Biochemical Consequences of Reduced Metabolism

How is the depth-related reduction in metabolic rate achieved, and can biochemical analyses shed light on the causes of this pattern? One of the most energetically demanding cellular processes is protein synthesis, so one might expect that the amount of protein would be reduced in deep-living fishes, or that protein turnover rates would be greatly reduced. Consistent with this idea, deep-sea fishes have higher body water contents and reduced protein levels (Childress and Nygaard, 1973; Torres *et al.*, 1979; Siebenaller *et al.*, 1982; Siebenaller and Yancey, 1984; Childress *et al.*, 1990b; Donnelly *et al.*, 1990).

Reduced protein levels are not due to a generalized decrease in the amounts of all proteins. Actin levels in skeletal muscle exhibit no depth-related changes (Swezey and Somero, 1982b; Siebenaller and Yancey, 1984), whereas several researchers have found a depth-related decline in activities of metabolic enzymes in fishes (Childress and Somero, 1979; Sullivan and Somero, 1980; Siebenaller and Somero, 1982; Torres and Somero, 1988a; Vetter *et al.*, 1994). This pattern is tissue specific; brain enzyme levels do not change with depth (Sullivan and Somero, 1980). Because skeletal muscle accounts for ~70% of the total body mass in fishes, the reduction in muscle metabolism implied by the biochemical differences should provide a substantial contribution to reduced overall metabolism.

The reduction of muscle enzyme levels is consistent with the hypothesis that there has been relaxed selection for locomotory capability in deep-sea pelagic species, due to the lack of light (Childress, 1995). One might then expect to find reduced enzyme levels in species from the deep-sea hydrothermal vents. This is not the case; the Galapagos vent zoarcid, *Thermarces andersoni,* from 2600 m, has muscle LDH and pyruvate kinase activities in the range of values for surface-living fishes, and higher than any other species below 200 m (Hand and Somero, 1983). This has been taken as evidence for the food limitation hypothesis for reduced metabolic rates, because species from these highly productive deep-sea habitats presumably have access to plenty of food. Alternatively, one could argue that vent species are exposed to greater predation and/or stronger currents, either of which should select for greater locomotory capacity.

Another quantitatively important component of metabolism is osmoregulation, which may account for over one-fourth of metabolism in fishes (Febry and Lutz, 1987). If this is the case, then deep-living fishes whose metabolic rates are less than 10% of those in shallow waters must have greatly reduced osmoregulatory costs. This has not been achieved by the evolution of osmoconformity, because deep-sea fishes have plasma ionic compositions similar to shallow-living species (Blaxter *et al.*, 1971; Shelton *et al.*, 1985). Evidence that deep-sea fishes do have reduced osmoregulatory costs comes from the observation that levels of the primary osmoregulatory enzyme in marine teleosts, gill Na^+,K^+-ATPase, decrease significantly with depth (Gibbs and Somero, 1990b). As in the case of the metabolic enzymes, two vent species had Na^+/K^+-ATPase activities similar to shallow-water fishes.

The depth-related patterns in enzyme levels just considered have all involved comparisons among species. In many cases, the choice of species may have been inappropriate; Torres *et al.* (1979) essentially compared myctophids to other midwater fishes, whereas Gibbs and Somero (1990b) included fishes from extremely different habitats and life-styles (active pelagic, midwater, benthic, hydrothermal vents). An alternative approach to the study of depth-related biochemical changes is the comparison of individuals from the same species collected at different depths. These studies are complicated by body size, because many fishes undergo ontogenetic vertical migrations (Stein and Pearcy, 1982; Wakefield and Smith, 1990), and many enzymes exhibit significant scaling relationships (Somero and Childress, 1980, 1990; Sullivan and Somero, 1983). However, intraspecific analyses may be able to distinguish between the food limitation and relaxed locomotory selection hypotheses, which make different predictions for intraspecific comparisons. The former predicts that deeper-living individuals will have lower metabolic rates and enzyme levels, as a direct physiological response to low food. On the other hand, if individuals are not food-limited, relaxed selection for locomotory capacity would result in genetic differences among species from different depths, but not acclimatory changes within species.

Intraspecific changes in enzyme levels have been examined on only a few occasions. Siebenaller (1984a) found that pelagic juveniles of *Sebastolobus altivelis* had higher levels of metabolic enzymes than did adults, consistent with their higher mass-specific metabolic rates (Smith and Brown, 1983). These findings are of little value for understanding depth-related changes, however, because these life stages differ so much in their body size, behavior, diet, and other factors. Additionally, scaling relationships may change as a result of metamorphosis (Kaupp and Somero, 1989). Two studies have found evidence consistent with the food limitation hypothesis. Gibbs and Somero (1990b) found significant depth-related decreases in gill Na^+,K^+-ATPase activities in two of four species of benthic and benthopelagic fishes. Vetter *et al.* (1994) found lower levels of metabolic enzymes in Dover sole collected on the continental slope off Southern California (depths greater than 400 m) than in individuals collected on the shelf (<200 m). Intraspecific studies cannot negate the hypothesis that relaxed selection for locomotory capacity contributes to interspecific differences in metabolic rates, but the limited biochemical information is consistent with a role for food limitation in reduced metabolic rates of deep-living fishes.

One important question that has not been addressed is the physiological mechanisms by which enzyme activities and protein levels are regulated. Protein levels depend on rates of protein synthesis and degradation, which have not been examined in deep-sea fishes. Pressure can have substantial effects on interactions between DNA and DNA-binding proteins (Mozhaev *et al.*, 1996), suggesting that pressure may perturb gene expression. Recent studies with bacteria have supported this idea (Bartlett *et al.*, 1995). Application of high pressure results in altered expression patterns of numerous proteins in *Escherischia coli* (Welch *et al.*, 1993) and *Methanococcus thermolithotrophicus* (Jaenicke *et al.*, 1988). Of more interest to deep-sea researchers is that the transcription of a specific gene, *ompH*, is affected by pressure in a barophilic isolate, and is maximal at the strain's pressure optimum (Bartlett *et al.*, 1989). Although the gene product, an outer membrane porin, is not required for survival at high pressure (Bartlett and Chi, 1994), this work does demonstrate the potential significance of pressure-dependent changes in gene regulation, and that certain genes may be expressed only at high pressures.

C. Summary

Metabolic rates of deep-sea fishes decrease with depth, especially in the pelagic realm. This pattern is probably the result of several interacting factors. Metabolic and biochemical analyses suggest that direct effects of temperature, physiological acclimation to low food levels, and genetic adaptation to these and low light levels (resulting in relaxed selection for locomotory capabilities) are responsible.

V. FUTURE DIRECTIONS: PHYLOGENETIC AND MOLECULAR APPROACHES

The phylogenetic limitations of most deep-sea fish studies have been pointed out several times. Any of the figures in this review could serve to illustrate this problem. For example, depth-related trends in membrane lipid composition (Fig. 4) were found using different, unrelated species at each depth, and early work on metabolic rates was biased by the high proportion of myctophids among shallow-living species (Fig. 9). Such broad comparisons ignore the fact that species are not evolutionarily independent of one another, and that the choice of organisms studied can significantly affect the outcome of an analysis. Comparative studies of *Sebastolobus* (Figs. 2, 3, 7, and 8) suffer the opposite problem; they have consisted of multiple two-species comparisons of the type criticized by Garland and Adolph (1994).

Actin (Fig. 1), Na^+,K^+-ATPase (Fig. 5), and dehydrogenases in general provide examples of an intermediate approach. A correspondence between

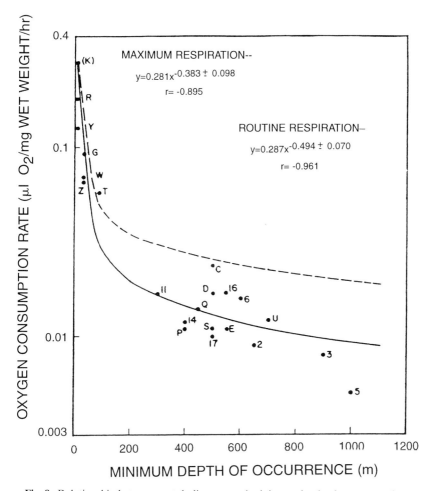

Fig. 9. Relationship between metabolic rates and minimum depth of occurrence in eastern Pacific midwater fishes. Each symbol indicates a different species. Oxygen consumption in the shallow-occurring group, which includes five vertically migrating myctophids, was measured at 10°C. Metabolism in the other species was measured at 5°C. Reprinted from *Deep-Sea Res.* **26**; J. J. Torres, B. W. Belman, and J. J. Childress. Oxygen consumption rates of midwater fishes as a function of depth of occurrence, 185–197. Copyright 1979, with permission from Elsevier Science Ltd, The Boulevard, Langford Lane, Kidlington OX5 1GB, UK.

broad patterns across taxa and similar relationships within smaller units such as genera can be taken as evidence supporting convergent evolution of pressure-adapted proteins. These studies have not been performed with the statistical methods recommended by Garland and Adolph (1994).

(Many were performed before phylogenetically based methods of analysis were invented.) This does not imply that deep-sea fishes are not adapted to their environment, only that conclusions regarding the adaptive significance of specific characters are limited. Detailed phylogenetic information is lacking for nearly all deep-sea fishes, so this situation is unlikely to improve in the near future.

Some hypotheses about adaptation to the deep sea may never be testable in a rigorous phylogenetic context, because so few groups of organisms have invaded this habitat. One would like to have an easily collected group of organisms, with well-defined phylogenetic relationships, whose members span the range of the environmental factor of interest (e.g., pressure). Unfortunately, appropriate taxa may not exist, or the evolutionary relationships of the group's members may directly correspond to environmental differences. For example, a deep-sea genus might contain two clades, each of whose members lived at similar depths. Thus, any putative pressure-adaptive differences in enzyme properties could not be distinguished from phylogenetic effects. This should not deter researchers from continuing to study deep-sea organisms; they simply need to choose their study subjects wisely and be more aware of potential artifacts. Better phylogenetic information is a must for such work, and molecular systematics studies will surely play an important role in the future.

Technological developments in the laboratory should provide another productive avenue for future research. It has become possible to perform almost any biochemical or biophysical measurement at high pressure, but few of these techniques have been applied to proteins from deep-sea organisms. One limitation of some techniques (e.g., NMR, infrared spectroscopy) is the large quantity of sample required. These problems can be overcome by cloning the gene for a given protein and then mass-producing it in the laboratory. A commercial application has already been found in the production of DNA polymerase from hydrothermal vent bacteria.

The techniques of molecular biology can also be used to gain a better understanding of protein function in the deep sea. One can determine which amino acid residues have been changed in the course of evolution, and use site-directed mutagenesis and other molecular techniques to test hypotheses about the functional significance of specific structural differences. This approach has been successful in the context of temperature adaptation (Powers *et al.*, 1991; Somero, 1995), and similar studies have been initiated for deep-sea proteins (G. N. Somero, personal communication). Molecular biology has already begun to play a significant role in understanding microbial adaptations to the deepsea (Bartlett *et al.*, 1995), and will certainly become just as important in the study of fishes.

ACKNOWLEDGMENTS

I thank G. N. Somero, the other members of the High Pressure Zone, A. A. Yayanos, and C. Phaenoides for their discussions over the years. Funding for manuscript preparation was provided by a generous grant from the Ubu Endowment.

REFERENCES

Adachi, S., and Morishima, I. (1989). The effects of pressure on oxygen and carbon monoxide binding kinetics for myoglobin. *J. Biol. Chem.* **264,** 18896–18901.
Airriess, C. N., and Childress, J. J. (1994). Homeoviscous properties implicated by the interactive effects of pressure and temperature on the hydrothermal vent crab *Bythograea thermydron. Biol. Bull.* **187,** 208–214.
Avrova, N. F. (1984). The effect of natural adaptations of fishes to environmental temperature on brain ganglioside fatty acid and long chain base composition. *Comp. Biochem. Physiol. B* **78,** 903–909.
Bailey, T. G., Torres, J. J., Youngbluth, M. J., and Owen, G. P. (1994). Effect of decompression on mesopelagic gelatinous zooplankton: A comparison of *in situ* and shipboard measurements of metabolism. *Mar. Ecol. Prog. Ser.* **113,** 13–27.
Bartlett, D. H., and Chi, E. (1994). Genetic characterization of *ompH* mutants in the deep-sea bacterium *Photobacterium* species strain SS9. *Arch. Microbiol.* **162,** 323–328.
Bartlett, D. H., Wright, M., Yayanos, A. A., and Silverman, M. (1989). Isolation of a gene regulated by hydrostatic pressure in a deep-sea bacterium. *Nature (London)* **342,** 572–574.
Bartlett, D. H., Kato, C., and Horikoshi, K. (1995). High pressure influences on gene and protein expression. *Res. Microbiol.* **146,** 697–706.
Behan, M. K., Macdonald, A. G., Jones, G. R., and Cossins, A. R. (1992). Homeoviscous adaptation under pressure: The pressure dependence of membrane order in brain myelin membranes of deep-sea fish. *Biochim. Biophys. Acta* **1103,** 317–323.
Belman, B. W., and Gordon, M. S. (1979). Comparative studies on the metabolism of shallow-water and deep-sea fishes. 5. The effects of temperature and hydrostatic pressure on oxygen consumption in the mesopelagic zoarcid, *Melanostigma pammelas. Mar. Biol.* **50,** 275–281.
Blaxter, J. H. S., Wardle, C. S., and Roberts, B. L. (1971). Aspects of the circulatory physiology and muscle systems of deep-sea fish. *J. Mar. Biol. Assoc. U.K.* **51,** 991–1006.
Braganza, L. F., and Worcester, D. L. (1986). Structural changes in lipid bilayers and biological membranes caused by hydrostatic pressure. *Biochemistry* **25,** 7484–7488.
Ceuterick, F., Peeters, J., Heremans, K., de Smedt, H., and Olbrechts, H. (1978). Effect of high pressure, detergents and phospholipase on the break in the Arrhenius plot of *Azotobacter* nitrogenase. *Eur. J. Biochem.* **87,** 401–407.
Childress, J. J. (1975). The respiratory rates of midwater crustaceans as a function of depth of occurrence and relation to the oxygen minimum layer off Southern California. *Comp. Biochem. Physiol. A* **50,** 787–799.
Childress, J. J. (1995). Are there physiological and biochemical adaptations of metabolism in deep-sea animals? *Trends Ecol. Evol.* **10,** 30–36.
Childress, J. J., and Nygaard, M. H. (1973). The chemical composition of midwater fishes as a function of depth of occurrence off Southern California. *Deep-Sea Res.* **20,** 1093–1109.

Childress, J. J., and Somero, G. N. (1979). Depth-related enzymic activities in muscle, brain and heart of deep-living pelagic marine teleosts. *Mar. Biol.* **52,** 273–283.

Childress, J. J., and Thuesen, E. V. (1993). Effects of hydrostatic pressure on metabolic rates of six species of deep-sea gelatinous zooplankton. *Limnol. Oceanogr.* **38,** 665–670.

Childress, J. J., and Thuesen, E. V. (1995). Metabolic potentials of deep-sea fishes: A comparative approach. Environmental and ecological biochemistry. *Biochem. Mol. Biol. Fishes* **5,** 175–196.

Childress, J. J., Cowles, D. L., Favuzzi, J. A., and Mickel, T. J. (1990a). Metabolic rates of benthic deep-sea decapod crustaceans decline with increasing depth primarily due to the decline in temperature. *Deep-Sea Res.* **37,** 929–949.

Childress, J. J., Price, M. H., Favuzzi, J., and Cowles, D. (1990b). The chemical composition of midwater fishes as a function of depth of occurrence off the Hawaiian Islands: Food availability as a selective factor? *Mar. Biol.* **105,** 235–246.

Chong, P. L.-G., Fortes, P. A. G., and Jameson, D. M. (1985). Mechanisms of inhibition of (Na,K)-ATPase by hydrostatic pressure studied with fluorescent probes. *J. Biol. Chem.* **260,** 14484–14490.

Cossins, A. R., ed. (1994). "Temperature Adaptation of Biological Membranes." Portland, London.

Cossins, A. R., and Macdonald, A. G. (1984). Homeoviscous theory under pressure: II. The molecular order of membranes from deep-sea fish. *Biochim. Biophys. Acta* **776,** 144–150.

Cossins, A. R., and Macdonald, A. G. (1986). Homeoviscous adaptation under pressure: III. The fatty acid composition of liver mitochondrial phospholipids of deep-sea fish. *Biochim. Biophys. Acta* **860,** 325–335.

Cossins, A. R., and Macdonald, A. G. (1989). The adaptations of biological membranes to temperature and pressure: Fish from the deep and cold. *J. Bioenerg. Biomembranes* **21,** 115–135.

Cowles, D. L., Childress, J. J., and Wells, M. E. (1991). Metabolic rates of midwater crustaceans as a function of depth of occurrence off the Hawaiian Islands: Food availability as a selective factor? *Mar. Biol.* **110,** 75–83.

Dahlhoff, E., and Somero, G. N. (1991). Pressure and temperature adaptation of cytosolic malate dehydrogenases of shallow- and deep-living marine invertebrates: Evidence for high body temperatures in hydrothermal vent animals. *J. Exp. Biol.* **159,** 473–487.

Dahlhoff, E., Schneidemann, S., and Somero, G. N. (1990). Pressure-temperature interactions on M_4-lactate dehydrogenases from hydrothermal vent fishes: Evidence for adaptation to elevated temperatures by the zoarcid *Thermarces andersoni*, but not the bythitid, *Bythites hollisi*. *Biol. Bull.* **179,** 134–139.

DeLong, E. F., and Yayanos, A. A. (1985). Adaptation of the membrane lipids of a deep-sea bacterium to changes in hydrostatic pressure. *Science* **228,** 1101–1103.

DeLong, E. F., and Yayanos, A. A. (1986). Biochemical function and ecological significance of novel bacterial lipids in deep-sea prokaryotes. *Appl. Environ. Microbiol.* **51,** 730–737.

Donnelly, J., and Torres, J. J. (1988). Oxygen consumption of midwater fishes and crustaceans from the eastern Gulf of Mexico. *Mar. Biol.* **97,** 483–494.

Donnelly, J., Torres, J. J., Hopkins, T. L., and Lancraft, T. M. (1990). Proximate composition of Antarctic mesopelagic fishes. *Mar. Biol.* **106,** 13–23.

Febry, R., and Lutz, P. (1987). Energy partitioning in fish: The activity-related cost of osmoregulation in a euryhaline cichlid. *J. Exp. Biol.* **128,** 63–85.

Garland, T., Jr., and Adolph, S. C. (1994). Why not to do two-species comparative studies: Limitations on inferring adaptation. *Physiol. Zool.* **67,** 797–828.

Garland, T., Jr., and Carter, P. A. (1994). Evolutionary physiology. *Annu. Rev. Physiol.* **56,** 579–621.

Gibbs, A. (1995). Temperature, pressure and the sodium pump: The role of homeoviscous adaptation. Environmental and ecological biochemistry. *Biochem. Mol. Biol. Fishes* **5**, 197–212.

Gibbs, A., and Somero, G. N. (1989). Pressure adaptation of Na$^+$,K$^+$-ATPase in gills of marine teleosts. *J. Exp. Biol.* **143**, 475–492.

Gibbs, A., and Somero, G. N. (1990a). Pressure adaptation of teleost gill Na$^+$,K$^+$-adenosine triphosphatase: Role of the lipid and protein moieties. *J. Comp. Physiol. B* **160**, 431–439.

Gibbs, A., and Somero, G. N. (1990b). Na$^+$–K$^+$-adenosine triphosphatase activities in gills of marine teleost fishes: Changes with depth, size and locomotory activity level. *Mar. Biol.* **106**, 315–321.

Gross, M., Auerbach, G., and Jaenicke, R. (1993). The catalytic activities of monomeric enzymes show complex pressure dependence. *FEBS Lett.* **321**, 256–260.

Hand, S. C., and Somero, G. N. (1983). Energy metabolism pathways of hydrothermal vent animals: Adaptations to a food-rich and sulfide-rich deep-sea environment. *Biol. Bull.* **165**, 167–183.

Harper, A. A., Macdonald, A. G., Wardle, C. S., and Pennec, J.-P. (1987). The pressure tolerance of deep-sea fish axons: Results of *Challenger* cruise 6B/85. *Comp. Biochem. Physiol. A* **88**, 647–653.

Hazel, J. R. (1995). Thermal adaptation in biological membranes: Is homeoviscous adaptation the explanation? *Annu. Rev. Physiol.* **57**, 19–42.

Hazel, J. R., and Williams, E. E. (1990). The role of alterations in membrane lipid composition in enabling physiological adaptation of organisms to their physical environment. *Prog. Lipid Res.* **29**, 167–227.

Heezen, B. C. (1957). Whales entangled in deep sea cables. *Deep-Sea Res.* **4**, 105–115.

Hennessey, J. P., and Siebenaller, J. F. (1985). Pressure inactivation of tetrameric lactate dehydrogenase homologues of confamilial deep-living fishes. *J. Comp. Physiol. B* **155**, 647–652.

Heremans, K. (1982). High pressure effects on proteins and other biomolecules. *Annu. Rev. Biophys. Bioeng.* **11**, 1–21.

Heremans, K., and Wuytack, F. (1980). Pressure effect on the Arrhenius discontinuity in Ca^{2+}-ATPase from sarcoplasmic reticulum. *FEBS Lett.* **117**, 161–163.

Jaenicke, R. (1983). Biochemical processes under high hydrostatic pressure. *Naturwissenschaften* **70**, 332–341.

Jaenicke, R., Bernhardt, G., Lüdemann, H.-D., and Stetter, K. O. (1988). Pressure-induced alterations in the protein pattern of the thermophilic archaebacterium *Methanococcus thermolithotrophicus*. *Appl. Environ. Microbiol.* **54**, 2375–2380.

Jannasch, H. W., Marquis, R. E., and Zimmerman, A. M., eds. (1987). "Current Perspectives in High Pressure Biology." Academic Press, London, Orlando, Florida.

Johnson, F. H. and Schlegel, F. M. (1948). *J. Cell. Comp. Physiol.* **31**, 421-425.

Johnston, I. A., and Bennett, A. F., eds. (1996). Animals and temperature: Phenotypic and evolutionary adaptation. *Soc. Exp. Biol. Semin. Ser.* **59**.

Kamimura, K., Fuse, H., Takimura, O., and Yamaoka, Y. (1993). Effects of growth pressure and temperature on fatty acid composition of a barotolerant deep-sea bacterium. *Appl. Environ. Microbiol.* **59**, 924–926.

Kaneshiro, S. M., and Clark, D. S. (1995). Pressure effects on the composition and thermal behavior of lipids from the deep-sea thermophile *Methanococcus jannaschii*. *J. Bacteriol.* **177**, 3668–3672.

Kaupp, S. E., and Somero, G. N. (1989). Empirically determined metabolic scaling in larval and juvenile marine fish. *Am. Zool.* **29**, 55A.

Klein, R. A. (1982). Thermodynamics and membrane processes. *Q. Rev. Biophys.* **15**, 667–757.

Kornblatt, M. J., Kornblatt, J. A., and Hui Bon Hoa, G. (1993). The role of water in the dissociation of enolase, a dimeric enzyme. *Arch. Biochem. Biophys.* **306**, 495–500.

Low, P. S., and Somero, G. N. (1975a). Pressure effects on enzyme structure and function under *in vitro* and simulated *in vivo* conditions. *Comp. Biochem. Physiol. B* **52**, 67–74.

Low, P. S., and Somero, G. N. (1975b). Activation volumes in enzymic catalysis: Their sources and modification by low molecular weight solutes. *Proc. Natl. Acad. Sci. U.S.A.* **72**, 3014–3018.

Macdonald, A. G. (1984). The effects of pressure on the molecular structure and physiological functions of cell membranes. *Philos. Trans. R. Soc. London B* **304**, 47–68.

Macdonald, A. G., and Cossins, A. R. (1985). The theory of homeoviscous adaptation of membranes applied to deep-sea animals. Physiological adaptations of marine animals. *Soc. Exp. Biol. Symp.* **39**, 301–322.

Macdonald, A. G., and Gilchrist, I. (1980). Effects of hydraulic decompression and compression on deep sea amphipods. *Comp. Biochem. Physiol. A* **67**, 149–153.

Macdonald, A. G., Gilchrist, I., and Wardle, C. S. (1987). Effects of hydrostatic pressure on the motor activity of fish from shallow water and 900 m depths: Some results of *Challenger* cruise 6B/85. *Comp. Biochem. Physiol. A* **88**, 543–547.

McFall-Ngai, M., and Horowitz, J. (1990) A comparative study of the thermal stability of the vertebrate eye lens: Antarctic fish to the desert iguana. *Exp. Eye Res.* **50**, 703–709.

Morild, E. (1981). The theory of pressure effects on enzymes. *Adv. Protein Chem.* **34**, 93–166.

Mozhaev, V. V., Heremans, K., Frank, J., Masson, P., and Balny, C. (1996). High pressure effects on protein structure and function. *Proteins.* **24**, 81–91.

Murray, T. F., and Siebenaller, J. F. (1993). Differential susceptibility of guanine nucleotide-binding proteins to pertussis toxin-catalyzed ADP-ribosylation in brain membranes of two congeneric marine fishes. *Biol. Bull.* **185**, 346–354.

Mylvaganam, S. E., Bonaventura, C., Bonaventura, J., and Getzoff, E. D. (1996). Structural basis for the Root effect in haemoglobin. *Nature Struct. Biol.* **3**, 275–283.

Neer, E. J., Lok, J. M., and Wolf, L. G. (1984). Purification and properties of the inhibitory guanine nucleotide regulatory unit of brain adenylate cyclase. *J. Biol. Chem.* **259**, 14222–14229.

Noble, R. W., Kwiatowski, L. D., De Young, A., Davis, B. J., Haedrich, R. L., Tam, L., and Riggs, A. F. (1986). Functional properties of hemoglobins from deep-sea fish: Correlations with depth distribution and presence of a swimbladder. *Biochim. Biophys. Acta* **870**, 552–563.

Ogunmola, G. B., Kauzmann, W., and Zipp, A. (1976). Volume changes in binding of ligands to methemoglobin and metmyoglobin. *Proc. Natl. Acad. Sci. U.S.A.* **73**, 4271–4273.

Pelster, B., and Scheid, P. (1992). Countercurrent concentration and gas secretion in the fish swim bladder. *Physiol. Zool.* **65**, 1–16.

Pennec, J.-P., Wardle, C. S., Harper, A. A., and Macdonald, A. G. (1988). Effects of high hydrostatic pressure on the isolated hearts of shallow water and deep sea fish; results of *Challenger* cruise 6B/85. *Comp. Biochem. Physiol. A* **89**, 215–218.

Phleger, C. F., and Laub, R. J. (1975). Skeletal fatty acids in fish from different depths off Jamaica. *Comp. Biochem. Physiol. B* **94**, 329–334.

Powers, D. A., Lauerman, T., Crawford, D., and DiMichele, L. (1991). Genetic mechanisms for adapting to a changing environment. *Annu. Rev. Genet.* **25**, 629–659.

Priede, I. G., Smith, K. L., Jr., and Armstrong, J. D. (1990). Foraging behavior of abyssal grenadier fish: Inferences from acoustic tagging and tracking in the North Pacific Ocean. *Deep-Sea Res.* **37**, 81–101.

Priede, I. G., Bagley, P. M., Armstrong, J. D., Smith, K. L., Jr., and Merrett, N. R. (1991). Direct measurement of active dispersal of food-falls by deep-sea demersal fishes. *Nature (London)* **351**, 647–649.

Rashin, A. A., Iofin, M., and Honig, B. (1986). Internal cavities and buried waters in proteins. *Biochemistry* **25**, 3619–3625.
Ruan, K., and Weber, G. (1989). Hysteresis and conformational drift of pressure-dissociated glyceraldehydephosphate dehydrogenase. *Biochemistry* **28**, 2144–2153.
Ruan, K. and Weber, G. (1993). Physical heterogeneity of muscle glycogen phosphoyrylase revealed by hydrostatic pressure dissociation. *Biochemistry* **32**, 6295–6301.
Sanders, N. K., and Childress, J. J. (1990). Adaptations to the deep-sea oxygen minimum layer: Oxygen binding by the hemocyanin of the bathypelagic mysid, *Gnathophausia ingens* Dohrn. *Biol. Bull.* **178**, 286–294.
Shelton, C., Macdonald, A. G., Pequeaux, A, and Gilchrist, I. (1985). The ionic composition of the plasma and erythrocytes of deep-sea fish. *J. Comp. Physiol. B* **155**, 629–633.
Siebenaller, J. F. (1978). Genetic variability in deep-sea fishes of the genus *Sebastolobus*. In "Marine Organisms: Genetics, Ecology and Evolution" (B. Battaglia and J. Beardmore, eds.), pp. 95–122. Plenum, New York.
Siebenaller, J. F. (1984a). Analysis of the biochemical consequences of ontogenetic vertical migration in a deep-living teleost fish. *Physiol. Zool.* **57**, 598–608.
Siebenaller, J. F. (1984b). Pressure-adaptive differences in NAD-dependent dehydrogenases of congeneric marine fishes living at different depths. *J. Comp. Physiol. B* **154**, 443–448.
Siebenaller, J. F., and Murray, T. F. (1990). A_1 adenosine receptor modulation of adenylyl cyclase of a deep-living teleost fish, *Antimora rostrata*. *Biol. Bull.* **178**, 65–73.
Siebenaller, J. F., and Murray, T. F. (1994a). The effects of hydrostatic pressure on pertussis toxin-catalyzed ribosylation of guanine nucleotide-binding proteins from two congeneric marine fish. *Comp. Biochem. Physiol. B* **108**, 423–430.
Siebenaller, J. F., and Murray, T. F. (1994b). The effects of hydrostatic pressure on the low-K_m GTPase in brain membranes from two congeneric marine fishes. *J. Comp. Physiol. B* **163**, 626–632.
Siebenaller, J. F., and Murray, T. F. (1995). The effects of pressure on G protein-coupled signal transduction. *In* Environmental and ecological biochemistry. *Biochem. Mol. Biol. Fishes* **5**, 147–174.
Siebenaller, J. F., and Somero, G. N. (1978). Pressure-adaptive differences in lactate dehydrogenases of congeneric marine fishes living at different depths. *Science* **201**, 255–257.
Siebenaller, J. F., and Somero, G. N. (1979). Pressure-adaptive differences in the binding and catalytic properties of muscle-type (M_4) lactate dehydrogenase of shallow- and deep-living marine fishes. *J. Comp. Physiol.* **129**, 295–300.
Siebenaller, J. F., and Somero, G. N. (1982). The maintenance of different enzyme activity levels in congeneric fishes living at different depths. *Physiol. Zool.* **55**, 171–179.
Siebenaller, J. F., and Somero, G. N. (1989). Biochemical adaptation to the deep sea. *In Crit. Rev. Aquat. Sci.* **1**, 1–25.
Siebenaller, J. F., and Yancey, P. H. (1984). Protein composition of white skeletal muscle from mesopelagic fishes having different water and protein contents. *Mar. Biol.* **78**, 129–137.
Siebenaller, J. F., Somero, G. N., and Haedrich, R. L. (1982). Biochemical characteristics of macrourid fishes differing in their depths of distribution. *Biol. Bull.* **163**, 240–249.
Siebenaller, J. F., Hagar, A. F., and Murray, T. F. (1991). The effects of hydrostatic pressure on A_1-adenosine receptor signal transduction in brain membranes of two congeneric marine fishes. *J. Exp. Biol.* **159**, 23–43.
Silva, J. L., and Weber, G. (1993). Pressure stability of proteins. *Annu. Rev. Phys. Chem.* **44**, 89–113.
Silvius, J. R., and McElhaney, R. N. (1981). Non-linear Arrhenius plots and the analysis of reaction and methanal rates in biological membranes. *J. Theor. Biol.* **88**, 135–152.

Smith, K. L., Jr. (1978). Metabolism of the abyssopelagic rattail *Coryphaenoides armatus* measured *in situ*. *Nature (London)* **274,** 362–364.
Smith, K. L., Jr. (1983). Metabolism of two dominant epibenthic echinoderms measured at bathyal depths in the Santa Catalina Basin. *Mar. Biol.* **72,** 249–256.
Smith, K. L., Jr., and Brown, N. O. (1983). Oxygen consumption of pelagic juveniles and demersal adults of the deep-sea fish *Sebastolobus altivelis*, measured at depth. *Mar. Biol.* **76,** 325–332.
Smith, K. L., Jr., and Hessler, R. R. (1974). Respiration of benthopelagic fishes: *In situ* measurements at 1230 meters. *Science* **184,** 72–73.
Smith, K. L., Jr., and Laver, M. B. (1981). Respiration of the bathypelagic fish *Cyclothone acclinidens*. *Mar. Biol.* **61,** 216–266.
Somero, G. N. (1992). Adaptations to high hydrostatic pressure. *Annu. Rev. Physiol.* **54,** 557–577.
Somero, G. N. (1995). Proteins and temperature. *Annu. Rev. Physiol.* **57,** 43–68.
Somero, G. N., and Childress, J. J. (1980). A violation of the metabolism-scaling paradigm: Activities of glycolytic enzymes in muscle increase in larger-size fish. *Physiol. Zool.* **53,** 322–337.
Somero, G. N., and Childress, J. J. (1990). Scaling of ATP-supplying enzymes, myofibrillar proteins and buffering capacity in fish muscle—Relationship to locomotory habit. *J. Exp. Biol.* **149,** 319–333.
Stamatoff, J., Guillon, D., Powers, L., Cladis, P., and Aadsen, D. (1978). X-Ray diffraction measurements of dipalmitoylphosphatidylcholine as a function of pressure. *Biochem. Biophys. Res. Commun.* **85,** 724–728.
Stein, D. L. (1985). Towing large nets by single warp at abyssal depths: Methods and biological results. *Deep-Sea Res.* **32,** 183–200.
Stein, D. L., and Pearcy, W. G. (1982). Aspects of reproduction, early life history, and biology of macrourid fishes off Oregon, U.S.A. *Deep-Sea Res.* **29,** 1313–1329.
Sullivan, K. M., and Somero, G. N. (1980). Enzyme activities of fish skeletal muscle and brain as influenced by depth of occurrence and habits of feeding and locomotion. *Mar. Biol.* **60,** 91–99.
Sullivan, K. M., and Somero, G. N. (1983). Size- and diet-related variations in enzymic activity and tissue composition in the sablefish, *Anoplopoma fimbria*. *Biol. Bull.* **164,** 315–326.
Swezey, R. R., and Somero, G. N. (1982a). Polymerization thermodynamics and structural stabilities of skeletal muscle actins from vertebrates adapted to different temperatures and hydrostatic pressures. *Biochemistry* **21,** 4496–4503.
Swezey, R. R., and Somero, G. N. (1982b). Skeletal muscle actin content is strongly conserved in fishes having different depths of distribution and capacities of locomotion. *Mar. Biol. Lett.* **3,** 307–315.
Swezey, R. R., and Somero, G. N. (1985). Pressure effects on actin self-assembly: Interspecific differences in the equilibrium and kinetics of the G to F transformation. *Biochemistry* **24,** 852–860.
Thuesen, E. V., and Childress, J. J. (1993). Enzymatic activities and metabolic rates of pelagic chaetognaths: Lack of depth-related declines. *Limnol. Oceanogr.* **38,** 935–948.
Thuesen, E. V., and Childress, J. J. (1994). Oxygen consumption rates and metabolic enzyme activities of oceanic California medusae in relation to body size and habitat depth. *Biol. Bull.* **187,** 84–98.
Tiku, P. E., Gracey, A. Y., Macartney, A. I., Beynon, R. J., and Cossins, A. R. (1996). Cold-induced expression of Δ_9-desaturase in carp by transcriptional and posttranslational mechanisms. *Science* **271,** 815–818.

Torres, J. J., and Somero, G. N. (1988a). Metabolism, enzymic activities and cold adaptation in Antarctic mesopelagic fishes. *Mar. Biol.* **98,** 169–180.

Torres, J. J., and Somero, G. N. (1988b). Vertical distribution and metabolism in Antarctic mesopelagic fishes. *Comp. Biochem. Physiol. B* **90,** 521–528.

Torres, J. J., Belman, B. W., and Childress, J. J. (1979). Oxygen consumption rates of midwater fishes as a function of depth of occurrence. *Deep-Sea Res.* **26,** 185–197.

Unno, M., Ishimori, K., and Morishima, I. (1990). High-pressure laser photolysis study of hemoproteins. Effects of pressure on carbon monoxide binding dynamics for R- and T-state hemoglobins. *Biochemistry* **29,** 10199–10205.

Unno, M., Ishimori, K., Morishima, I., Nakayama, T., and Hamanoue, K. (1991). Pressure effects on carbon monoxide rebinding to the isolated α and β chains of human hemoglobin. *Biochemistry* **30,** 10679–10685.

Vetter, R. D., Lynn, E. A., Garza, M., and Costa, A. S. (1994). Depth zonation and metabolic adaptation in Dover sole, *Microstomus pacificus,* and other deep-living flatfishes: factors that affect the sole. *Mar. Biol.* **120,** 145–159.

Wakefield, W. W., and Smith, K. L., Jr. (1990). Ontogenetic vertical migration in *Sebastolobus altivelis* as a mechanism for transport of particulate organic matter at continental slope depths. *Limnol. Oceanogr.* **35,** 1314–1328.

Wann, K. T., and Macdonald, A. G. (1980). The effects of pressure on excitable cells. *Comp. Biochem. Physiol. A* **66,** 1–12.

Weber, G. (1992). "Protein Interactions." Chapman & Hall, New York and London.

Weber, G., and Drickamer, H. G. (1983). The effect of high pressure upon proteins and other biomolecules. *Q. Rev. Biophys.* **16,** 89–112.

Welch, T. J., Farewell, A., Neidhardt, F. C., and Bartlett, D. H. (1993). Stress response in *Escherichia coli* induced by elevated hydrostatic pressure. *J. Bacteriol.* **175,** 7170–7177.

Wilson, R. R., Siebenaller, J. F., and Davis, B. J. (1991). Phylogenetic analysis of species of three subgenera of *Coryphaenoides* (Teleostei: Macrouridae) by peptide mapping of homologs of LDH-A_4. *Biochem. Syst. Ecol.* **19,** 277–287.

Wirsen, C. O., Jannasch, H. W., Wakeham, S. G., and Canuel, E. A. (1987). Membrane lipids of a psychrophilic and barophilic deep-sea bacterium. *Curr. Microbiol.* **14,** 319–322.

Yang, T., and Somero, G. N. (1993). Effects of feeding and food deprivation on oxygen consumption, muscle protein concentration and activities of energy metabolism enzymes in muscle and brain of shallow-living (*Scorpaena guttata*) and deep-living (*Sebastolobus alascanus*) scorpaenid fishes. *J. Exp. Biol.* **181,** 213–232.

Yang, T., Lai, N. C., Graham, J. B., and Somero, G. N. (1992). Respiratory, blood, and heart enzymatic adaptations of *Sebastolobus alascanus* (Scorpaenidae; Teleostei) to the oxygen minimum zone: A comparative study. *Biol. Bull.* **183,** 490–499.

Yayanos, A. A. (1981). Reversible inactivation of deep-sea amphipods (*Paracella capresca*) by a decompression from 601 bars to atmospheric pressure. *Comp. Biochem. Physiol. A* **69,** 563–565.

Yayanos, A. A. (1995). Microbiology to 10,500 meters in the deep sea. *Annu. Rev. Microbiol.* **49,** 777–805.

7

PRESSURE EFFECTS ON SHALLOW-WATER FISHES

PHILIPPE SÉBERT

I. Introduction
II. The Fish as a Model
   A. Why the Fish?
   B. What Type of Fish?
III. Methods
IV. Effects of Short-Term Pressure Exposure
   A. Whole Animal
   B. Organs and Tissues
   C. Factors Interacting with Hydrostatic Pressure
   D. Mechanisms: The Membrane Hypothesis
V. Acclimatization of Fish to Hydrostatic Pressure
   A. Behavior and Oxygen Consumption
   B. Metabolism
   C. Tissue Composition
   D. Membrane Fluidity and Composition
   E. Structural Changes
VI. Comparison of Shallow-Water Fishes and Deep-Water Fishes
   A. Oxygen Consumption
   B. Muscle Biochemistry
   C. Comparing Shallow-Water and Deep-Water Fishes
VII. Conclusion
    References

I. INTRODUCTION

The reader may be surprised to find a chapter concerning shallow-water fishes in a book dedicated to the physiology of deep-sea fishes. However, the explanation is relatively clear when one considers the fact that several species experience a deep-sea environment during their life cycle. Additionally, the successful acclimatization of shallow-water fishes to high pressure (see Section V) leads us to suppose that fishes living strictly in shallow water may potentially live in deeper waters and thus could provide an interesting method of studying the evolution processes. Finally, as outlined in the following section, the fish can be used as a model to study the specific effects of pressure.

The term *deep-sea fish* means that such animals live at depth [it is also true for some freshwater species; see Gordon (1970)], i.e., they are submitted to an important environmental factor, namely, hydrostatic pressure (HP). Unfortunately, HP is not the only environmental factor present at depth; others include decreased (or absence of) light, temperature, oxygenation, and food availability (see Chatpers 1 and 4). Pressure and temperature are perhaps more important than the other factors because they are the two main thermodynamic parameters that affect living processes. For example, the equations concerning the rate of enzyme reactions and thus the functioning of the organisms, $kp = k_0 e^{-(P\Delta V/RT)}$ (Johnson and Eyring, 1970) or the Clausius–Clapeyron relationship, $dT/dP = \Delta V(T/\Delta H)$, both show that increase in pressure, P, and temperature, T, can act in opposite directions. Thus, fishes living at depth are submitted to high pressure and low temperature, which could have similar metabolic effects.

Deep-sea fishes have adapted to depths greater than abyssal plain, i.e., 5000 m depth (500 atm or 50 MPa pressure) and to temperatures just above 0°C. In addition, certain fish have developed adaptations necessary for their diurnal and ontogenic vertical migrations, which result in simultaneous T and P changes, with the relative variations of P having a greater magnitude. Beyond these ecophysiological considerations, the reader must keep in mind that human hyperbaric physiology attracts a lot of attention (experimental deep diving or simply diving as a sport) and therefore it is necessary to know the specific effects of pressure.

II. THE FISH AS A MODEL

A. Why the Fish?

The study of a specific environmental factor requires a model. Why the fish? The first reason could be that 60% of all vertebrate species are fishes. Thus by merely regarding the number of species, a "typical" vertebrate would be a fish (Bone et al., 1992). Additionally, the fish has often been used as a model in development and clinical genetic studies (Powers, 1989; Ekker and Akimenko, 1991; Brenner et al., 1993; Kahn, 1994), and what follows shows that they are also a useful model for the study of the physiological effects of pressure.

When air breathers (mammals) are studied under conditions of high pressure they inhale gas mixtures consisting of oxygen and one or two "inert gases" (He, N_2, H_2). Consequently, air breathers under these conditions are submitted simultaneously to HP per se and to an increase in inert gas partial pressures (IGP). There is no physiological method to differentiate between HP and IGP effects in air breathers. However, a solution to this problem was suggested by Fenn (1967): "To resolve this problem of the possible practical role of hydrostatic pressure in diving, it seems necessary to use fish or mammals inhaling water instead of gas." Although some experiments have been performed using mammals breathing in a liquid medium having a high O_2 affinity, such as fluorocarbons (Kylstra et al., 1967; Lundgren and Örnhagen, 1976), this does not seem to be an ideal solution because it is doubtful that mammalian lungs filled with liquid are in optimal physiological condition. In contrast, a fish model to study pressure effects (Barthélémy, 1985) has four significant advantages: (1) fishes are vertebrates whose anatomo-functional organization can be compared to mammals, (2) fishes are ectotherms and consequently enable interactions between metabolic changes and pressure to be studied, as well as the temperature/pressure interactions, (3) fishes breathe water, the density of which is little modified by pressure, and the drastic modifications in ventilatory mechanics due to the inhalation of gases by mammals under pressure are thus avoided, and (4) the major probable reason for using the fish as a model is the possibility of dissociating HP effects from IGP effects, which may help in understanding the observed effects in mammals under pressure (HP + IGP).

B. What Type of Fish?

Little work has been devoted to the effects of pressure on true shallow-water fishes, i.e., fishes living near the surface. In fact, we consider that from a physiological point of view, fishes caught at depths less than 10–20 m are not deep-sea fishes, yet they still live at pressures two to three times that of atmospheric pressure. Thus, a distinction must be made between shallow waters (where pressure is very low when compared to the greatest depths encountered in the sea) and fishes living in shallow waters (where pressure can be very high when compared to atmospheric pressure). Gordon (1970) set the limit between low and high pressure at 5 atm (40 m). This chapter concerns mainly fishes living at the surface and experimentally submitted to high pressure. Fishes living at "low pressure" are considered as a reference for their congeners living at depth. The last section of the chapter presents some comparisons and speculations on the state of affairs for deep-sea fishes.

Since the pioneering work of the French physiologist Regnard (1884, 1885) on the pressure tolerance of both *Pleuronectes platessa* and *Carassius auratus*, several studies have been conducted (see Gordon, 1970, for review). Eels (*Anguilla anguilla*) and goldfish (*Carassius auratus*) are probably the shallow-water fishes most often used because they have an inherent high tolerance to HP. They are able to survive at 101 ATA (1 ATA = 1 atmosphere absolute = 0.1 MPa = 1 bar) for at least 10 h in confined conditions (see Sébert and Macdonald, 1993). This ensures that the fish is studied in good physiological condition when pressure is applied for only 2 to 3 h. However, knowing the relatively low tolerance of rainbow trout (*Oncorhynchus mykiss*) to pressure, it is sometimes interesting to use such a fish to obtain magnified responses. Concerning deep-water fishes, there is no specific model, but some species are currently studied mainly from the genus *Coryphaenoides*. Other species studied include *Mora moro*, *Bathysaurus mollis*, and *Antimora rostrata*. The genus *Sebastolobus* is also very interesting because it exists as shallow-water and deep-water species.

III. METHODS

Techniques used both in the laboratory and *in situ* when studying deep-sea fishes are discussed in Chapter 9 of this book. Here, we consider only the specific methods used in the study of shallow-water fishes.

Studying hydrostatic pressure necessitates the total absence of gas pockets in the aquaria (Fig. 1A). Thus the experimental tank must be completely filled with water at atmospheric pressure. Compression is ensured with a hydraulic pump (see Kynne, 1970; Avent *et al.*, 1972; Theede, 1972) or via a gas-proof soft rubber membrane, in which case the experimental tank is placed in a hyperbaric chamber compressed with gases (Barthélémy *et al.*, 1971). It must be stressed that in such aquaria, the fishes are in confinement and this limits the duration of pressure exposure. It is necessary to use a high-pressure water circulation system in order to maintain the fishes under pressure for several days or weeks (Sébert *et al.*, 1990). In contrast, the study of a combination of hydrostatic and gas pressure effects requires contact between the water phase and the gas phase and/or gas bubbling through the water contained in the experimental aquarium (Fig. 1B).

Due to the techniques used, the number of available direct measurements on whole animals is reduced. It is possible to perform measurements on organs or tissues of fishes exposed to high pressure, but it must be

7. PRESSURE EFFECTS ON SHALLOW-WATER FISHES

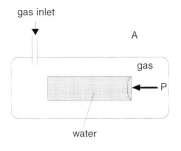

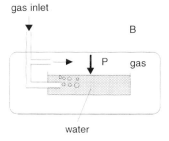

Fig. 1. Experimental set-up to study fishes under pressure. (A) Hydrostatic pressure. The aquarium in the hyperbaric chamber is completely filled with water; a gas-proof soft rubber membrane transmits the pressure P without modifying water gas content. (B) Hydrostatic pressure and gas pressure. The aquarium is open to the gas phase of the hyperbaric chamber.

remembered that currently such tissue samples can be taken only after decompression.

IV. EFFECTS OF SHORT-TERM PRESSURE EXPOSURE

A. Whole Animal

1. BEHAVIOR AND SURVIVAL TIMES

In 1885, the French physiologist Regnard was the first to report observations on aquatic animals experimentally subjected to high pressures. His observations were later confirmed by Fontaine (1928), Ebbecke (1944), and Nishiyama (1965). Broadly speaking, the effects of pressure are excitatory and induce abnormal activity. Generally (see Sébert and Macdonald,

1993), the only symptoms observed up to 20 atm are active swimming, often upward, which may be a response to a change in buoyancy (see Chapter 5). When pressure is further increased, movements become progressively less well coordinated and more jerky. Frequent loss of equilibrium, violent seizures or convulsions, and sometimes writhing movements are often observed as well. Other manifestations such as color changes can be observed (Nishiyama, 1965; Barthélémy and Belaud, 1972). Subsequently the fish becomes motionless at the bottom of the aquarium and a further pressure increase kills the animal. Generally, the fish recovers "normal" activity and physiological functions within some hours after decompression. It is important to stress, however, that the above-described symptoms are dependent on the species, on the temperature (Brauer et al., 1974), and, for a given species, on the compression rate and protocol. The faster the compression rate, the lower the pressure at which the first symptoms appear and the greater their intensity.

Complementary studies have been performed concerning the role of compression rate, oxygen partial pressure, metabolic rate (modifying T_w), and the like in the observed pressure effects. In some experiments, oxygen was made more available to trout (artificial ventilation, increasing P_{wO_2} to 2 ATA) but there was no change in compression symptoms or survival times (Barthélémy et al., 1981; Sébert et al., 1987). Similarly, the responses to HP changed neither in cold water (4°C), which decreases the metabolic rate, nor after the removal of the swim bladder or the implantation of a catheter to equilibrate the pressure between the ambient medium and the abdominal cavity (see Barthélémy et al., 1981). This last observation shows that hyperexcitability is not simply a mechanical response to body compression.

By comparison with mammals, motor behavior under HP is generally characterized by a threshold pressure for tremors (P_t) followed by a threshold pressure for convulsions (P_c) (see Brauer et al., 1974; Sébert and Macdonald, 1993). These symptoms are species dependent. Thus both fishes and mammals show varying degrees of hyperbaric tolerance and this affects survival time.

The pioneering work concerning pressure tolerance of the fish was carried out by Regnard (1884), who studied *Pleuronectes platessa* and *Carassius auratus*. Both species died at 300 ATA. Ebbecke (1944), studying tiny individuals of *Gobus*, *Pleuronectes*, and *Spinochia*, observed that death began at approximately 200 ATA, and all animals were dead at 500 ATA. He concluded that surface-dwelling fishes are unable to cope with pressure at 2000 m and more. Naroska (1968) compressed aquatic animals for 1 h and determined the LD_{50} after 24 h recovery at normal atmospheric pressure. The principal feature of his experiments was that vertebrates have a considerably lower pressure tolerance than invertebrates. For example, the LD_{50} for *P. platessa* is 150 ATA, whereas *Mytilus edulis* survives at a pressure above 800 atm. Pressure tolerance also decreases with salinity in many ectotherms (Ponat, 1967; Flügel, 1972). However, these studies of pressure tolerance were obtained under varying conditions, without necessarily specifying the duration of exposure, the rate of compression, the hydrostatic or gas pressures, or the temperature, and did not measure survival time directly.

In order to evaluate correctly the time a fish may survive a given pressure, it is necessary to expose it to an experimental pressure that must be maintained until its death. An example of pressure tolerance in terms of survival time is given in Fig. 2. A drawback of these experiments is the animal confinement. Thus, hypoxia (which can be limited by increasing P_{wO_2} before compression) and hypercapnia appear progressively, together with an accumulation of metabolites such as ammonia, which is extremely toxic for fishes (Smart, 1978). These physical changes in water quality can considerably modify (decrease) pressure tolerance. By resolving such technical problems (Sébert *et al.*, 1990) and maintaining normoxic, normocapnic conditions, without an accumulation of metabolic wastes, the fish can survive at high pressure for several days (eels at 41 ATA) (Johnstone *et al.*, 1989) or even several weeks at 101 ATA (eels, trout, goldfish) (Simon *et al.*, 1989a; Sébert *et al.*, 1991; Sébert and B. Simon, unpublished data, 1990). Thus, experiments that simply study survival times become techni-

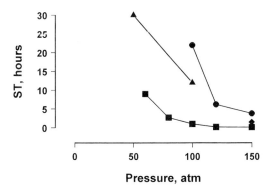

Fig. 2. Survival time (ST) of fishes under pressure. ■, Trout under hydrostatic pressure. ●, Trout under hydrostatic and helium pressure. ◆, Anaesthetized trout (Nembutal) under hydrostatic pressure. ▲, Goldfish under hydrostatic pressure. Results for trout from Barthélémy *et al.* (1981); results for goldfish from Sébert and Simon (unpublished data, 1990). Experimental temperature ranges from 15° to 17°C.

cally very complex when pressure is varied. This may explain why relatively little work has been devoted to pressure effects in whole fishes (see Gordon, 1970) and why such studies originate from a very small number of research teams. In addition to water quality considerations, survival times can be greatly increased, under confinement conditions, when fish are anaesthetized or when inert gases are dissolved in water (see Fig. 2).

2. Nervous System, Ventilation, and Circulation

In 1930, Fontaine wrote that the death of animals under pressure was not due to a pressure effect on the cell in general, but that death was the consequence of specific pressure effects on both the nervous and muscular systems (contraction), leading to a decrease in oxygen availability to the cell. There is evidence for pressure effects on the central nervous system (CNS) component. This in itself affects behavior and muscular activity, but there are additional pressure effects.

In the eel (Belaud et al., 1976a) slow electroencephalogram (EEG) waves with a frequency of 5–6/sec appeared at 31 ATA and disappeared at 101 ATA. Rapid waves emanating from both the telencephalum and cerebellum increased in frequency. At 151 ATA, EEG activity disappeared after about 15 min. Similar experiments performed in trout (Barthélémy et al., 1981) showed anomalies in spontaneous electrical activity from opticum tectum above 51 ATA. Slow large waves (as in the eel), which originally appeared as separate bursts, became predominant at higher pressures. The EEG activity progressively disappeared within the first 10 min at 151 ATA. Concomitantly, the magnitude of visual evoked potentials (VEPs) decreased between 1 and 101 ATA, and the latency period decreased. Between 101 and 151 ATA, the decrease in VEP magnitude was associated with an increase in latency duration: the VEP disappeared after 5 min of HP exposure. In both the eel and trout, a biphasic change in behavior correlated well the biphasic change in EEG activity.

There is limited information on ventilation patterns of the fish under pressure. In trout, the first symptoms of pressure effects appear at about 21 ATA, when breathing frequency increases. At 81 ATA there is a progressive disorganization in the magnitude and pattern of opercular movements without any change in ventilatory frequency (Barthélémy et al., 1981). Above 131 ATA, ventilatory frequency becomes irregular, until finally, at 151 ATA, frequency and tidal volume decrease to a complete ventilatory arrest. This description of ventilatory disorganization holds for the eel (Balouet et al., 1973) but the pressure threshold (which varies with compression rate) (Belaud and Barthélémy, 1973) and the intensity of the changes differ depending on the protocol used.

Cardiovascular changes also occur with high pressure. Increasing pressure above 21–31 ATA induces tachycardia in the eel (Belaud et al., 1976b). This cardiac response has previously been shown in other teleosts by Drapers and Edwards (1932) and Naroska (1968), and was confirmed for the eel by Sébert and Barthélémy (1985a). It seems that the pressure effects on cardiac function are biphasic: an excitation below 101 ATA and an inhibition above this pressure (Flügel and Schlieper, 1970). An investigation of the blood pressure in the dorsal aorta and in the mesenteric vein of the eel has been carried out using differential electronic manometers. During compression, there is arterial hypotension that disappears within 20 min at 101 ATA. At the same time, an observed venous hypertension disappears rapidly and tends to change into hypotension after 1 h under pressure (Belaud and Barthélémy, 1973). These responses suggest decreased systemic resistance and possibly reduced cardiac output as a result of venous pooling of blood.

At atmospheric pressure, it is well known that cardiac and ventilatory functions are highly dependent on blood catecholamines (CAs) (Peyraud-Waitzenegger et al., 1980). Furthermore, when eels are submitted to 101 ATA, there is a large increase of CAs in the blood (Sébert et al., 1986) (Table I), with a smaller increase of CA content in the brain and heart (Sébert et al., 1984). Thus the possibility exists that CAs are responsible for the observed cardiovascular changes under HP. It is evident that any CA-mediated cardiovascular effect would be by HP modifying the receptor site (e.g., its structure, ligand affinity, and membrane environment) (see Macdonald, 1984, for review). For example, an experiment performed by Sébert and Barthélémy (1985a) on the eel and using different agonists and antagonists has shown that HP is capable of reversing CA effects. Phentolamine induces tachycardia at 1 ATA (see Fig. 3). When used under pressure (which induces tachycardia), phentolamine induces bradycardia. More experiments of this nature are needed because it is clearly unwise to assume that receptor-mediated control mechanisms always operate the same under HP as they do at atmospheric pressure.

3. OXYGEN CONSUMPTION

Oxygen consumption M_{O_2}, is a good index of aerobic metabolism and thus of energy metabolism. The oxygen consumption of several species has been studied: *Pleuronectes platessa, Ammodytes lanceolatus, Gobius minutus* (Fontaine, 1928, 1929a,b), *Cottus kessleri* (Roer et al., 1984), *Platichthys flesus* (Naroska, 1968), *Anguilla anguilla* (Belaud and Barthélémy, 1973) [see Sébert and Barthélémy (1985b) for yellow eels; these authors also have unpublished data for the silver eel, 1986], *Salmo gairdneri* (Sébert et al., 1987; Cann-Moisan et al., 1988), and *Carassius auratus* (Sébert

Table I
Plasma Contents

Substance[a]	Species	1 ATA	Pressure[b]	Conditions	Reference
CPK activity (μmol min^{-1} liter^{-1})	Trout	563 514	558 13201	24 h postdecompression at 3 and 4 ATA	Bark and Smith (1982)
NE (pmol ml^{-1})	Eel (15°C)	5.7 ± 1.7	14.7 ± 1.7	No difference after 1 or 3 h at 101 ATA	Sébert et al. (1986)
E (pmol ml^{-1})	Eel (15°C)	11.1 ± 1.6	20.2 ± 2.1		—
Soluble proteins (mg 100 ml^{-1})	Eel (17°C)	3.28 ± 0.22	3.62 ± 0.21	After 3 h at 101 ATA	Simon (1990)
Lactates (mM)	Eel (17°C)	1.65 ± 0.42	1.02 ± 0.43	After 3 h at 101 ATA	Simon (1990)
Glucose (g liter^{-1})	Eel (17°C)	1.14 ± 0.18	1.52 ± 0.20	After 3 h at 101 ATA	Simon (1990)
Total FFA (μM)	Eel (17°C)	501 ± 22	265 ± 61	After 3 h at 101 ATA	Simon (1990)
P_{aO_2} (torr)	Trout (13°C)	155 ± 1.5	162 ± 12.8	After 0.6 h at 101 ATA	Sébert et al. (1987)
C_{aO_2} (mM)	Trout (13°C)	3.0 ± 0.28	2.9 ± 0.39	After 0.6 h at 101 ATA	Sébert et al. (1987)

[a] CPK, Creatine kinase; NE, norepinephrine; E, epinephrine; FFA, free fatty acids.
[b] Samples obtained postdecompression.

7. PRESSURE EFFECTS ON SHALLOW-WATER FISHES

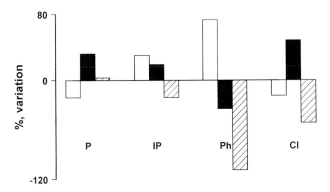

Fig. 3. Pressure and drug interactions on eel heart rate. P, Propanolol; IP, isoproterenol; Ph, phentolamine; Cl, clonidine. Open bars: treated at 1 ATA (percentage variation compared with untreated fishes at 1 ATA). Solid bars: treated at 101 ATA (percentage variation compared with untreated fishes at 101 ATA). Hatched bars: absolute percentage difference (treated–untreated) at 101 ATA. In order to allow for seasonal variation in catecholamine reactivity, Ph and Cl experiments were performed at 15°C (winter) and P and IP experiments were performed at 20°C (summer). Data from Sébert and Barthélémy (1985a).

and Simon, unpublished data, 1990). All these studies have shown that compression induces a large increase in oxygen consumption, concomitantly with periods of high motor activity, depending on the species, temperature, and salinity (Sébert, 1993). Furthermore, the higher the HP, the higher the $\dot{M}_{O_2}$ (Fontaine, 1928).

The rate of compression directly affects the increase in $\dot{M}_{O_2}$. In the yellow eels (weighing about 100 g), which have an $\dot{M}_{O_2}$ value of about 1 mmol h^{-1} kg^{-1} at 1 ATA ($T_w = 17°C$), a compression rate of 10 atm min^{-1} up to 101 ATA increases $\dot{M}_{O_2}$ to 6 mmol h^{-1} kg^{-1}, whereas a compression rate of 2 atm min^{-1} increases $\dot{M}_{O_2}$ to only 3.5 mmol h^{-1} kg^{-1} (see Sébert and Barthélémy, 1985b; Simon et al., 1989a). Similarly, compressing smaller eels (weighing about 2 g) at a rate of 10 atm min^{-1} to 101 ATA also increases $\dot{M}_{O_2}$ sixfold (from 4.2 mmol h^{-1} kg^{-1} to 24 mmol h^{-1} kg^{-1}) (P. Sébert, unpublished data, 1992).

All the previously cited authors have interpreted the $\dot{M}_{O_2}$ increase as a consequence of the large increase in motor activity of the fish during the compression period (see Sébert and Macdonald, 1993 for review); thus, although it has been shown that HP increases $\dot{M}_{O_2}$, one must remember that it is more a compression effect (ΔP versus time) than a pressure effect (ΔP). In fact, when pressure is maintained for some hours (Fontaine, 1929b; Naroska, 1968; Sébert and Barthélémy, 1985b) or weeks (Simon et al.,

1989a; P. Sébert and B. Simon, unpublished data on goldfish, 1990), $\dot{M}_{O_2}$ decreases despite the fact that the fishes continue to be as active as control fishes at atmospheric pressure. Such an observation implies that maintaining HP resets aerobic metabolism. (HP effects on anaerobic metabolism are alluded to in Section IV,B,6).

B. Organs and Tissues

Measurements of concentrations of various substances in organs and/or tissues are generally performed on tissue samples taken from fishes first exposed to high pressure, and then decompressed. Sometimes, tissues (muscle, nerve, heart) have been sampled at atmospheric pressure, and then compressed to perform measurements.

1. Isolated Muscle and Nerve

Due to its particular anatomical orientation, the myotomes of fish skeletal muscle cannot be studied with a conventional nerve–muscle preparation, as in land animals. Instead, a flat strain device is "pinned" onto an isolated block of muscle, allowing the muscle to be electrically stimulated, and different parameters are studied. When isolated muscle is pressurized, an increase in maximum twitch tension and in the time taken to reach peak tension is noted (Wardle, 1985; Wardle *et al.*, 1987). In contrast, half-relaxation time is not modified by pressure, nor is the tetanic tension modified at high stimulus frequency (Tetteh-Lartey, 1985). Additionally, Harper *et al.* (1987) have shown a decrease in compound action potential and conduction velocity of the vagus nerve in the cod.

2. Isolated Heart

Cardiac muscle is interesting because it has the unique property of initiating its own excitatory impulse. Some studies have been performed on the hearts of *Fundulus* embryos (Drapers and Edwards, 1932), myocardial strips (Edwards and Cattel, 1928), and isolated heart [see Ebbecke (1935) on catshark and Gennser *et al.* (1990), Belaud *et al.* (1976b), Pennec *et al.* (1988) on the eel]. Although Pennec *et al.* (1988) found a decrease in beat frequency (in yellow eels adapted to seawater and with nonlinear compression), results generally show a pressure-induced increase in isometric twitch tension and in beat frequency. However, it seems that when pressure is maintained on eel atria, the excitatory effect disappears (Gennser *et al.*, 1990). As shown by Belaud *et al.* (1976b), the cardiac response (tachycardia or bradycardia) is dependent not only on the temperature and the milieu used, but also on the mode of compression (linear or pulses). The results obtained in the isolated fish heart are similar to what is observed in the hearts of mammals. A comparison of cardiac responses from intact animals and isolated hearts shows that pressure acts both through extrinsic control mechanisms and through a direct action on the cardiac cells.

3. ION MOVEMENTS IN ISOLATED GILL AND RED BLOOD CELLS

When isolated gill preparations are incubated in artificial seawater and then submitted to hydrostatic pressure (Péqueux and Gilles, 1986), various changes occur in tissue Na^+, K^+, and Cl^- contents, depending on the experimental pressure. Na^+ content varies much more rapidly than Cl^- and K^+: the threshold pressures at which the increases are observed are higher for Cl^- than for the other ions. These results raise the possibility that pressure could act selectively on the various transport mechanisms given that Na^+ and Cl^- movements are, to some extent, independent processes in the teleostean gills.

It has been shown that pressure inhibits Na^+, K^+-ATPase. However, the increase in gill Na^+ content observed under pressure is considered to be due only to a pressure effect on the Na^+ passive diffusion from the environment toward body fluids (maybe via an enhanced Na^+ permeability) (see Péqueux and Gilles, 1986). Pressure decreases gill K^+ content, which can also be due to a change in K^+ permeability (Péqueux, 1981; Péqueux and Gilles, 1986).

As a general rule, Na^+, K^+-ATPase is marginally inhibited by moderate pressure steps (about 100–200 ATA). If pressure is increased, there is a substantial decrease in enzyme activity irrespective of species and/or the organ (Péqueux and Gilles, 1978; Pfeiler, 1978; Roer and Péqueux, 1985; Gibbs and Somero, 1989). Thus, it can be suggested that, at the experimental pressures normally used, i.e., 100–200 ATA, the changes in gill ionic contents are due to an action of HP on ion transport channels. The observed increases in permeability are presumably a result of these channels, increasing their probability of being in an open state at high pressure.

Some studies on erythrocyte ion transfers have also been performed (see review from Macdonald and Shelton, 1985; Shelton and Macdonald, 1987). It seems that generally, fish erythrocyte behavior under pressure resembles that seen in comparable studies in human red blood cells in which pressure of about 30–150 atm inhibits active Na^+ efflux (Goldinger et al., 1980) and slightly increases net K^+ passive efflux (see Péqueux and Gilles, 1986). However, there are many disparities in the results that may be due to differences in the methods used for the isolation and treatment of the samples and/or to differences in their stability level, especially when active transport is considered.

In conclusion, in shallow-water fishes exposed to pressures of about 50 to 100 ATA, changes in ion contents are principally due to changes in permeability and not to changes in active transport. The effects are mainly an increase in Na^+ content (by decreasing Na^+ efflux) and a decrease in K^+ content (by increasing K^+ efflux). An interestingly review by Hall et al. (1993) discusses membrane transport under pressure in an erythrocyte model.

4. Brain

In the brains of eels exposed to hydrostatic pressure for 6 h, Sébert et al. (1995a) found a 124% increase in malondialdehyde (MDA) levels at 51 ATA and a 290% increase at 101 ATA. Because MDA is generally considered to be an indicator of lipoperoxidation and thus membrane impairment, its increase could explain, at least in part, the excitation periods that are observed during and after animal compression.

Neurotransmitters have been measured in the brain of the yellow eel (see Table II). Generally 101 ATA pressure does not induce any great changes in brain catecholamine, serotonin, glycine, or glutamic acid content. However, these observations are limited to one temperature, and an interaction seems to exist between temperature and pressure: negative for CAs

Table II
Neurotransmitter Contents in Eel Brain[a]

Neurotransmitter[b]	T_w(°C)	1 ATA	101 ATA (PD) (3 h)	Reference
NE	15	1.40 ± 0.072	1.42 ± 0.085	Sébert et al. (1986)
E	15	0.24 ± 0.010	0.28 ± 0.009	Sébert et al. (1986)
DA	15	0.83 ± 0.041	1.01 ± 0.032	Sébert et al. (1986)
DOPAC	15	0.27 ± 0.016	0.30 ± 0.022	Sébert et al. (1986)
5-HT	14	0.89 ± 0.095	1.06 ± 0.136	Sébert et al. (1985a)
5-HIAA	14	0.21 ± 0.011	0.29 ± 0.028	Sébert et al. (1985a)
Gly	16	1394 ± 58.6	1390 ± 75.5	Sébert et al. (1985b)
Gln	16	14377 ± 439.2	15834 ± 674.5	Sébert et al. (1985b)
GABA	17	2560 ± 99.0	2530 ± 76.0	Barthélémy et al. (1991)
MDA	16	6.1 ± 0.5	24.2 ± 6.5	Sébert et al. (1995a)

[a] Values, expressed in nmol g^{-1}, are mean ± SEM. Compression rate, 10 atm min^{-1}. PD, Samples obtained postdecompression.

[b] Abbreviations: NE, norepinephrine; E, epinephrine; DA, dopamine; DOPAC, 3,4-dihydroxyphenylacetic acid; 5-HT, 5-hydroxytryptamine; 5-HIAA, 5-hydroxyindolacetic acid; Gly, glycine; Gln, glutamine; GABA, γ-aminobutyric acid; MDA, malondialdehyde.

(Sébert et al., 1984) and positive for indolamines (Sébert et al., 1985a). Further experiments need to be carried out in order to discover whether these interactions are related to changes in receptor–ligand affinity.

5. BLOOD

Hydrostatic pressure effects on the blood of fishes (or other species) are not clear. Certain studies show a decrease in fish hemoglobin oxygen-binding affinity (Brunori et al., 1978; Wells, 1975) and a change in its absorption spectrum (Gibson and Carey, 1975) at pressures up to 100 ATA. In contrast, Johnson and Schlegel (1948) found no HP effect. Some studies show that mammalian hemoglobin increases its oxygen affinity (Kiesow, 1974; Reeves and Morin, 1986); however, these studies were performed under high inert gas pressures that can oppose or even exceed HP effects, depending on the gas (Wells, 1975). In the same manner, the study of the hemoglobin molecule of different fish living at different depths shows some differences, but there is no unequivocal answer to the physiological significance of these differences (Diprisco and Tamburrini, 1992). Gordon (1970), reporting several Russian studies, noted that changes in blood oxyhemoglobin and blood sugars of pressure-exposed fish varied seasonally. However, at the pressures commonly used, the changes in Hb affinity were very small. This can explain why Sébert et al. (1987) failed to find any changes in the arterial oxygen content of trout exposed to 101 ATA when compared to fishes at atmospheric pressure. This observation means that despite the changes in ventilation, circulation, and ion transfers reported elsewhere, oxygen movement across the gills is adequate to saturate the Hb.

For reference, Table I lists several other substances that have been measured in blood or plasma.

6. MUSCLE BIOCHEMISTRY

Table III lists compounds measured in the white muscle of fishes exposed to HP. It is interesting to note that the trout, which is very dependent on the oxygenation of the ambient medium, is quite sensitive to pressure and exhibits pronounced metabolic alterations.

Recent studies on eels have shown that short-term (3 h) exposure to 101 ATA induces substantial modifications in certain enzyme activities in muscle (white and red). The most important of these changes is a decrease in cytochrome c oxidase together with an increase in glycolytic enzyme activities, but without great changes in the enzymes participating in the Krebs' cycle (Simon et al., 1992). Thus, there is an increase in the pyruvate kinase/cytochrome oxidase (PK/COX) ratio, which is used as a biochemical index of anaerobic versus aerobic capacities for metabolism (see Crockett and Sidell, 1990). At the same time, glycogen stores decrease and fatty

Table III
White Muscle Contents[a]

Component	1 ATA	101 ATA (PD)	Fish	Reference[b]
Fatty acids (μmol g_{ww}^{-1})	1.70 ± 0.27	2.12 ± 0.36	Eel	1
Glycogen (μg g_{ww}^{-1})	0.37 ± 0.04	0.13 ± 0.02*	Eel	1
COX (μmol substrate min^{-1} kg$_{ww}^{-1}$)	0.025 ± 0.066	0.012 ± 0.005*	Eel	1
MDH (μmol substrate min^{-1} kg$_{ww}^{-1}$)	43.6 ± 6.0	59.0 ± 5.2	Eel	1
IDH (μmol substrate min^{-1} kg$_{ww}^{-1}$)	0.74 ± 0.06	0.79 ± 0.08	Eel	1
CS (μmol substrate min^{-1} kg$_{ww}^{-1}$)	1.29 ± 0.16	1.22 ± 0.10	Eel	1
LDH (μmol substrate min^{-1} kg$_{ww}^{-1}$)	376 ± 43	529 ± 25*	Eel	1
PK/COX	7238 ± 2116	20983 ± 9328	Eel	1
ATP (nmol g_{ww}^{-1})	3388 ± 250	2967 ± 204	Eel	2
	3386 ± 472	1164 ± 206*	Trout	3
ADP (nmol g_{ww}^{-1})	690 ± 48	682 ± 91	Eel	2
	878 ± 73	508 ± 52*	Trout	3
AMP (nmol g_{ww}^{-1})	49 ± 14	111 ± 22*	Eel	2
	130 ± 20	141 ± 11	Trout	3
Energy charge	0.90 ± 0.01	0.89 ± 0.01	Eel	2
Energy charge	0.85 ± 0.02	0.74 ± 0.03*	Trout	3
IMP (nmol g_{ww}^{-1})	1691 ± 415	6509 ± 709*	Trout	3
NAD (nmol g_{ww}^{-1})	218 ± 15	233 ± 15	Trout	3
NADH (nmol g_{ww}^{-1})	61 ± 11	62 ± 7	Trout	3
NADP (nmol g_{ww}^{-1})	6 ± 2	2 ± 1	Trout	3
NADPH (nmol g_{ww}^{-1})	10 ± 2	8 ± 3	Trout	3

[a] Values are mean ± SEM; Fish were exposed at 101 ATA hydrostatic pressure for a short period. PD, Samples obtained postdecompression. Temperatures: reference 1, 17.5°C; reference 2, 13.5°C; reference 3, 12°C.

[b] 1, Sébert et al. (1993a); 2, Sébert et al. (1987); 3, Cann-Moisan et al. (1988).

* Significant difference ($P < 0.05$ or better).

acids accumulate in white muscle. In other words, a decrease in aerobic energy production is compensated for by an increase in anaerobic energy production. However, the metabolic compensation is far from complete. The ATP tissue contents (except for brain) decrease by 15% in the eel

(Sébert et al., 1987) and by 65% in the trout (Cann-Moisan et al., 1988), and energy charge decreases.

The decrease in oxygen consumption described in Section IV,A,3 could be the consequence of an impairment in oxygen transfer from the ambient medium to the cell. However, because arterial blood is fully oxygenated under pressure (Table I), it is most likely that the decrease in $\dot{M}_{O_2}$ observed under pressure is not due to a decrease in O_2 availability, but to either reduced delivery (see Section II,A,2) or a decrease in O_2 use at the cell level, i.e., an alteration in aerobic metabolism. This latter hypothesis is in agreement with the results observed (Table III), i.e., accumulation of fatty acids (specific substrate), and a decrease in the activity of cytochrome oxidase and ATP content, leading to a decrease in energy charge. In fact, the changes reported concerning the effects of hydrostatic pressure on energy metabolism are in agreement with the hypothesis that HP could induce a state resembling histotoxic hypoxia (Sébert et al., 1993a). The origin of this hypoxia will be discussed in section IV,D. A partial compensation for the energy production is achieved by the activation of the anaerobic pathway (increased glycogen use and LDH activity).

C. Factors Interacting with Hydrostatic Pressure

1. ANAESTHESIA

Wann and Macdonald (1988) have considered in detail the interactions of high pressure and general anaesthetics. Studies of interactions have mainly been performed on mammals, because these interactions are of special interest in humans. In trout, the depression of ventilation, EEG, and VEP by anaesthetics is cancelled out under HP. Additionally, anaesthetic administration before compression increases survival time under pressure (Fig. 2) and reduces restlessness during compression in relation to untreated fishes. Similarly, HP (101 ATA) decreases the duration of anaesthesia. Also, the higher the water temperature, the shorter the recovery time (Belaud et al., 1976c, 1977), an observation that may be explained by the effect of temperature on metabolism and elimination of anaesthetics. Clearly, HP and anaesthetic drugs thus act in opposite directions on the trout body as a whole (Barthélémy et al., 1981).

2. INERT GASES

Several reviews have been devoted to the biological effects of inert gases (Varene and Valiron, 1980) and their interactions with hydrostatic pressure (Brauer et al., 1982).

In 1981, Beaver and Brauer failed to show any effect of hyperbaric Heliox (helium and oxygen mixtures) on the convulsion threshold pressure of *Symphurus plagiusa.* In contrast, Barthélémy *et al.* (1981) show that helium and nitrogen mixtures oppose the HP effect in trout (Fig. 2). Like anaesthetic drugs, helium and nitrogen (in certain conditions of administration of N_2) can extend the survival times under HP. Survival times are maximal for a P_{N_2} of 41–61 ATA under HP of 101 to 151 ATA (when N_2 is introduced at 51 ATA). Also the changes in EEG, behavior, and ventilation induced by HP decrease when N_2 is introduced in water.

Similar results have been obtained with isolated eel atria. The introduction of N_2 at pressure has been shown to reduce the HP effects on twitch tension (Gennser and Karpe Förnhagen, 1990). Barthélémy *et al.* (1988) and Simon *et al.* (1989b) have reconsidered the experiments of Sébert and Barthélémy (1985a,b) on the eel, saturating the water with nitrogen at 71 ATA. These results show additional evidence for the opposing effects of N_2 and HP. Whereas $\dot{M}_{O_2}$ decreases continuously when eels are submitted to 101 ATA, metabolism reaches a steady state within 90 min of exposure to HP and nitrogen. The time period needed to observe maximal and stable opposite effects corresponds to the time calculated as necessary to saturate the major tissues of the eel with nitrogen (Belaud and Barthélémy, 1979). Some results of Simon *et al.* (1989b) have also shown that when eels are exposed to HP and high water P_{N_2}, the cardiac effects of catecholaminergic drugs are different from those observed in conditions of high pressure only. Thus, under conditions of HP it is inappropriate to consider N_2 as a truly inert gas.

3. TEMPERATURE

As early as 1958, Marsland wrote that "pressure has begun to take its place with temperature as a fundamental factor governing physiological processes." Temperature is a very important factor because (1) of its thermodynamic interactions with pressure and (2) fishes are ectotherms.

Many of the pressure effects described in fishes can be compared to those observed when water temperature is modified at atmospheric pressure (see Shaklee *et al.,* 1977; Crawshaw, 1979; Hazel, 1979; Walesby and Johnston, 1980; White and Somero, 1982; Cameron, 1984; Heisler, 1984; Hazel and Carpenter, 1985; Hazel and Zerba, 1986). This may increase our understanding of the corresponding pressure effects, because all thermodynamic equations (see Balny *et al.,* 1989) show that pressure (or pressure variations) and temperature can have opposite effects. However, adaptation to low temperature does not necessarily mean preadaptation to high pressure (Somero, 1992a,b).

Pressure/temperature interactions and their consequences on deep-living fishes have been considered by Sébert (1993). Pressure/temperature interactions clearly affect the behavior of fishes. For example, using hydrostatically driven gradient tube systems, Brauer et al. (1985) reported an increase in the temperature preferendum averaging +4°C per 100 atm for *Chasmodes bosquianus*. Thus, it appears that fishes seek a water temperature capable of opposing the pressure effects. In fact, such a $\Delta T/\Delta P$ value corresponds closely to the characteristics of nonisotropic membrane systems and may be relevant for the performance of excitable cells (Brauer et al., 1985). More recently, Sébert et al. (1995b) have shown that when temperature increases (+5°C) together with pressure (101 ATA), the resulting oxygen consumption increase observed during compression is lower than that when pressure acts alone. Consequently Q_{10} is much greater than 2 at the beginning of a 3-h temperature exposure at 1 ATA than at the end ($Q_{10} = 2.3$), but it is always less than 1.5 (starting from 0.35) when under pressure. In contrast, a study concerning measurements of $\dot{M}_{O_2}$, in both winter and summer showed no differences in pressure effects (P. Sébert and B. Simon, unpublished data, 1995). In addition, the threshold pressure for convulsions is independent of test temperature but varies with acclimatization temperature. This variation is species dependent (Beaver and Brauer, 1981). This absence of pressure/temperature interactions *in vivo*, in contrast to the thermodynamic predictions and *in vitro* observations (see Macdonald, 1984; Brauer et al., 1985), probably reflects the nature of the *in vivo* protocols, which tend to vary greatly.

D. Mechanisms: The Membrane Hypothesis

The basis for all HP effects is the change in system volume that accompanies a physiological or biochemical process (Somero, 1992a). This may lead to major changes in both the total amount of metabolic flux and the relative activities of different pathways (Somero and Hand, 1990). Furthermore, HP is capable of modifying the physicochemical state of membranes and their major components, the phospholipids (Clausius–Clapeyron relation). The effects of pressure on the molecular structures and physiological functions of cell membranes have been extensively reviewed by Macdonald (1984).

Pressure can decrease membrane fluidity by acting directly on phospholipids and thus indirectly (at least in the "physiological" pressure range) on membrane proteins (enzymes, ionic channels, receptors). Pressures of 50 ATA or less are sufficient to perturb many membrane-localized functions (Somero, 1991).

HP also modifies the structure and dynamic properties of macromolecules, including enzymes (Johnson et al., 1974; Kunugi, 1992; Finch and Kiesow, 1979). Pressure, as well as temperature can act directly on proteins (see Macdonald, 1984; Balny et al., 1989; Balny and Masson, 1993; Somero,

1995) and thus modify allosteric configuration of enzymes and the rate of enzyme reactions (see Johnson and Eyring, 1970; Low and Somero, 1975). Most transporters show some degree of pressure sensitivity. To operate, these transporters require conformational changes; conformational changes may be inhibited by pressure either directly, via the protein, or indirectly, via the lipid environment, in such a way that the protein is constrained during the translocation process (Hall *et al.*, 1993). Pressure can also dissociate peripheral protein–membrane complexes (Plager and Nelsestuen, 1992).

As reported by Péqueux (1981), "little can be said about the molecular aspect of pressure-induced disturbance. Nevertheless, several pieces of evidence prompt us to explain such effects in terms of phase transition in the lipidic components of the membrane." It is true that alterations in cellular and/or subcellular membranes can explain most of the behavioral and physiological changes observed under pressure [alternative explanations are presented by Macdonald (1987)]. The experimental data support the membrane hypothesis: (1) COX activity correlates with membrane fluidity (Vik and Capaldi, 1977); (2) HP decreases the activity of certain membrane-bound enzymes (Section IV,B,6); (3) HP modifies the effects of catecholaminergic drugs, which act through membrane-bound receptors (Section IV,A,2); (4) HP opposes the effects of anesthesia and N_2 (Section IV,C); (5) HP induces an increase in the tissue content of malondialdehyde (Sébert *et al.*, 1995a) [MDA is a product of lipid peroxidation, a process known to cause a deleterious decrease in mitochondrial membrane fluidity (Chen and Yu, 1994; Esterbauer *et al.*, 1990, 1991; Rifkind *et al.*, 1993)]; and (6) when HP is maintained for weeks, membrane fluidity (measured at 1 ATA) is higher in pressure-exposed fishes than in control fishes. This implies that if fluidity is restored during a long period under pressure, it must have decreased during the first hours under pressure (see Section V).

Clearly there is much experimental support for the hypothesis that HP acts at the membrane level, and a modification in membrane lipid order (fluidity) is the paradigm most widely invoked to explain the observed effects. However, there are also data indicating that other features of membrane organization (such as acyl chain length, balance between conically and cylindrically shaped phospholipids, and existence of discrete membrane domains) can influence membrane function (Lee, 1991; Hazel, 1995). But, whatever the mechanism, the cell membrane appears to be a major target for the HP effect.

The results described in this section have concerned fishes exposed for some hours to HP. This, in fact, is a feature of most hyperbaric studies. It is likely that the observed effects are mainly the consequences of pressure variations during the compression phase (ΔP versus time), because the time under constant pressure is relatively short. The next section addresses the following questions: What are the effects of long-term exposure to HP? Are shallow-water fish able to acclimatize to high hydrostatic pressure?

V. ACCLIMATIZATION OF FISH TO HYDROSTATIC PRESSURE

High-pressure aquaria have been used in a few cases to study the acclimatization of aquatic animals to high hydrostatic pressure. The laboratory and *in situ* methods of studying deep-sea fishes are described in Chapter 9 of this book [see also Sébert and Macdonald (1993) for general techniques for studying fishes under pressure]. However, at the present time, it appears that only the system of Sébert *et al.* (1990) has been used in a prolonged manner to study the acclimatization of fishes or other aquatic animals to pressure, and thus most results concerning fish acclimatization to HP originate from this research team.

A. Behavior and Oxygen Consumption

Water circulation during compression does not modify the behavior of fishes. The symptoms described for short-term exposure (Section IV,A,1) are evident at the beginning of long-term pressure exposure but disappear within a few days. Behavior is then similar to that of fishes at atmospheric pressure.

Heart rate exhibits slight variations, with mean values about 5% higher than those found in control fish (P. Sébert, unpublished data, 1988). When $\dot{M}_{O_2}$ is measured in yellow eel exposed to 101 ATA for 30 days (normoxic conditions), the maximum value (which is dependent on the compression rate) is observed at the end of the compression period and it then decreases exponentially with time (Fig. 4). The time constant is 1.4 days. This means that $\dot{M}_{O_2}$ reaches a steady-state value averaging 65% of the control value within 7 days (Simon *et al.*, 1989a). In such acclimatized fishes, Sébert *et al.* (1995b) increased water temperature by 5°C and observed that after some hours in warm water, Q_{10} was lower than 1.2, thus showing that pressure-acclimatized shallow-water fishes exhibit only slight thermal sensitivity.

When the migratory pattern suggested by Tesch (1978) is reproduced for silver eels in the laboratory (a diurnal vertical migration of about 60 atm), the relationship $\dot{M}_{O_2}$ versus time is not very different from that observed for yellow eels under sustained pressure (Sébert, 1993). It is interesting to note that animals that are decompressed after 30 days under

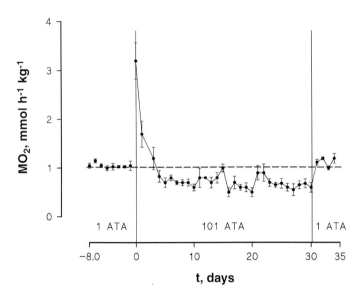

Fig. 4. The oxygen consumption of *Anguilla anguilla* over 30 days at 101 ATA at 17°C. Each point is the mean (± SEM) obtained from four separate experiments on different fishes. Data from Simon *et al.* (1989a).

pressure are nearly motionless, not reactive, and exhibit some loss of equilibrium. Such an observation can be compared with the reported behavior exhibited by fishes trawled from the lake or sea bottoms to the surface (Brauer *et al.*, 1984; Macdonald *et al.*, 1987). However, when the previously pressure-acclimatized eels are recompressed some days after decompression, normal behavior is restored (Simon *et al.*, 1989a). During this second compression, $\dot{M}_{O_2}$ increases less and rapidly reaches a steady state, in contrast with naive fishes (Fig. 5). Preliminary experiments have also been performed by Johnstone *et al.* (1989) using 72 h confinement at 40 atm. They have shown that such "training" significantly increases the mean onset pressure for convulsions by about the same pressure (40 ATA).

All of the above information clearly shows that a shallow-water fish (the eel) is able to acclimatize to high pressure, and that during this acclimatization physiological changes occur that makes the fishes less sensitive to further compression. Acclimatization of trout and goldfish for at least 21 days at 101 ATA has also been successful, provided that care was taken to compress the fish slowly (B. Simon and P. Sébert, unpublished data, 1996).

B. Metabolism

Figure 4 shows that during the acclimatization of the eel to HP, metabolic acclimatization involves perfect, and perhaps overcompensation (as

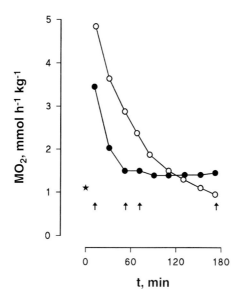

Fig. 5. The decline in the rate of oxygen consumption in *Anguilla anguilla* ($T_w = 17°C$) previously pressurized and decompressed as in Fig. 4 (●) and in naive specimens (○). The fishes undergoing compression after previous exposure to 101 ATA reach a steady-state $\dot{M}_{O_2}$ before the previously untreated fishes. The arrows indicate a significant difference at $P < 0.05$; the star indicates control $\dot{M}_{O_2}$ for 1 ATA. Data from Simon *et al.* (1989a).

defined by Prosser, 1991) of, energy production. After 1 month under pressure (Fig. 6), muscle contents of ATP, ADP, and AMP are restored to normal values (sometimes increased) and enzyme activities measured at 1 ATA return to or are higher than the values observed in nonexposed fishes. The maintenance of aerobic energy production and energy charge, despite a decrease in $\dot{M}_{O_2}$, suggests that oxidative phosphorylation efficiency improves during pressure acclimatization (Simon *et al.*, 1992). The process of pressure acclimatization is not a question of adapting to the energetic state observed after some hours under HP, but rather a return to the state observed before compression.

The metabolic changes induced by long-term exposure to 101 ATA are in agreement with an increased use of pyruvate synthesis as a substrate for aerobic and anaerobic pathways. As the enzyme activities of the TCA cycle increase and as the PK/COX ratio decreases, it is suggested that the aerobic pathway predominates. Such observations are, in general, consistent with those made regarding cold acclimatization (see Tyler and Sidell, 1984) and therefore further support the thermodynamic similarity between high pressure and low temperature. Likewise, kinetic studies of NADP-isocitrate dehydrogenase (IDH) by Simon *et al.* (1997) show that specific activity

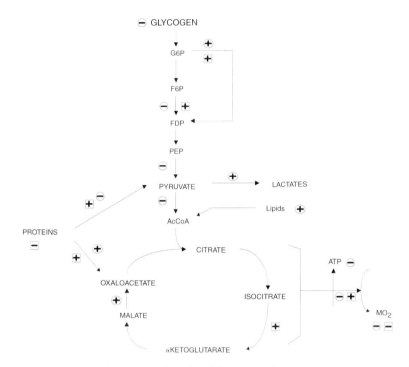

Fig. 6. Effects of short-term (circles) and long-term (4 weeks, squares) exposure to 101 ATA on the energy metabolism in the white muscle of *Anguilla anguilla* compared to values in fishes maintained at 1 ATA. The positive or negative sign in the symbol indicates the direction of change. From Sébert and Macdonald (1993). Fish *In* "Effects of High Pressure on Biological Systems" (A. G. Macdonald, ed.), pp. 147–196. Copyright 1993 Springer-Verlag.

increases in pressure- and cold-acclimatized eels without any difference in activation energy. Concomitantly, the study of IDH (but also lactate and malate dehydrogenases and hexokinase) isoenzymes has shown that pressure acclimatization induces an increase in the intensity, i.e., an increase in enzyme activity, without any modifications in the proportion of isoenzymes (Fig. 7).

C. Tissue Composition

Various substances have been measured in gill, blood, and muscle of eels acclimatized for 30 days at 101 ATA hydrostatic pressure (Table IV). Increases in plasma Na^+, Mg^{2+}, and (mainly) Cl^- contents were found, whereas in muscle and gill, only Na^+ and Cl^- (mainly Cl^-) were increased under constant HP. Concomitantly with the changes in ion contents, there was a decrease in the maximum activities of gill Na^+, K^+-ATPase and

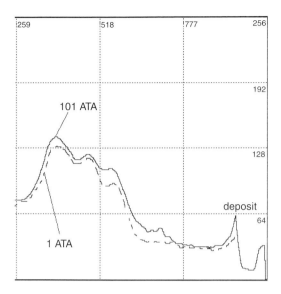

Fig. 7. Densitometry data for malate dehydrogenase in eel white muscle. Each peak corresponds to one isoenzyme and the area under the peak is proportional to isoenzyme activity. Each profile is the mean of five eels. From B. Simon (unpublished data, 1996).

Mg^{2+}-ATPase. These results are in agreement with a Na^+ balance impairment in the tissues studied. No variation in neurotransmitters such as gamma amino butyric acid (GABA) and glutamine (Glu) in the brain of eels acclimatized to high pressure was found (Barthélémy et al., 1991). In contrast, female silver eels kept in a cage at a depth of 450 m for 3 months show a slightly but significantly higher gonadosomatic ratio when compared to control fishes at 1 ATA. Interestingly, the pituitary gonadotropin content was found to be 27 times higher. It has been suggested that HP could be one of the factors contributing to the stimulation of the gonadotropic function and the onset of puberty in European eels (Fontaine et al., 1985). It is known that silver eels migrate without feeding and that white muscle protein content decreases (−30%) under pressure. It would be interesting to compare the loss of muscle proteins and the gain in gonad mass of the same fish under pressure.

D. Membrane Fluidity and Composition

The fluidity of the bilayer estimated from measurements with the fluorescent probe diphenylhexatriene (DPH) at normal atmospheric pressure increases within the animal's normal depth range. Concomitantly, the main

Table IV
Tissue Contents in Pressure-Acclimatized Fishes[a]

Measure	Water	Gill	Blood	White muscle
Na^{+b}	1	59 (1)	151 (1)	22 (1)
Cl^{-b}	1	31 (1)	73 (1)*	8 (1)*
K^{+b}	0.9	63 (1)	7 (1)	89 (1)
Ca^{2+b}	1	—	7 (1)	—
Mg^{2+b}	0.6	—	5 (1)	—
H_2O (%)	100	76 (1)	—	70 (1)
Proteins[c]	—	—	31 (1)	25 (2)*
Hematocrit (%)	—	—	27 (1)	—
Lactates[d]	—	—	0.7 (3)	42 (2)
Gly, Glu[e]	—	—	1.5 (3)	0.3 (2)
Fatty acids (μmol g_{ww}^{-1})	—	—	—	1.9 (2)

[a] $T_w = 17°C$. Samples obtained postdecompression. Source of each measurement is given in parentheses: 1, Sébert et al. (1991); 2, Simon et al. (1992); 3, Simon (1990).
[b] Units: mEq liter^{-1} (liquid) or mEq g_{ww}^{-1} (tissue).
[c] Units: g liter^{-1} (blood) or mg g_{ww}^{-1} (tissue).
[d] Units: mM (blood) or μmol g_{ww}^{-1} (tissue).
[e] Units: g liter^{-1} (blood) or μmol g_{ww}^{-1} (tissue). Gly, Glycogen; Glu, glucose.
* Significantly different from control fish.

phospholipids in liver mitochondria show a decrease in their fatty acid saturation ratio. Such results are consistent with the homeoviscous theory, although compensation for the manner in which pressure reduces bilayer fluidity is less than perfect (see Cossins and Macdonald, 1984, 1986, 1989; Macdonald and Cossins, 1985). Nevertheless, Behan et al. (1992) have shown that brain myelin membranes have a similar order at the respective ambient pressure and temperature of the species concerned. Some evidence exists that shows that membrane structure affects gene transcription activity, and that membrane lipid biosynthesis is perhaps controlled by a negative-feedback loop based on lipid order (Maresca and Cossins, 1993). In fact, alterations in membrane lipid composition may relate to the conservation of dynamic membrane properties rather than to the fine tuning of lipid order (Hazel, 1995). In yellow freshwater eels acclimatized to 101 ATA for 1 month, similar results have been obtained (Table V). The fluidity of gill membrane fragments, as measured at 1 ATA, increased (Sébert, et al., 1993b).

Studies on goldfish brain membranes have shown that HP orders the bilayer by an amount equivalent to a cooling of 15°C per 1000 ATA (Chong

Table V
Gill membrane in Pressure-Acclimatized Fishes[a]

Measure[b]	1 ATA	101 ATA
Anisotropy	0.220 ± 0.001	0.215 ± 0.001*
M + P (%)	67.8 ± 1.4	73.2 ± 1.4*
Unsaturation index	191 ± 8	197 ± 7
Saturation ratio	0.48 ± 0.03	0.37 ± 0.03*

[a] The values are means ± SEM at $T_w = 17°C$. Data modified from Sébert et al. (1993b).

[b] Anisotropy is related to membrane order: a decrease in anisotropy means a decrease in membrane order, i.e., an increase in fluidity. The samples at 101 ATA were obtained postdecompression. M + P, sum of mono- and polyunsaturated fatty acids.

* Statistical significance ($P < 0.05$ or better).

and Cossins, 1983). The difference in anisotropy (fluidity) of gill extracts between control and pressure-acclimatized eels was approximately 0.005 (Table V), which is thus equivalent to a cooling of approximately 0.5°C. In other words, the change in DPH anisotropy observed after 30 days exposure to 101 atm offsets about 33% of the pressure-induced ordering. This value is within the range of homeoviscous efficiencies observed in various membrane preparations of different fish species in response to temperature acclimatization (Cossins, 1983; Cossins and Macdonald, 1989; Lee and Cossins, 1990; Hazel 1995). Again there is general support for the membrane hypothesis presented in Section IV,D.

Lipid analysis of the gill membrane using gas–liquid chromatography has shown that there is a higher unsaturation index in HP-exposed fishes than in control fishes. This increase has been shown to be due to an increase in polyunsaturated fatty acids (Sébert et al., 1993b). In mitochondria-rich fractions from the liver, the decrease in phosphatidylcholine in favor of phosphatidylethanolamine (modifying membrane phase behavior and/or membrane order) is also believed to compensate for the loss in fluidity induced by pressure at the beginning of the acclimatization period (Sébert et al., 1994). Thus it appears from these experiments that homeoviscous regulation, described for temperature acclimatization (see White and Somero, 1982) and in deep-water fishes (see above), can also be observed in shallow-water fishes exposed to high pressure for a long period of time. As stated in Section IV, homeoviscous adaptation has limitations as an adaptative paradigm because several experimental data argue against a role for fluidity (Lee, 1991; Hazel, 1995). Nevertheless, whatever the exact mechanism may be, important changes that occur at the membrane level allow the fishes to acclimatize to HP.

E. Structural Changes

The preceding sections have discussed the many changes observed in muscle metabolism during the pressure acclimatization process. For example, it appears that there are differences in the relative participation of aerobic and anaerobic pathways in energy production. Muscle represents about 60% of the eel weight. Muscle mass consists of about 75% white fast fibers and 25% red slow fibers (Cornish and Moon, 1985). Simon *et al.* (1991) have shown that red muscle is unaltered in protein content or fiber cross-sectional area. In contrast, white muscle undergoes a 32% decrease in protein content (Table IV) that is compensated in mass by an increase in water content. Concomitantly, fiber composition is modified: there is an increase in small-diameter fibers at the expense of large fibers, leading to an overall 16% decrease in mean fiber area (Fig. 8). It is not clear whether a relationship exists between the decrease in fiber area and the decrease in protein content, although similar observation have been made in muscles of deep-sea fishes.

Apart from the morphological changes at muscle level, gills have also been studied. As mentioned previously, changes in gill membrane composi-

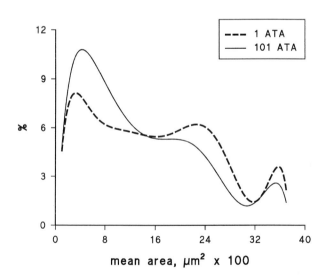

Fig. 8. Area frequency distribution of eel white muscle fibers at 1 and 101 ATA. Data adapted from Simon *et al.* (1991).

7. PRESSURE EFFECTS ON SHALLOW-WATER FISHES

Table VI
Gill Histology in Pressure-Acclimatized Eel[a]

Measure	1 ATA	101 ATA
Number of mucus cells (afferent side)	17.8 ± 1.1	4.4 ± 0.6
Number of mucus cells (efferent side)	15.3 ± 0.7	6.5 ± 0.4
Number of Cl^- cells/mm^2	798 ± 174	3095 ± 403
Fractional area ($\mu m^2/mm^2$)	6804 ± 1316	46194 ± 4470

[a] Data from Dunel-Erb et al. (1996). The values are mean ± SEM at T_w = 20°C (N = 5). For mucus cells, values are given for 12 cross-sections of filament per fish. The samples from pressure-acclimatized eels were obtained postdecompression.

tion and fluidity are accompanied by modifications in ion tissue contents and ATPase activities. Thus, the gill seems to be a pressure-sensitive tissue, most likely because of its important role in ionic and osmotic regulation of the ambient and internal media. Dunel-Erb et al. (1996) studied gill epithelium of freshwater yellow eels acclimatized to HP and found a significant decrease in the number of mucus cells and a large increase in density and in fractional area of chloride cells on the apical surface (Table VI). To explain these results, Dunel-Erb et al. (1996) suggest that acclimatization to high pressure mimics the environmental conditions of migration for silver eels. These conditions include a rise in salinity, which requires NaCl excretion and thus Na^+,K^+-ATPase activity. The increase in density and size of chloride cells could compensate for the impairment of ATPase by high pressure (Section V,C).

VI. COMPARISON OF SHALLOW-WATER FISHES AND DEEP-WATER FISHES

Deep-sea fish physiology has been reviewed by Hochachka (1975), Torres et al. (1979), Somero et al. (1983), Siebenaller and Somero (1989), Siebenaller (1991), Somero (1990, 1991, 1992a,b), and Sébert and Macdonald (1993). The following discussions report the principal results and compare them with what is known about shallow-water species under pressure. Most of the results concern oxygen consumption and muscle biochemistry.

A. Oxygen Consumption

The physiological state of fishes trawled from the bottom of the sea has often been questioned. It is certain that such fish suffer from trawling and hydrostatic decompression, but many experiments have been performed on tissue and organ samples removed from moribund fishes and on whole animals that "seem normal" at the surface (see Sébert and Macdonald, 1993). When oxygen consumption is measured at depth (between 1230 and 3650 m) using a slurp gun respirometer, a value averaging 0.13 mmol min^{-1} kg^{-1} is obtained for an environmental temperature of approximately 3°C without a clear relationship with depth (Smith and Hessler, 1974; Smith, 1978; Smith and Brown, 1983; Smith and Baldwin, 1983). In contrast, when M_{O_2} is measured at 1 ATA on fish trawled from bottom to surface, the metabolic rates observed (1) are much lower than in shallow-water fishes and (2) decrease as the capture depth increases (see Torres et al. 1979; Siebenaller and Somero, 1989). These low metabolic rates are generally interpreted to reflect a reduced locomotor activity. In fact, when deep-water fishes are recompressed to their depth of habitat, it appears that pressure has little effect on $\dot{M}_{O_2}$ (Gordon et al., 1976; Belman and Gordon, 1979; Roer et al., 1984). Although it is evident that $\dot{M}_{O_2}$ decreases under pressure, as for shallow-water fishes, the differences between shallow-water and deep-water fishes have perhaps been overestimated. Figure 9 illustrates this point. In general, a decrease in $\dot{M}_{O_2}$ with depth correlates with a decrease in the activity of the enzymes involved in aerobic and anaerobic pathways.

B. Muscle Biochemistry

Deep-water fishes seem to have a low energy requirement, which can account for a decrease in activity rather than a pressure and/or a temperature effect. Many studies have been performed on the adjustments of enzyme activities in the white muscle of deep-water fishes. The most studied enzymes are lactate dehydrogenase, malate dehydrogenase, glyceraldehyde 3-phosphate dehydrogenase, pyruvate kinase, creatine phosphokinase, isocitrate dehydrogenase, citrate synthase, and cytochrome c oxidase, which allow estimations of fluxes through anaerobic and aerobic pathways. Some studies have also examined enzymes that are not directly involved in energy production, such as acetylcholinesterase or Na$^+$,K$^+$-ATPase (see Somero, 1992a).

There are four major findings: (1) Deep-water fish enzymes under pressure have a higher structural stability compared to enzymes of shallow-water fishes (Siebenaller, 1991; Davis and Siebenaller, 1992). The increased structural stability of the deep-water fish proteins reduces protein turnover, an energetically wasteful process in the food-poor deep sea (Siebenaller, 1991). (2) Enzyme activities are lower in deep-water fish than in congeneric shallow-water fishes and conform with $\dot{M}_{O_2}$ decreasing at depth. (3) Enzyme

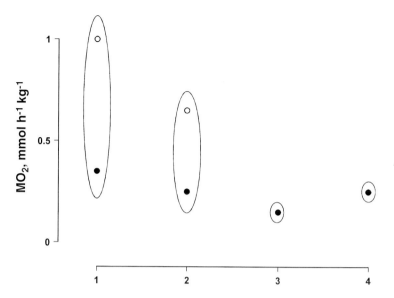

Fig. 9. Oxygen consumption of fishes at different pressures. ○ Ambient temperature 17°C; ●, $\dot{M}_{O_2}$ values computed at 3°C using $Q_{10} = 2$. 1, Shallow-water fishes at 1 ATA; 2, shallow-water fishes at 101 ATA (Sébert, 1993). 3, Deep-sea fishes living at about 101 ATA and 3–5°C; *in situ* measurements (Smith and Hessler, 1974; Smith, 1978; Smith and Brown, 1983). 4, Deep-sea fishes living at about 101 ATA and 3–5°C; measurements at 1 ATA (Torres *et al.*, 1979).

kinetics (measuring V_{max} and sometimes the Michaelis–Menten constant, K_m) of deep-water fishes are only slightly affected by pressure changes, in contrast to what is observed in shallow water fishes (Fig. 10) (Gibbs and Somero, 1989). The difference in pressure sensitivity may result partly from small conformational changes (Murray and Siebenaller, 1993; Siebenaller and Murray, 1994). Thus, it seems that the advantage of deep-water fishes in possessing enzymes relatively unaffected by pressure is somewhat weakened by the inconvenience that these enzymes have lower catalytic efficiencies (Somero, 1990, 1991). (4) Deep-water fishes have a lower protein content than the equivalent white muscle of shallow-water fishes (Blaxter *et al.*, 1971; Whitt and Prosser, 1971; Childress and Nygaard, 1973; Torres *et al.*, 1979; Sullivan and Somero, 1980; Siebenaller *et al.*, 1982; Somero *et al.*, 1983; Yancey *et al.*, 1992).

C. Comparing Shallow-Water and Deep-Water Fishes

When comparing shallow-water and deep-water fishes, it is important to remember that at depth, except for pressure, most environmental factors

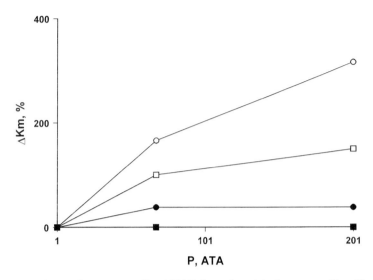

Fig. 10. Effects of pressure on K_m of NAD-dependent dehydrogenases. The effects are expressed as percentages of the values measured at 1 ATA. Measurements were performed on skeletal muscle from two congeneric fish: *Sebastolobus alascanus* (a shallow-water fish, open symbols) and *Sebastolobus altivelis* (a deep-water fish, closed symbols). ○, ●, Malate dehydrogenase-2; □, ■, glyceraldehyde-3-phosphate dehydrogenase. Data adapted from Siebenaller (1984).

have a lower intensity than at surface: this is true for temperature, pH, oxygen, and light, but also biomass and predators. Is there a causal relationship with the lower metabolism of deep-water fishes? The answer is probably yes, but the extent of the relationship is not clear.

Table VII lists similarities and differences between shallow-water fishes acclimatized to high pressure and deep-water fishes. It is interesting to note that for deep-water fishes there is a correlation between $\dot{M}_{O_2}$ and enzyme activities that is not the case for shallow-water fishes. For deep-water fishes, it must be understood as a balance between energy production and requirements, i.e., energy production is lowered because locomotor activity (and thus energy requirement) decreases. The fishes are adapted to their complex environment and not only to pressure and low temperature. For shallow-water fishes, the decrease in $\dot{M}_{O_2}$ is considered as a thermodynamic effect of pressure (such as a decrease in temperature). In these conditions, maintaining enzyme activities and energy production despite the $\dot{M}_{O_2}$ decrease implies that there is an improvement in enzyme function efficiency (Simon et al., 1989a, 1992). As previously reported, pressure can affect membrane fluidity and thus the related functions together with a potential direct effect

Table VII
Comparison of Shallow-Water and Deep-Water Fishes[a]

Common	Different
$\dot{M}_{O_2}$ decrease	Activities of enzymes involved in energy production (decrease in deep-water fishes, normal in shallow-water fishes)
Decrease in muscle protein content	
Increase in muscle water content	
Higher membrane fluidity at 1 ATA	Motor activity at depth? (normal in shallow-water fishes, decrease for deep-water fishes)
Lower saturation ratio (membrane lipids)	
Decrease in NA$^+$/K$^+$-ATPase activity	
Decrease in activity after decompression	

[a] The comparison concerns the sense of variation, not its intensity. Control fish: congeneric shallow-water fishes for deep-water fishes (*in situ* temperature); the same fish at 1 ATA for shallow-water fishes ($T_w = 17°C$).

on proteins (receptors, ion channels, and enzymes). In this way, the overall physiology of the organism is altered, and to survive, the fish must optimize its response using the less impaired functions in order to balance the most impaired ones. Thus, shallow-water fishes acclimatize to pressure effects. In contrast, the physiology of deep-water fishes is presumed to be optimum at depth (like shallow water fishes at surface): such fishes are adapted to HP. This fundamental difference reflects the difference between what is called acclimatization and adaptation.

It is evident that environmentally induced variations can take place only within the limits set for an animal by its genotype (Prosser, 1991). It has been suggested that deep-water fish adaptation to the deep environment is partly ensured by selecting pressure-adapted proteins with a reduced volume change, which may require pressure regulation of gene expression (Somero, 1990, 1992a,b). Such a selection may involve genetic regulation, i.e., genes encoding proteins may be expressed only at a given pressure, as has been shown in deep-sea bacteria (Bartlett *et al.*, 1989, 1993; Somero, 1991). The efficiency of deep-water fishes to select pressure-resistant proteins would be extremely reduced if protein function was impeded by an altered membrane environment, unless, of course, proteins are adapted to function in such conditions. However, this is not the case, and membranes may adapt to pressure in the sense that their optimal fluidity is partially or fully restored.

The compensatory adjustment of membrane fluidity to maintain an optimal state when disturbed has been termed "homeoviscous adaptation." This type of adaptation involves changes in lipid saturation (double-bond content), acyl chain length, and sometimes phospholipid head-group compositions and their consequent effect on enzyme kinetics. Homeoviscous adaptation has already been shown for temperature effects and can be observed at the cellular level (see White and Somero, 1982; Bols *et al.*, 1992 for review). Evidence has been provided for deep-living species such as bacteria (Delong and Yayanos, 1985; Kamimura *et al.*, 1992) and fishes (Cossins and Macdonald, 1984, 1986; Gibbs and Somero, 1990), whose lipid metabolism is known to be modified (Patton, 1975; Phleger, 1975; Phleger and Laub, 1989).

The adaptation of molecular and cellular structures to pressure does not imply that deep-water fishes have lower metabolic rates. Torres and Somero (1988) think that it is unlikely that pressure contributes significantly to the decline of metabolism with depth. In contrast, low temperature at depth may be considered as a contributing factor, although Siebenaller and Somero (1989) have stated that "adaptations that confer tolerance to high pressures and low temperatures on deep-water fishes do not appear to play important roles in establishing the low rates of metabolism observed." Shallow-water fishes requires a higher muscular activity and thus a higher metabolic rate to avoid predators and to accommodate their extensive horizontal movement for foraging. Such a locomotor pattern in deep-water fishes would be of little interest due to the lower biomass existing at depth. They must perform vertical migrations in order to intercept food gradients, and thus problems of buoyancy are raised (Childress and Nygaard, 1973; Tytler and Blaxter, 1973, 1977). Deep-water fishes have thus sacrificed muscular strength and mobility in exchange for lower metabolic rates, although the contractile machinery is maintained at a similar level among fishes, as shown, for example, by the constancy of muscle actin content (Swezey and Somero, 1982). The muscle concentrations of enzymes involved in energy production are thus adjusted according to the locomotory needs of the whole animal. In contrast, it has been suggested (see above) that shallow-water fishes acclimatized to high pressure adapt their metabolism to the effects of pressure possibly by increasing the efficiency of chemical reactions inside the cell. In fact, long-term HP exposure can elicit or favor adaptive changes in the association states of subunits of multimeric enzymes and in the binding states of enzymes to other enzymes (multienzyme associations, or "metabolons") or to structural proteins, a process termed *compartmentation*. Such a compartmentation, where enzymes are physically associated, not only improves metabolic flux (metabolite channeling) but also causes marked alterations in both steady-state and transient kinetics of the participating enzymes (see Somero and Hand, 1990). Thus, during long-term HP exposure, if the changes in membrane fluidity explain why COX activity is restored, the compartmentation of several enzymes could explain why ATP generation in maintained despite the $\dot{M}_{O_2}$ decrease (Simon et al., 1992). In fact, when two dehydrogenases, E1 and E2, are coupled, the K_m for NADH of E2 is generally lower when the substrate for E2 is E1–NADH rather than free NADH (see Somero and Hand, 1990). This adaptive decrease in K_m (also observed for cold adaptation) could compensate for the known pressure-induced increase of K_m observed in shallow-water fishes (see Somero et al., 1983 for review). It can therefore be concluded that when shallow-water fishes are exposed to HP for long periods, the observed acclimatization processes allow optimization of energetic fluxes, so that, at pressure, the fish has an energy production quite similar to that observed at 1 ATA. In contrast, deep-water fish metabolism is adapted not to the pressure effects but to the various factors present in the deep environment involving resistance and capacity adaptations as defined by Somero (1992a). Shallow-water fishes acclimatized to HP are a useful model in the study of specific pressure effects and may provide the missing link between surface and deep-water fishes. This model may also help in understanding the evolution of metabolism, but also, with deep-water fishes in illustrating the range of evolutionary possibilities.

Are the shallow-water and deep-water fish genetic pools different from each other? Is pressure able to induce new gene expression in pressure-acclimatized shallow-water fishes? Is the gene expression different in congeneric fishes living at depth or at surface? Only genetic studies will give complete answers to these questions. However, indirect arguments can be obtained by studying, for example, the isoenzymes or the ion channels, which are the direct gene expression.

VII. CONCLUSION

In this chapter, fishes have been regarded mainly as a model for high-pressure physiology. By using fishes, it is possible to understand some basic mechanisms whereby pressure can alter the functioning of a living organism, irrespective of its complexity. The possibility of maintaining shallow-water fishes under pressure for a long period now exists and provides great potential in the field of integrated physiology, especially in muscle energetics. The comparison of pressure-acclimatized fishes and deep-living species provides an opportunity to understand phenotypic and genotypic adaptations to the environment. If pressure-acclimatized shallow-water fishes can be considered as the "missing link" between surface and deep-sea fishes, more research will be needed to elucidate whether pressure is capable of inducing genetic regulation in such species. In other words, it would be fascinating to know if fishes living only at surface have lost their capacities to adapt to pressure, perhaps even to deep environment. The answer to such a question, which is just one example, is important in understanding some of the processes involved in species evolution. In this sense, the fact that trout are able to survive for at least 21 days at 101 ATA (1000 m) is extremely interesting. Likewise, it would be interesting to achieve the acclimatization of deep-water fishes to atmospheric pressure in order to compare physiological changes with those observed during the acclimatization of shallow-water fishes to high pressure. Such an experiment requires sampling fishes at depth in a hyperbaric chamber in order to control the decompression. Without taking into account important technical problems, such acclimatization should be possible some time in the future and the involved physiological changes should be not too different from those described in Section V.

ACKNOWLEDGMENTS

The author thanks B. Simon for numerous and helpful discussions. Many thanks are also expressed to S. Madec for technical assistance in the preparation of the manuscript.

REFERENCES

Avent, R. M., Menzies, R. J., and Phillips, D. (1972). An observational hydrostatic pressure vessel for the study of behavior and metabolism of whole animals. In "Barobiology and the Experimental Biology of the Deep Sea" (R. W. Brauer, ed.), pp. 372–382. University of North Carolina, Chapel Hill.

Balny, C., and Masson, P. (1993). Effects of high pressure on proteins. Food Rev. Int. **9,** 611–628.

Balny, C., Masson, P., and Travers, F. (1989). Some recent aspects of the use of high pressure for protein investigations in solution. High. Press. Res. **2,** 1–28.

Balouet, G., Barthélémy, L., and Belaud, A. (1973). Etude, à partir d'un vertébré aquatique, des effets spécifiques de la pression per se. Proc. 1st Annu. Sci. Meeting EUBS, Försvarmedicin **9,** 483–488.

Bark, D. H., and Smith, L. S. (1982). Creatine phosphokinase activities in rainbow trout, Salmo gairdneri, associated with rapid decompression. Enzyme **27,** 156–162.

Barthélémy, L. (1985). Le poisson, modèle scientifique en hyperbarie. Bull. Inst. Oceanogr. Monaco 4(suppl.), 9–31.

Barthélémy, L., and Belaud, A. (1972). Constations physiologiques et physiopathologiques faites sur un poisson (Anguilla anguilla L.) en conditions hyperbares. Med. Sub. Hyp. **8,** 33.

Barthélémy, L., Belaud, A., Bellet, M., and Peyraud, C. (1971). Etude, chez le poisson, des effets de la pression soit en tant que force appliquée, soit associée à une augmentation de la masse des gaz dissous. C. R. Seances Soc. Biol. **165,** 1754–1756.

Barthélémy, L., Belaud, A., and Saliou, A. (1981). A study of the specific action of per se hydrostatic pressure on fish considered as a physiological model. In "Proceedings of the Seventh Symposium on Underwater Physiology, Undersea Medical Society" (A. J. Bachrach and M. M. Matzen, eds.), pp. 641–649. Undersea Medical Society, Bethesda, Maryland.

Barthélémy, L., Sébert, P., and Simon, B. (1988). Opposite effects of hydrostatic pressure (HP) and water nitrogen partial pressure (Pw_{N_2} #71 ATA) on eel oxygen consumption. *J. Physiol.* (*London*) **406**, 95P.

Barthélémy, L., Cann-Moisan, C., Simon, B., Caroff, J., and Sébert P. (1991). Concentrations encéphaliques d'acide gamma-aminobutyrique (GABA) et de glutamine chez un poisson (*Anguilla anguilla* L.) soumis à une pression hydrostatique de 101 ATA. *Med. Sub. Hyp.* **1**, 35–47.

Bartlett, D. H. l, Wright, M. E., Yayanos, A. A., and Silverman, M. (1989). Isolation of a gene regulated by hydrostatic pressure in a deep-sea bacterium. *Nature* (*London*) **342**, 572–574.

Bartlett, D. H., Chi, E., and Wright, M. E. (1993). Sequences of the *ompH* gene from the deep-sea bacterium *Photobacterium* SS9. *Gene* **131**, 125–128.

Beaver, R. W., and Brauer, R. W. (1981). Pressure/temperature interaction in relation to development of high pressure convulsions in ectotherm vertebrates. *Comp. Biochem. Physiol.* **69A**, 665–674.

Behan, M. K., Macdonald, A. G., Jones, G. R., and Cossins, A. R. (1992). Homeoviscous adaptation under pressure: The pressure dependence of membrane order in brain myelin membrane of deep-sea fish. *Biochim. Biophys. Acta* **1103**, 317–323.

Belaud, A., and Barthélémy, L. (1973). Effects of hydrostatic pressure (31 to 101 ATA) on eel (*Anguilla anguilla* L.). *IRCS Cardiovasc. Syst.* **10**, 11-1–15.

Belaud, A., and Barthélémy, L. (1979). Influence of body temperature on nitrogen transport and decompression sickness in fish. *Aviat. Space Environ. Med.* **50**, 672–677.

Belaud, A., Mabin, D., Barthélémy, L., and Peyraud, C. (1976a). Activité électroencéphalographique d'un poisson (*Anguilla anguilla* L.) soumis à diverses conditions hyperbares. *J. Physiol.* (*Paris*) **72**, 639–652.

Belaud, A., Barthélémy, L., Lesaint, J., and Peyraud, C. (1976b). Trying to explain an effect of per se hydrostatic pressure on heart rate in fish. *Aviat. Space Environ. Med.* **47**, 252–257.

Belaud, A., Barthélémy, L., Peyraud, C., and Chouteau, J. (1976c). Evidence of a pressure reversal of anesthesia in fish. *IRCS* **4**, 45.

Belaud, A., Barthélémy, L., and Peyraud, C. (1977). Temperature and per se hydrostatic pressure reversal of pentobarbital anesthesia in fish. *J. Appl. Physiol.* **42**, 329–334.

Belman, B. W., and Gordon, M. S. (1979). Comparative studies on the metabolism of shallow-water and deep-sea marine fishes. V. Effects of temperature and hydrostatic pressure on oxygen consumption in the mesopelagic Zoarcid *Melanostigma pammelas*. *Mar. Biol* **50**, 275–281.

Blaxter, J. H. S., Wardle, C. S., and Roberts, B. L. (1971). Aspects of the circulatory physiology and muscle systems of deep-sea fish. *J. Mar. Biol. Assoc. U.K.* **51**, 991–1006.

Bols, N. C., Mosser, D. D., and Steels, G. B. (1992). Temperature studies and recent advances with fish cells *in vitro*. *Comp. Biochem. Physiol.* **103A**, 1–14.

Bone, Q., Marshall, N. B., and Blaxter, J. H. S. (1992). "Biology of Fishes." Blackie, London.

Brauer, R. W., Beaver, R. W., Hogue III, C. D., Ford, B., Goldman, S. M., and Venters, R. T. (1974). Intra- and interspecies variability of vertebrate high-pressure neurological syndrome. *J. Appl. Physiol.* **37**, 844–851.

Brauer, R. W., Hogan, P. M., Hugon, M., Macdonald, A. G., and Miller, K. W. (1982). Patterns of interactions of effects of light metabolically inert gases with those of hydrostatic pressure as such—A review. *Undersea Biomed. Res.* **9**, 353–396.

Brauer, R. W., Sidelyova, V. G., Dail, M. B., Galazii, G. I., and Roer, R. D. (1984). Physiological adaptation of cottoid fishes of Lake Baïkal to abyssal depth. *Comp. Biochem. Physiol.* **77A,** 699–705.

Brauer, R. W., Jordan, M. R., Miller, C. G., Johnson, E. D., Dutcher, J. A., and Sheeman, M. E. (1985). Interaction of temperature and pressure in intact animals. *In* "High Pressure Effects on Selected Biological Systems" (A. J. R. Péqueux and R. Gilles, eds.), pp. 3–28. Springer-Verlag, Berlin.

Brenner, S., Elgar, G., Sanford, R., Macrae, A., Venkatesh, B., and Aparicio, S. (1993). Characterisation of the pufferfish (*Fugo*) genome as a compact model vertebrate genome. *Nature (London)* **366,** 265–268.

Brunori, M., Coletta, M., Giardina, B., and Wyman, J. (1978). A macromolecular transducer illustrated by trout hemoglobin. IV. *Proc. Natl. Acad. Sci. U.S.A.* **75,** 4310–4312.

Cameron, J. N. (1984). Acid–base status of fish at different temperature. *Am. J. Physiol.* **246,** R452–R459.

Cann-Moisan, C., Sébert, P., Caroff, J., and Barthélémy, L. (1988). Effects of hydrostatic pressure (HP = 101 ATA) on nucleotides and pyridine dinucleotides of tissue contents in trout. *Exp. Biol.* **47,** 239–242.

Chen, J. J., and Yu, B. P. (1994). Alterations in mitochondrial membrane fluidity by lipid peroxidation products. *Free Radicals Biol. Med.* **17,** 411–418.

Childress, J. J., and Nygaard, M. H. (1973). The chemical composition of midwater fishes as a function of depth of occurence off southern California. *Deep-Sea Res.* **20,** 1093–1109.

Chong, P. L., and Cossins, A. R. (1983). A differential polarized fluorimetric study of the effects of high hydrostatic pressure upon the fluidity of cellular membranes. *Biochemistry* **22,** 409–415.

Cornish, I., and Moon, T. W. (1985). Glucose and lactate kinetics in american eel *Anguilla rostrata. Am. J. Physiol.* **249,** R67–R72.

Cossins, A. R. (1983). Adaptation of biological membrane to temperature. *In* "Cellular Acclimatisation to Environment Change" (A. R. Cossins and P. Sheterline, eds.), pp. 3–32. Cambridge Univ. Press, Cambridge.

Cossins, A. R., and Macdonald, A. G. (1984). Homeoviscous theory under pressure. II. The molecular order to membranes from deep sea fish. *Biochim. Biophys. Acta* **776,** 144–150.

Cossins, A. R., and Macdonald, A. G. (1986). Homeoviscous theory under pressure. III. The fatty acid composition of liver mitochondrial phospholipids of deep sea fish. *Biochim. Biophys. Acta* **860,** 325–335.

Cossins, A. R., and Macdonald, A. G. (1989). The adaptation of biological membranes to temperature and pressure: Fish from the deep and cold. *J. Bioeng. Biomemb.* **21,** 115–135.

Crawshaw, L. I. (1979). Responses to rapid temperature change in vertebrate ectotherms. *Am. Zool.* **19,** 225–237.

Crockett, E. L., and Sidell, B. D. (1990). Some pathways of energy metabolism are cold adapted in antarctic fishes. *Physiol. Zool.* **63,** 472–488.

Davis, B. J., and Siebenaller, J. F. (1992). Proteolysis at pressure and HPLC peptide mapping of M4 lactate dehydrogenase homologs from marine fishes living at different depths. *Int. J. Biochem.* **24,** 1135–1139.

Delong, F., and Yayanos, A. A. (1985). Adaptation of the membrane lipid of a deepsea bacterium to changes in hydrostatic pressure. *Science* **228,** 1101–1103.

Diprisco, G., and Tamburrini, M. (1992). The hemoglobins of marine and freshwater fish: The search for correlations with physiological adaptation. *Comp. Biochem. Physiol.* **102B,** 661–671.

Drapers, J. W., and Edwards, D. J. (1932). Some effects of high pressure on developing marine forms. *Biol. Bull. Mar. Lab. Woods Hole* **63,** 99–107.

Dunel-Erb, S., Sébert, P., Chevalier, C., Simon, B., and Barthélémy, L. (1996). Morphological changes induced by acclimation to high pressure in the gill epithelium of freshwater yellow eel (*Anguilla anguilla* L.). *J. Fish Biol.* **48,** 1018–1022.

Ebbecke, U. (1935). Uber die wirkang hoher drucke auf herzschlag und elektrokardiogram. *Arch. Ges. Physiol.* **236,** 416–426.

Ebbecke, U. (1944). Lebensvorgänge unter der Einwirkung hoher Drücke. *Ergebn. Physiol.* **45,** 34–183.

Edwards, D. J., and Cattel, K. (1928). The stimulating action of hydrostatic pressure on cardiac function. *Am. J. Physiol.* **84,** 472–484.

Ekker, M., and Akimenko, M. A. (1991). Embryology and genetics of the zebrafish, *Brachydanio zerio*. *Med. Sci.* **7,** 553–560.

Esterbauer, H., Eckl, P., and Ortner, A. (1990). Possible mutagens derived from lipids and lipids precursors. *Mutat. Res.* **238,** 233.

Esterbauer, H., Schaur, R. J., and Zollner, H. (1991). Chemistry and biochemistry of 4-hydroxynomenal, malondialdehyde and related aldehydes. *Free Radicals Biol. Med.* **11,** 81–128.

Fenn, W. O. (1967). Possible role of hydrostatic pressure in diving. In "Underwater Physiology, Proceedings of the Third Symposium on Underwater Physiology" (C. J. Lambertsen, ed.), pp. 395–403. Williams & Wilkins, Baltimore, Maryland.

Finch, E. D., and Kiesow, L. A. (1979). Pressure, anaesthetics and membrane structure: A spin probe study. *Undersea Biomed. Res.* **6,** 41–45.

Flügel, H. (1972). Adaptation and acclimatization to high pressure environments. In "Barobiology and the Experimental Biology of the Deep Sea" (R. W. Brauer, ed.), pp. 69–88. University of North Carolina, Chapel Hill.

Flügel, H., and Schlieper, C. (1970). The effects of pressure on marine invertebrates and fishes. In "High Pressure Effects on Cellular Processes" (A. M. Zimmerman, ed.), pp. 211–234. Academic Press, New York.

Fontaine, M. (1928). Les fortes pressions et la consommation d'oxygène de quelques animaux marins. Influences de la taille de l'animal. *C. R. Seances Soc. Biol.* **99,** 1789–1790.

Fontaine, M. (1929a). De l'augmentation de la consommation d'O des animaux marins sous l'influence des fortes pressions, ses variations en fonction de l'intensité de la compression. *C. R. Acad. Sci.* **188,** 460–461.

Fontaine, M. (1929b). De l'augmentation de la consommation d'oxygène des animaux marins sous l'infuence des fortes pressions, ses variations en fonction de la durée de la compression. *C. R. Acad. Sci.* **188,** 662–663.

Fontaine, M. (1930). Recherches expérimentales sur les réactions des êtres vivants aux fortes pressions. *Ann. Inst. Oceanogr. Monaco* **8,** 1–99.

Fontaine, Y. A., Dufour, S., Alinat, J., and Fontaine, M. (1985). L'immersion prolongée en profondeur stimule la fonction hypophysaire gonadotrope de l'anguille européenne (*Anguilla anguilla* L.) femelle. *C. R. Acad. Sci.* **300,** 83–87.

Gennser, M., and Karpe Förnhagen, H. C. (1990). Effects of hyperbaric pressure and temperature on atria from ectotherm animals (*Rana pipiens* and *Anguilla augilla*). *Comp. Biochem. Physiol.* **95A,** 219–228.

Gibbs, A., and Somero, G. N. (1989). Pressure adaptation of Na^+/K^+-ATPase in gills of marine teleosts. *J. Exp. Biol.* **143,** 475–492.

Gibbs, A., and Somero, G. N. (1990). Pressure adaptation of teleost gill Na^+/K^+ adenosinetriphosphatase: Role of the lipid and protein moieties. *J. Comp. Physiol.* **160B,** 431–439.

Gibson, Q. H., and Carey, F. G. (1975). Effects of pressure on the absorption spectrum of some heme compounds. *Biochem. Biophys. Res. Commun.* **67,** 747–751.

Goldinger, J. M., Kang, B. S., Choo, Y. E., Paganelli, C. V., and Hong, S. K. (1980). Effect of hydrostatic pressure on ion transport and metabolism in human erythrocytes. *J. Appl. Physiol.* **49,** 224–231.

Gordon, M. S. (1970). Hydrostatic pressure. *In* "Fish Physiology" (W. S. Hoar, and D. J. Randall, eds.), Vol. 4, pp. 445–464. Academic Press, New York and London.

Gordon, M. S., Belman, B. W., and Chow, P. H. (1976). Comparative studies on the metabolism of shallow-water and deepsea marine fishes. IV; Patterns of Aerobic metabolism in the mesopelagic deep-sea Fangtooth fish *Anoplogaster cornuta. Mar. Biol.* **35,** 287–293.

Hall, A. C., Pickles, D. M., and Macdonald, A. G. (1993). Aspects of eukaryotic cells. *In* "Effects of High Pressure on Biological Systems" (A. G. Macdonald, ed.), pp. 29–85. Springer-Verlag, Berlin.

Harper, A. A., Macdonald, A. G., Wardle, C. S., and Pennec, J. P. (1987). The pressure tolerance of deep-sea fish axons. Result of *Challenger* cruise 6B/85. *Comp. Biochem. Physiol.* **88A,** 647–653.

Hazel, J. R. (1979). Influence of thermal acclimation on membrane lipid composition of rainbow trout liver. *Am. J. Physiol.* **236,** R91–R101.

Hazel, J. R. (1995). Thermal adaptation in biological membranes: Is homeoviscous adaptation the explanation? *Annu. Rev. Physiol.* **57,** 19–42.

Hazel, J. R., and Carpenter, R. (1985). Rapid changes in the phospholipid composition of gill membranes during thermal acclimation of the rainbow trout, *Salmo gairdneri. J. Comp. Physiol.* **155B,** 597–602.

Hazel, J. R., and Zerba, E. (1986). Adaptation of biological membranes to temperature: Molecular species compositions of phosphatidylcholine and phosphatidylethanolamine in mitochondrial and microsomal membranes of liver from thermally-acclimated rainbow trout. *J. Comp. Physiol.* **156B,** 665–674.

Heisler, N. (1984). Role of ion transfer processes in acid–base regulation with temperature changes in fish. *Am. J. Physiol.* **246,** R441–R451.

Hochachka, P. W. (1975). Biochemistry at depth. "Pressure Effects on Biochemical Systems of Abyssal and Midwater Organisms: The 1973 Kona Expedition of the Alpha-Helix." Pergamon, Oxford.

Johnson, F. H., and Eyring, H. (1970). The kinetic basis of pressure effects in biology and chemistry. *In* "High Pressure Effects on Cellular Processes" (A. M. Zimmerman, ed.), pp. 1–44. Academic Press, New York.

Johnson, F. H., and Schlegel, M. C. (1948). Hemoglobin oxygenation in relation to hydrostatic pressure. *J. Cell. Comp. Physiol.* **31,** 421–425.

Johnson, F. H., Eyring, H., and Stover, B. (1974). "The Theory of Rate Processes in Biology and Medicine." Wiley, New York.

Johnstone, D. F., Macdonald, A. G., Mojsiewicz, W. R., and Wardle, C. S. (1989). Preliminary experiments in the adaptation of the European eel (*Anguilla anguilla*) to high hydrostatic pressure. *J. Physiol. (London)* **417,** 87P.

Kahn, P. (1994). Zebrafish hit the big time. *Science* **264,** 904–905.

Kamimura, K., Fuse, H., Takimura, O., Yamaoka, Y., Ohwada, K., and Hashimoto, J. (1992). Pressure-induced alteration in fatty acid composition of barotolerant deep-sea bacterium. *J. Oceanogr.* **48,** 93–104.

Kiesow, L. A. (1974). Hyperbaric inert gases and the hemoglobin–oxygen equilibrium in red blood cells. *Undersea Biomed. Res.* **1,** 29–43.

Kunugi, S. (1992). Enzyme reactions under high pressure and their applications. *Ann. N.Y. Acad. Sci.* **672,** 293–304.

Kylstra, J. A., Nantz, R. Crowe, J., Wagner, W., and Saltzman, H. A. (1967). Hydraulic compression of mice to 166 atmospheres. *Science* **158,** 793–794.

Kynne, O. (1970). "Marine Ecology." Wiley(Interscience), London.
Lee, A. G. (1991). Lipids and their effects on membrane proteins: Evidence against a role for fluidity. *Prog. Lipid Res.* **30,** 323–348.
Lee, J. A. C., and Cossins, A. R. (1990). Temperature adaptation of biological membranes: Differential homeoviscous responses in brush-border and basolateral membranes of carp intestinal mucosa. *Biochim. Biophys. Acta* **1026,** 195–203.
Low, P. S., and Somero, G. N. (1975). Pressure effects on enzyme structure and function *in vitro* and under simulated *in vivo* conditions. *Comp. Biochem. Physiol.* **52B,** 67–74.
Lundgren, C. E. G., and Örnhagen, H. C. (1976). Hydrostatic pressure tolerance in liquid breathing mice. *In* "Proceedings of the Fifth Symposium on Underwater Physiology" (C. J. Lambertsen, ed.), pp. 397–404. Fed. Am. Soc. Exp. Biol., Bethesda, Maryland.
Macdonald, A. G. (1984). The effects of pressure on the molecular structure and physiological functions of cell membranes. *Philos. Trans. R. Soc.* (*London*) **B304,** 47–68.
Macdonald, A. G. (1987). The role of membrane fluidity in complex processes under high pressure. *In* "Current Perspectives in High Pressure Biology" (H. W. Jannash, R. E. Marquis, and A. M. Zimmerman, eds.), pp. 207–223. Academic Press, London.
Macdonald, A. G., and Cossins, A. R. (1985). The theory of homeoviscous adaptation of membranes applied to deep sea animals. *In* "Physiological Adaptations of Marine Animals" (M. Laverack, ed.), pp. 301–322. *Comparative Biology,* Cambridge.
Macdonald, A. G., and Shelton, C. J. (1985). Ionic regulation under high pressure: Some observations of comparative and theoretical interest. *In* "High Pressure on Selected Biological Systems" (A. J. R. Péqueux and R. Gilles, eds.), pp. 51–67. Springer-Verlag, Berlin.
Macdonald, A. G., Gilchrist, I., and Wardle, C. S. (1987). Effects of hydrostatic pressure on the motor activity of fish from shallow water and 900 m depths; some results of *Challenger* Cruise 6B/85. *Comp. Biochem. Physiol.* **88A,** 543–547.
Maresca, B., and Cossins, A. R. (1993). Fatty feedback and fluidity. *Nature* (*London*) **365,** 606–607.
Marsland, D. A. (1958). Cells at high pressure. *Sci. Am.* **199,** 36–43.
Murray, T. F., and Siebenaller, J. F. (1993). Differential susceptibility of guanine nucleotide-binding proteins to pertussis toxin-catalyzed ADP-ribosylation in brain membranes of two congeneric marine fishes. *Biol. Bull.* **185,** 346–354.
Naroska, V. (1968). Vergleichende Untersuchugen über den Einfluss des hydrostatischen Druckes auf überlebensfähigkeit und Stoffwechsel intensität mariner Everteraten und Teleoteer. *Kieler Meeresforsch.* **24,** 95–123.
Nishiyama, T. (1965). A preliminary note on the effect of hydrostatic pressure on the behavior of some fish. *Bull. Fac. Fish. Hokkaido Univ.* **15,** 213–214.
Patton, J. S. (1975). The effect of pressure and temperature on phospholipid and triglyceride fatty acids of fish white muscle: A comparison of deepwater and surface marine species. *Comp. Biochem. Physiol.* **52B,** 105–110.
Pennec, J. P., Wardle, C. S., Harper, A. A., and Macdonald, A. G. (1988). Effects of high hydrostatic pressure on the isolated hearts of shallow water and deep sea fish; results of *Challenger Cruise,* 6B/85. *Comp. Biochem. Physiol.* **89A,** 215–218.
Péqueux, A. (1981). Effects of high hydrostatic pressures on Na^+ transport across isolated gill epithelium of sea water-acclimated eels *Anguilla anguilla. In* "Underwater Physiology" (A. J. Bachrach and M. M. Matzen, eds.), Vol. 7, pp. 601–609. Undersea Medical Society, Bethesda, Maryland.
Péqueux, A., and Gilles, R. (1978). Effects of high hydrostatic pressures on the activity of the membrane ATPases of some organs implicated in hydromineral regulation. *Comp. Biochem. Physiol.* **55A,** 103–108.

Péqueux, A., and Gilles, R. (1986). Effects of hydrostatic pressure on ionic and osmotic regulation. *In* "Diving in Animals and Man" (A. O. Brubakk, J. W. Kanwisher, and G. Sundnes, eds.), pp. 161–189. Tapir, Trondheim, Norway.

Peyraud-Waitzenegger, M., Barthélémy, L., and Peyraud, C. (1980). Cardiovascular and ventilatory effects of catecholamines in unrestrained eels (*Anguilla anguilla* L.). A study of seasonal changes of reactivity. *J. Comp. Physiol.* **138,** 367–375.

Pfeiler, E. (1978). Effects of hydrostatic pressure on (Na^+ + K^+-ATPase and Mg^{2+}-ATPase in gills of marine teleost fish. *J. Exp. Zool.* **205,** 393–402.

Phleger, C. F. (1975). Lipid synthesis by *Antimora rostrata*, an abyssal codling from the Kona coast. *Comp. Biochem. Physiol.* **52B,** 97–99.

Phleger, C. F., and Laub, R. J. (1989). Skeletal fatty acids in fish from depths off Jamaica *Comp. Biochem. Physiol.* **94B,** 329–334.

Plager, D. A., and Nelsestuen, G. L. (1992). Dissociation of peripheral protein–membrane complexes by high pressure. *Protein Sci.* **1,** 530–539.

Ponat, A. (1967). Untersuchungen zur zellulären Druckresistenz Verschiedener Evertebrates der Nord-und Ostsee. *Kieler Meeresforsch.* **23,** 21–47.

Powers, D. A. (1989). Fish as model systems. *Science* **246,** 352–358.

Prosser, C. L. (1991). Introduction: Definition of comparative physiology: Theory of adaptation. *In* "Environmental and Metabolic Animal Physiology" (C. L. Prosser, ed.), pp. 1–11. Wiley-Liss, New York.

Reeves, R. B., and Morin, R. A. (1986). Pressure increases oxygen affinity of whole blood and erythrocytes suspensions. *J. Appl. Physiol.* **61,** 486–494.

Regnard, P. (1884). Effet des hautes pressions sur les animaux marins. *C. R. Seances Soc. Biol.* **36,** 394–395.

Regnard, P. (1885). Phénomènes objectifs que l'on peut observer sur les animaux soumis aux hautes pressions. *C. R. Seances Soc. Biol.* **37,** 510–515.

Rifkind, J. L., Abugo, O., Levy, A., Monticone, R., and Heim, J. (1993). Formation of free radicals under hypoxia. *In* "Surviving Hypoxia" (P. W. Hochachka, P. L. Lutz, T. Sick, M. Rosenthal, and G. Van den Thillart, eds.), pp. 509–525. CRC Press, Boca Raton, Florida.

Roer, R. D., and Péqueux, A. J. R. (1985). Effects of hydrostatic pressure on ionic and osmotic regulation. *In* "High Pressure Effects on Selected Biological Systems" (A. J. R. Péqueux and R. Gilles, eds.), pp. 31–49. Springer-Verlag, Berlin.

Roer, R. D., Sidelyova, V. G., Brauer, R. W., and Galazii, G. I. (1984). Effects of pressure on oxygen consumption in cottid fish fro Lake Baïkal. *Experientia* **40,** 771–773.

Sébert, P. (1993). Energy metabolism of fish under hydrostatic pressure: A review. *Trends Comp. Biochem. Physiol.* **1,** 289–317.

Sébert, P., and Barthélémy, L., (1985a). Hydrostatic pressure and adrenergic drugs (agonists and antagonists): Effects and interactions in fish. *Comp. Biochem. Physiol.* **82C,** 207–212.

Sébert, P., and Barthélémy, L. (1985b). Effects of high hydrostatic pressure per se, 101 atm on eel metabolism. *Resp. Physiol.* **62,** 349–357.

Sébert, P., and Macdonald, A. G. (1993). Fish. *In* "Effects of High Pressure on Biological Systems" (A. G. Macdonald, ed.), pp. 147–196. Springer-Verlag, Berlin.

Sébert, P., Lebras, Y., Barthélémy, L., and Peyraud, C. (1984). Effect of high hydrostatic pressure on catecholamine contents in tissues of the eel acclimated at two temperatures. *Aviat. Space Environ. Med.* **55,** 931–934.

Sébert, P., Barthélémy, L., and Caroff, J. (1985a). Serotonin levels in fish brain: Effects of hydrostatic pressure and water temperature. *Experientia* **41,** 1429–1430.

Sébert, P., Bigot, J. C., and Barthélémy, L. (1985b). Effects of hydrostatic pressure on amino-acid contents of eel brain. *IRCS Med. Sci.* **13,** 834–835.

Sébert, P., Barthélémy, L., and Caroff, J. (1986). Catecholamine content (as measured by the HPLC method) in brain and blood plasma of the eel: Effects of 101 ATA hydrostatic pressure. *Comp. Biochem. Physiol.* **84C**, 155–157.

Sébert, P., Barthélémy, L., Caroff, J., and Hourmant, A. (1987). Effects of hydrostatic pressures per se (101 ATA) on energetic processes in fish. *Comp. Biochem. Physiol.* **86A**, 491–495.

Sébert, P., Barthélémy, L., and Simon, B. (1990). Laboratory system enabling long-term exposure (>30d) to hydrostatic pressure (<101 atm) of fishes or other animals breathing water. *Mar. Biol.* **104**, 165–168.

Sébert, P., Péqueux, A., Simon, B., and Barthélémy, L. (1991). Effects of long term exposure to 101 ATA hydrostatic pressure on blood, gill and muscle composition and on some enzyme activities of the Fw eel (*Anguilla anguilla* L.) *Comp. Biochem. Physiol.* **98B**, 573–577.

Sébert, P., Simon, B., and Barthélémy, L. (1993a). Hydrostatic pressure induces a state resembling histotoxic hypoxia in fish. *Comp. Biochem. Physiol.* **105A**, 255–258.

Sébert, P., Cossins, A. R., Simon, B., Meskar, A., and Barthélémy, L. (1983b). Membrane adaptations in pressure acclimated freshwater eels. *Proc. 32nd Congress IUPS, Glasgow*, **177.2/P**, 121.

Sébert, P., Meskar, A., Simon, B., and Barthélémy, L. (1994). Pressure acclimation of the eel and liver membrane composition. *Experientia* **50**, 121–123.

Sébert, P., Menez, J. F., Simon, B., and Barthélémy, L. (1995a). Effects of hydrostatic pressure on malondialdehyde (MDA) brain contents in yellow freshwater eels. *Redox Report* **1**, 379–382.

Sébert, P., Simon, B., and Barthélémy, L. (1995b). Effects of temperature increase on oxygen consumption of yellow freshwater eels exposed to 101 ATA hydrostatic pressure. *Exp. Physiol.* **80**, 1039–1046.

Shaklee, J. B., Christiansen, J. A., Sidell, B. D., Prosser, C. L., and Whitt, G. S. (1977). Molecular aspects of temperature acclimation in fish: Contribution of changes in enzyme activities and isoenzyme patterns to metabolic reorganization in the green sunfish. *J. Exp. Zool.* **201**, 1–20.

Shelton, C. J., and Macdonald, A. G. (1987). Effect of high hydrostatic pressure on $^{86}Rb^+$ influx in the erythrocyte of the plaice. (*Pleuronetes platessa*). *Comp. Biochem. Physiol.* **88A**, 481–485.

Siebenaller, J. F. (1984). Pressure adaptative differences in NAD dependent dehydrogenases of congeneric marine fishes living at different depths. *J. Comp. Physiol.* **154B**, 443–448.

Siebenaller, J. F. (1991). Pressure as an environmental variable: Magnitude and mechanisms of perturbation. *In* "Biochemistry and Molecular Biology of Fishes" (P. W. Hochachka and T. P. Mommsen, eds.), Vol. 1, pp. 323–343. Elsevier, Amsterdam.

Siebenaller, J. F., and Murray, T. F. (1994). The effects of hydrostatic pressure on pertussis toxin-catalyzed ribosylation of guanine nucleotide-binding proteins from two congeneric marine fish. *Comp. Biochem. Physiol.* **108B**, 423–430.

Siebenaller, J. F., and Somero, G. N. (1989). Biochemical adaptation to the deepsea. *CRC Crit. Rev. Aquat. Sci.* **1**, 1–25.

Siebenaller, J. F., Somero, G. N., and Haedrich, R. L. (1982). Biochemical characteristics of macrourid fishes differing in their depths of distribution. *Biol. Bull.* **163**, 240–249.

Simon, B. (1990). Métabolisme énergétique de l'anguille (*Anguilla anguilla* L): effets d'expositions de courte durée (3 heures) et de longue durée (un mois) à 101 ATA de pression hydrostatique. Thèse Sciences, Brest, France, pp. 1–105.

Simon, B., Sébert, P., and Barthélémy, L. (1989a). Effects of long-term exposure to hydrostatic pressure per se (101 ATA) on eel metabolism. *Can. J. Physiol. Pharmacol.* **67**, 1247–1251.

Simon, B., Sébert, P., and Barthélémy, L. (1989b). Opposition des effets de la pression hydrostatique et de la pression partielle d'azote chez l'anguille. *Med. Sub. Hyp.* **8,** 77–93.

Simon, B., Sébert, P., and Barthélémy, L. (1991). Eel (*Anguilla anguilla* L.), muscle modifications induced by long-term exposure to 101 ATA hydrostatic pressure. *J. Fish Biol.* **38,** 89–94.

Simon, B., Sébert, P., Cann-Moisan, C., and Barthélémy, L. (1992). Muscle energetics in yellow freshwater eels (*Anguilla anguilla* L.) exposed to high hydrostatic pressure (101 ATA) for 30 days. *Comp. Biochem. Physiol.* **102B,** 205–208.

Simon, B., Sébert, P., and Barthélémy, L. (1997). Effects of cold and pressure acclimations on liver isocitrate dehydrogenase in the yellow eel (*Anguilla anguilla* L.). *Comp. Biochem. Physiol.* in revision.

Smart, G. R. (1978). Investigation of the toxic mechanisms of ammonia to fish—Gas exchange in rainbow trout (*Salmo qairdneri*) exposed to acutely lethal concentrations. *J. Fish Biol.* **12,** 93–104.

Smith, KI. L. (1978). Metabolism of the abyssopelagic rattail *Coryphaenoides armatus* measured *in situ*. *Nature (London)* **274,** 362–364.

Smith, K. L., and Baldwin, R. J. (1983). Deep sea respirometry: *In situ* techniques. *In* "Polarographic Oxygensensors: Aquatic and Physiological Applications" (E. Gnaiger and H. Forstner, Eds.), pp. 298–319. Springer-Verlag, Berlin.

Smith, K. L., and Brown, N. C. (1983). Oxygen consumption of pelagic juveniles and demersal adults of the deepsea fish *Sebastolobus altivelis*, measured at depth. *Mar. Biol.* **76,** 325–332.

Smith, K. L., and Hessler, R. R. (1974). Respiration of benthopelagic fishes: *In situ* measurements at 1230 meters. *Science* **184,** 72–73.

Somero, G. N. (1990). Life at low volume change: Hydrostatic pressure as a selective factor in the aquatic environment. *Am. Zool.* **30,** 123–135.

Somero, G. N. (1991). Hydrostatic pressure and adaptations to the deep sea. *In* "Environmental and Metabolic Animal Physiology" (C. L. Prosser, ed.), pp. 167–204. Wiley-Liss, New York.

Somero, G. N. (1992a). Biochemical ecology of deep sea animals. *Experientia* **48,** 537–543.

Somero, G. N. (1992b). Adaptations to high hydrostatic pressure. *Annu. Rev. Physiol.* **54,** 557–577.

Somero, G. N. (1995). Proteins and temperature. *Annu. Rev. Physiol.* **57,** 43–68.

Somero, G. N., and Hand, S. C. (1990). Protein assembly and metabolic regulation: Physiological and evolutionary perspectives. *Physiol. Zool.* **63,** 443–471.

Somero, G. N., Siebenaller, J. F., and Hochachka, P. W. (1983). Biochemical and physiological adaptations of deep-sea animals. *In* "Deep Sea Biology" G. T. Rowe, ed., pp. 261–330. Wiley, New York.

Sullivan, K. M., and Somero, G. N. (1980). Enzyme activities of fish skeletal muscle and brain as influenced by depth of occurence and habits of feeding and locomotion. *Mar. Biol.* **60,** 91–99.

Swezey, R. R., and Somero, G. N. (1982). Skeletal muscle actin content is strongly conserved in fishes having different depths of distribution and capacities of locomotion. *Mar. Biol. Lett.* **3,** 307–315.

Tesch, F. W. (1978). Telemetric observations on the spawning migration of the eel (*Anguilla anguilla*) west of the European continental shelf. *Environ. Biol. Fish* **3,** 203–209.

Theede, H. (1972). Design and performance characteristics of currently existing high pressure aquarium systems. *In* "Barobiology and the Experimental Biology of the Deep Sea" (R. W. Brauer, ed.), pp. 362–371. University of North Carolina, Chapel Hill.

Tetteh-Lartey, N. A. (1985). Effects of temperature and hydrostatic pressure on contraction properties *in vitro* of skeletal muscle from a teleost. Thesis, Univ. Aberdeen, Scotland.

Torres, J. J., and Somero, G. N. (1988). Vertical distribution and metabolism in antarctic mesopelagic fishes. *Comp. Biochem. Physiol.* **90B,** 521–528.

Torres, J. J., Belman, B. W., and Childress, J. J. (1979). Oxygen consumption rates of midwater fishes as a function of depth of occurrence. *Deep-Sea Res.* **26A,** 185–197.

Tyler, S., and Sidell, B. D. (1984). Changes in mitochondrial distribution and diffusion distances in muscle of goldfish upon acclimation to warm and cold temperatures. *J. Exp. Zool.* **232,** 1–9.

Tytler, P., and Blaxter, J. H. S. (1973). Adaptation by cod and saithe to pressure changes. *Neth. J. Sea Res.* **7,** 31–45.

Tytler, P., and Blaxter, J. H. S. (1977). The effects of swimbladder deflation on pressure sensitivity in the saithe *Pollachius virens. J. Mar. Biol. Assoc. U.K.* **57,** 1057–1064.

Varene, P., and Valiron, M. O. (1980). Effets biologiques des gaz inertes. *Bull. Eur. Physiopathol. Respir.* **16,** 79–109.

Vik, S. B., and Capaldi, R. A. (1977). Lipid requirement for cytochrome C oxydase activity. *Biochemistry* **16,** 5755–5759.

Walesby, N. J., and Johnston, I. A. (1980). Temperature acclimation in brook trout muscle: Adenine nucleotide concentration, phosphorylation state and adenylate energy charge. *J. Comp. Physiol.* **139,** 127–133.

Wann, K. T, and Macdonald, A. G. (1988). Actions and interactions of high pressure and general anaesthetics. *Prog. Neurobiol.* **30,** 271–307.

Wardle, C. S. (1985). Swimming activity in marine fish. *In* "Physiological Adaptations of Marine Animals" (M. Laverack, ed.), pp. 521–540. Comparative Biology, Cambridge.

Wardle, C. S., Tetteh-Lartey, N., Macdonald, A. G., Harper, A. A., and Pennec, J. P. (1987). The effect of pressure on the lateral swimming muscle of the european eel *Anguilla anguilla* and the deep sea eel. *Histiobranchus bathybius;* results of *Challenger* cruise 6B/85. *Comp. Biochem.* **88A,** 595–598.

Wells, J. M. (1975). Hydrostatic pressure and hemoglobin oxygenation. *In* "Proceedings of the Fifth Symposium on Underwater Physiology" (C. J. Lambertsen, ed.), pp. 443–448. Undersea Medical Society, Bethesda, Maryland.

White, F. N., and Somero, G. (1982). Acid–base regulation and phospholipid adaptations to temperature: Time courses and physiological significance of modifying the milieu for protein function. *Physiol. Rev.* **62,** 40–90.

Whitt, G. S., and Prosser, C. L. (1971). Lactate dehydrogenase isozymes, cytochrome oxidase activity, and muscle ions of the rattail (*Coryphaenoides* sp.) *Am. Zool.* **11,** 503–511.

Yancey, P. H., Kulongoski, T., Usibelli, M. D., Lawrence-Berrey, R., and Pedersen, A. (1992). Adaptations in mesopelagic fishes. II. Protein contents and various muscles and actomyosin contents and structure of swimming muscle. *Comp. Biochem. Physiol.* **103B,** 691–697.

8

SENSORY PHYSIOLOGY

JOHN MONTGOMERY AND NED PANKHURST

I. Introduction
II. Olfaction/Chemoreception
III. Vision
IV. Touch
V. Octavolateralis Systems
   A. Introduction
   B. Vestibular System/Hearing
   C. Mechanosensory Lateral Line
   D. Electrosense
VI. General Comments
   A. Orientation and Navigation
   B. Comparisons with Antarctic Fishes
   C. Deep-Sea Sensory Biology
   References

I. INTRODUCTION

For the purposes of this chapter the deep sea can considered to be those areas of the oceans where there is insufficient daylight for vision. For clear oceanic waters this occurs below 900–1000 m (Denton, 1990), but this condition will also occur in shallower water, for example, where water clarity is reduced, at high latitude during the winter, and in submarine caves and below ice cover. No absolute definition is satisfactory though, because animals migrate, encountering different photic environments at different times of the day or year, or during different phases of their life-history. This definition does, however, remove the temptation to dwell at length on the extreme specializations of the visual system seen in mesopelagic fish that enable them to use the available faint down-welling light (Locket, 1977), and on the use of biological light as camouflage in mesopelagic fishes (Denton *et al.,* 1972). Instead, it focuses our attention on visual systems targeting biological light, and on nonvisual sensory systems including olfaction, touch and the mechano- and electrosensory lateral-line organs.

The inaccessibility of the deep oceans, and the delicate nature of the ichthyofauna, often mean that live, or even well-preserved, specimens are hard to come by. Physiological studies are almost nonexistent, and most of what can be said is by inference from anatomy, and by analogy with more accessible species occupying low light environments, or possessing similar sensory specializations and structures. Many fish that are nocturnally active, that live in caves, or that inhabit the high-latitude polar seas also depend only on biological light and on nonvisual senses, and so provide us with useful insights into the sensory problems of the deep sea, and their possible solutions.

Although the 1000-m cut-off is useful for delimiting the deep-sea fauna, very few of the fishes will occupy this region throughout their life cycle. A common life-history strategy for deep-sea fishes is to have buoyant eggs and a period of larval growth in the photic zone. One of the extraordinary features of fishes (in comparison with other vertebrates) is the requirement to be viable individuals over a very wide range of sizes, from first feeding, often only a few millimetres, to adulthood. Reorganization of biological structure during ontogeny is not infinitely plastic, so it would not be surprising to see some features of adult structure that relate to larval or juvenile requirements.

The approach of this chapter will be to discuss first the general structure and function of fish sensory systems, with particular reference to any studies on deep-sea fishes. The physiological principles underlying sensory function and the processing of sensory information will not be addressed, but good general works covering these issues include Atema *et al.* (1988), Bullock and Heiligenberg (1986), Coombs *et al.* (1989), Montgomery (1988a), and Tavolga *et al.* (1981). The chapter concludes with a more speculative consideration of some aspects of the sensory milieu of the deep sea.

II. OLFACTION/CHEMORECEPTION

The olfactory receptor cells are located in an epithelium lining the floor of the nasal capsule; this epithelium is typically folded into olfactory lamellae (Caprio, 1988). The geometry and number of olfactory lamellae within the nasal capsule, and the organization of the sensory and nonsensory epithelia, vary widely among different fish species (Yamamoto, 1982). These differences can, within limits, provide information on the relative importance of olfaction.

Among deep-sea fishes the benthopelagic fauna (which live near the deep-sea floor and could be described as demersal rather than benthic) typically have moderately to well-developed olfactory organs. This is true of rattails (Macrouridae), deep-sea cods (Moridae), and brotulids, with the largest olfactory organs found in black squaloid sharks and synaphobranchid eels (Marshall, 1979). Perhaps the most obvious use for olfaction is scavenging. Dead organic material that drifts to the sea floor provides a food source that can be tracked down along the olfactory trails carried by the bottom currents. Predation on some forms of benthos can also be mediated by olfaction, and sifting through mouthfuls of sediment identifying small food items by taste can be another effective feeding strategy. Another obvious use for olfaction is finding a mate. Pheromonal communication is a common component of mate location and recognition systems in fish. It undoubtedly occurs in benthopelagic fish, but to an unknown extent. However, in one benthopelagic group, the halosaurs, it is clearly indicated by their sexually dimorphic olfactory organs. At maturity the male anterior naris becomes large and tubular and the olfactory lamellae become enlarged and lobulated so that, as Marshall (1979) describes it, the entire rosette comes to look like a sprig of broccoli.

Away from the seafloor, in the bathypelagic fauna, sexual dimorphism of olfactory organs is the rule. The black species of *Cyclothone* (family Gonostomatidae), which are the most numerous of the bathypelagic fish, and the ceratioid anglerfishes, which are the most speciose, have macrosomatic males and microsomatic females (Marshall, 1979). In contrast, minor groups such as gulper eels and snipe eels have small or regressed olfactory organs in both sexes.

The potential for olfactory location of mates has been modeled in the deep-sea hatchetfish *Argyropelecus hemigymnus* (Jumper and Baird, 1991). Although this is a mesopelagic species (200–600 m), the model would be equally applicable to bathypelagic fish. The essential features of the model are that the female drifting with the local current releases into the water a pulse of pheromone that spreads in a horizontal "patch." Males, assumed to be moving at random, encounter the patch and then search within the patch for the female.

The premise that the pheromones spread in a horizontal patch requires some explanation (see Westerberg, 1984). It is not intuitively obvious that the spread of a substance released in midwater should be anisotropic (i.e.. orders of magnitude faster in the horizontal direction than in the vertical). The reason for this is that the water column is stratified below the thermocline and can have extensive fine structure. High resolution temperature profiles show homogeneous layers alternating with layers wherein the temperature changes rapidly with depth. The layers of rapid temperature change also correspond with changes in salinity and current speed and direction. The thickness of both kinds of layers varies from a few meters down to 0.1 m. The processes that generate the fine structure are not fully understood, but the basic principal is that a well-mixed water mass will spread horizontally at its appropriate density level in the water column. So small packets of water will exist as "pancakes," and layers only 1 m thick have been traced horizontally for over 1000 m. Small-scale diffusion processes in this quite stratified regime are dominated by "shear diffusion." The different current directions and speeds in adjacent homogeneous layers produce thin sheets of tracer in the intermediate layer within the strong vertical density gradient.

The model developed by Jumper and Baird (1991) predicts that a horizontal patch of detectable pheromone will expand to a maximum range of almost 100 m in about 9 h, then fully dissipate in about 1 day. Using some reasonable assumptions about male mobility, and assuming that the fish are uniformly distributed in their habitat, the model also predicts that the mean time for detection of a female is only about 1 h. The potential importance of pheromone communication for mate location in the deep-sea is illustrated by the finding (Jumper and Baird, 1991) that without the pheromone patch, the time for detection by other sensory means increases to 8 days.

III. VISION

At depths of 200 to 1000 m visual function is a fascinating story of eyes operating at the limits of useful function utilizing ambient light. Large eyes, tubular eyes that allow a large eye in a small head, pure rod retinas, multiple banks of rods, very long rod outer segments, high convergence ratios of receptors to ganglion cells, visual pigments matched to environmental light, and reflective tapeta lucida all provide enhanced visual sensitivity (Munk, 1966; Locket, 1977; Best and Nicol, 1980, Pankhurst, 1987; Partridge *et al.*, 1988, 1989). The other major theme of this story is the use of biological light for midwater camouflage, and visual tricks that could be used to break the camouflage. Fish in midwater can be seen from below as silhouettes, appearing dark against the down-welling light. Within the mesopelagic zone the ambient light is dim enough that visibility from below can be reduced by the production of light on the ventral surface. This intricate camouflage mechanism is based on ventral photophores that, to be effective, have to produce light of the correct color, intensity, and angular distribution to match the background (Denton *et al.*, 1972). The paradoxical presence of yellow lenses in mesopelagic fish has been interpreted as a means of enhancing the contrast of the ventral bioluminescence against the slightly shorter wavelength space light (Muntz, 1976). If yellow lenses are not involved in breaking midwater camouflage it is difficult to explain their presence, because they would greatly reduce light intensity in an already intensity-limited environment. Another indicator that yellow lenses are involved with problems created by space light is that yellow lenses have not been found in fish living below the photic zone (Douglas and Thorpe, 1992; Douglas et al., 1995).

Below 1000 m the only light is bioluminescence. Vision appears to become less important—at least adaptations for extreme sensitivity become less evident: relative eye size becomes less (Fig. 1), and ocular degeneration is common in bathyal species that lack bioluminescent organs (Nichol, 1978). Not withstanding this, many of the demersal species living below 1000 m have well-developed visual systems (Douglas et al., 1995), and display many of the retinal modifications to enhance sensitivity found in mesopelagic fish (Munk, 1966; Locket, 1977). Given that the intensity of bioluminescence is often very low (commonly of the order of 10^{-6} mWcm^{-2}, or equivalent to the intensity of spacelight at 800 m in clear ocean water) (Nicol, 1978), then it is understandable that visual systems of bathyal fishes resemble those of mesopelagic fishes. This also raises the point that many

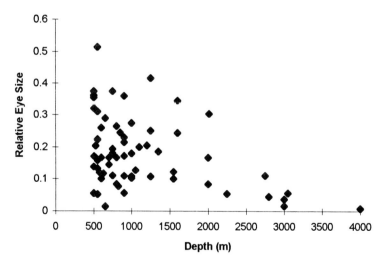

Fig. 1. Relative eye size as a function of depth of occurrence for fish, with a median depth range of greater than 500 m. Relative eye size is taken here as horizontal eye diameter divided by head length. [Data measured from illustrations in Paulin et al. (1989).] Each data point represents the illustrated representative species for one family and is plotted at the median depth range given for that family.

of the species typically described as mesopelagic or mesobenthic have lower limits to their depth distributions that extend well into the aphotic zone (e.g., Locket, 1980; Pankhurst, 1987). Ocular modifications interpreted as adaptive for low-intensity spacelight of the mesopelagic zone also equip these fishes for visual detection of bioluminescence in the bathyal zone.

Below the photic zone, bioluminescence falls into two main categories: blue to blue/green and red. Most species are blue emitting, with the wavelength of maximum emission (λ_{max}) falling between 450 and 490 nm, but with relatively broad emission curves (Herring, 1983). In a comparative study of the visual pigments of deep-sea fishes, Douglas et al. (1995) found that the majority of species had a single visual pigment, and in those species inhabiting depths >1100 m the action spectra of the pigments (λ_{max}) were centered between 475 and 485 nm. The narrower range of λ_{max} in visual pigments compared with bioluminescent emission spectra could be interpreted to mean that only sources within the range 457–485 nm are of interest. But the broad bandwidth of the bioluminescence means that a mismatch of the λ_{max} of emission and reception would not greatly reduce photon uptake. Bowmaker et al. (1994) make the point that the high axial density of visual pigment seen in deep-sea fish results in a wide flat absorption spectrum, resulting in nearly 100% absorption of photons even if the λ_{max} of the source and the pigment are not closely matched. They were led to this suggestion by the finding that the deepest fish in Lake Baikal have blue-sensitive rods with a λ_{max} of 480–500 nm. This is despite the lake water having a maximum transmission at 550–600 nm and there being no evidence for bioluminescence in this system. This mismatch prompts the consideration of alternative hypotheses for typical blue-sensitive photopigments found in deep-sea fishes. Perhaps at very low photon fluxes other factors such as reduced thermal noise override spectral matching. It is not known if blue photopigments have any advantage with respect to low thermal noise, but that visual sensitivity is noise limited has been clearly demonstrated in toads (Aho et al., 1988).

The most obvious use of blue bioluminescence is in communication. The pattern of bioluminescent flashing has been observed in two *Lampanyctus* species by Mensinger and Case (1990), who suggest that the distinct flash patterns may permit species recognition between species that otherwise have similar photophore arrays, and also have overlapping habitats. However, like all communication systems, the signals can be intercepted by predators, or subverted by them to attract prey. For example, the deep-sea anglerfishes use bioluminescence both for communication and as light lures. Of the 100 or so species of deep-sea anglers, there are but a few in which the females do not carry a light lure on the end of a modified fin ray (Marshall, 1979). The lures are highly species specific, but this does not

8. SENSORY PHYSIOLOGY

seem to be related to targeting different prey, because dietary studies have found no evidence of prey selection. Rather the specificity is thought to act as part of the mate recognition system. Parasitic males of the ceratiid anglerfishes and other groups do not seem to feed after metamorphosis, yet it is not until after metamorphosis that their visual systems reach full development. In the male ceratiids the eyes are very large, and in one group (the linophrynids) the eyes are tubular and look forward (Marshall, 1979). Given the lack of feeding in these males, the overriding function of their visual system must be mate localization and recognition.

A visual system geared to the interception of bioluminescent signals is found in *Bajacalifornia drakei* (family Alepocephalidae) studied by Locket (1985) (Fig. 2). This fish lives at depths between 700 and 1600 m and its main visual feature is that it has highly specialized foveas. The axes of vision to the foveas of the two eyes converge in front of the jaws. The foveas themselves are remarkable steep-sided (convexiclivate) structures with up to 28 superimposed banks of rods. The function of the fovea has been the matter of some speculation. Locket (1985) concludes that the convexiclivate fovea functions as a focus indicator, and that the two foveas allow the determination of direction and distance of the prey. Denton

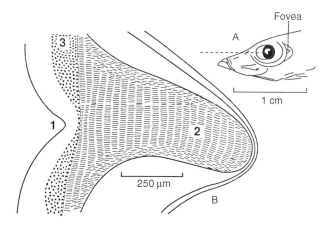

Fig. 2. The eye of *Bajacalifornia drakei*. (A) Head with tissues behind the eye cut away to show the location of the fovea. The axes through the center of the lens and the fovea cross in front of the fish. (B) Diagram of the fovea. The steep-sided pit in the retina (1) lies in front of the fovea, which contains up to 28 banks of ellipsiod outer-segment complexes. Rod nuclei (3) are piled up in the foveal shoulders. Data from Locket (1985), after Denton (1990). Light and vision at depths greater than 200 metres. *In* "Light and Life in the Sea" (P. J. Herring, A. K. Campbell, M. Whitfield, and L. Maddock, eds.), pp. 127–148, by permission of Cambridge University Press.

(1990) adds that although bioluminescent sources are relatively weak, they are concentrated on a small part of the retina and will be bright enough to allow accurate determination of the position of the source. Denton and Locket (1989) have also proposed that such multibank retinas may also have the capacity to provide information about source wavelength, based on the fact that vitread banks of photoreceptors (those adjacent to the vitreous humor) of *Diretmus argenteus* (family Diretmidae) act as effective band-pass filters, changing the spectral character of light reaching more sclerad photoreceptors.

The eye of *Bajacalifornia* also has a marked anterior aphakic space (a gap between the lens and the iris) (Fig. 2). Indeed, the presence of aphakic eyes is particularly common among the deeper living fish, which deal almost exclusively with bioluminescent lights. The merit of an aphakic eye is that it increases the capture of light for the retina from obliquely placed sources. For example, in *Bajacalifornia* the anterior aphakic space allows light from in front of the jaws to be collected by the whole of the lens, rather than from the just the external half, as would normally occur where the lens is closely surrounded by the pupil. Under conditions of high ambient light, the aphakic space would allow diffuse light onto the retina, which could compromise visual function, but this is unlikely to be a problem in the deep sea, where biological light sources are punctate in space and time.

The second category of bioluminescence is the orange or red light produced by three genera of fishes. Although these are strictly mesopelagic fishes, their vision is based on biological light and thus is appropriate to our discussion. The distinctive feature of these fish is that they both emit and perceive red light. These are the only known active visual systems in the animal kingdom, whereby light produced by the organism is used to investigate its surroundings. In *Malacosteus* the light is produced in a suborbital photophore. A filter over the organ absorbs most of the generated light so that the only light emitted is of a narrow waveband centred at about 700 nm (Denton *et al.*, 1985). Detection of the long-wavelength light emitted is aided by the possession of red-shifted visual pigments (λ_{max} of 514 and 556 nm) (Crescitelli, 1989). Natural daylight penetrating to these depths is devoid of red light, so red coloration is adopted as a common camouflage among midwater invertebrates and these organisms are insensitive to red light. The ability to illuminate red-colored prey with a red light that they cannot see would seem to confer an almost unfair advantage to the predator. However, prey detection may not be limited to red-colored animals. The highly reflective tapeta lucida in the eyes of deep-water fishes will also serve to reflect interrogating light back to the source animal. If the illuminated fish possess the typical deep-water visual pigment of λ_{max} 485 nm, then, although the possibility of detecting 700-nm light remains,

8. SENSORY PHYSIOLOGY

the probability of detecting sufficient photons for a response is remote. As a result, the illuminated fish may be unaware that they have been detected. Visual detection of this type has been demonstrated in the shallow-water nocturnal flashlight fish *Anomalops* (Howland *et al.*, 1992). O'Day and Fernandez (1974) also suggest that red bioluminescence could be a good means of intraspecific communication via a private wavelength free from interception by potential predators.

IV. TOUCH

Not much is known about touch or somatosensory systems in deep-sea fish. In general terms it would be surprising if, as the possibilities for vision declined, touch did not become relatively more important. Hints of this are seen in the elaborate extended fin rays of many species, such as tripod fishes, and in the common occurrence of mental (i.e., attached to the chin) barbels. The potential for tactile stimuli to play a role in prey detection is shown in antarctic fishes. Antarctic benthic feeders show stereotypical responses to prey touching their pelvic or anal fins (Janssen, 1992). In response to a touch, the fish repositions the head to above where the prey collided with the fin. This would bring the lateral-line sense organs into a position where they could detect hydromechanical stimuli from the prey (see below). Janssen *et al.* (1993) have also shown that in another antarctic fish, the plunderfish (Artedidraconidae), the mental barbel is used as a lure, and that touching the lure initiates a strike.

V. OCTAVOLATERALIS SYSTEMS

A. Introduction

The octavolateralis systems are a related group of senses. The majority of these senses are based around mechanosensory hair cells that, depending on the way in which they are built into the sense organ, can be used to encode angular and linear accelerations of the fish, gravity, acoustic stimuli, and water movements. Octavolateralis senses also include electroreception, which is found in almost all nonteleost fishes, but for the deep sea we need only consider electroreception in sharks, skates, rays, and chimaeras. The mechanosensory and electrosensory lateral-line systems can provide high-resolution information about the location and movement of animate objects close to the fish, so probably provide the best sensory alternative to vision when vision is not available. A review of elasmobranch sensory systems, including the octavolateralis system is given by Montgomery (1988a).

B. Vestibular System/Hearing

The otic capsule behind the eye houses the vestibular system. A collection of sensory systems that are all innervated by the VIIIth cranial nerve (Fig. 3). The two senses used in orientation are the semicircular canals, which encode angular rotations of the head, and the gravity-receptive otolith organs, predominantly the utriculus. The hair cell receptors of the semicircular canals are located in a discrete patch within a swelling of the canal called an ampulla. Differential motion between the canal and the internal fluid (endolymph) during head rotation creates the mechanical stimulus to the hair cells. There are three semicircular canals to encode head rotations in three-dimensional space. In otolith organs, a heavy otolith provides the mechanical stimulus to the hair cells during linear accelerations, and in the utricule encodes the orientation of the head with respect to gravity. In normal circumstances light is also used for orientation, and in the absence of this cue it is perhaps not surprising that bathypelagic fish have noticeably

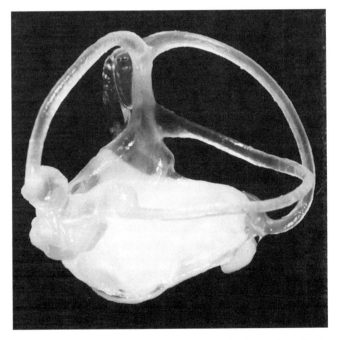

Fig. 3. The left vestibular labyrinth of an elasmobranch viewed from the left side. The three semicircular canals have within the ampullary swellings patches of hair cells that respond to fluid movements produced by angular rotations of the head. The three semicircular canals, mutually at right angles, encode head rotations in the three-dimensional space. The dense white areas are otoliths, which respond to linear accelerations and vibrations.

8. SENSORY PHYSIOLOGY

well-developed semicircular canals and utricular receptors (Marshall, 1979). For example, the dwarf male anglerfishes and *Cyclothone* spp. have over half the volume of the neurocranium devoted to the otic capsules, and the deep-water (1700–3700 m) *Acanthonus* has semicircular canals that are the largest, relative to body weight, of any vertebrate (Fine *et al.*, 1987).

The other main sense of the vestibular system is hearing, based principally on the saccular otolith. Movement of the fish in an acoustic field sets up the differential movement between the heavy otolith and the underlying sensory epithelium that is essential for mechanical stimulation of the hair cells. Detection of the pressure component of the acoustic field requires specialized connections between the swim bladder and the inner ear, a situation that has not been reported for deep-sea fishes. In bathypelagic fishes the sacculus is small, but in most benthopelagic fishes, including macrourids, deep-sea cods, and brotulids, it is very large. Size is not a necessary correlate of acoustic capability, but a larger sacculus would be expected to increase sensitivity by increasing the differential movement, and also by providing a larger sensory epithelium, allowing for a greater number of hair cells. All three of the above-mentioned families also have a means of sound production. In one ophidiid species, *Barathrodemus manatinus*, sexual dimorphism is apparent in the sound production mechanism, with only males possessing well-developed drumming muscles associated with the swim bladder (Carter and Musick, 1985). However, sound production and hearing are confined to species with large eyes found on the upper continental slope. Abyssal macrourids lack sound-production mechanisms and have small saccular otoliths (Marshall, 1979). For reasons unknown, it appears that sound production ceases at depth despite the continuing presence of swim bladders in the abyssal forms. Perhaps the decreasing elasticity of a gas-filled bladder at depth makes it difficult to vibrate, or increasing gas density makes swim bladders less efficient as sound radiators. Whatever the case, loss of sound production appears correlated with a reduction in size of the sacculus. Sound production and hearing seem coupled processes in deep-sea fishes. In a review of sound detection and processing by fish, Popper and Fay (1993) argue that the most general function of hearing is to identify and locate objects (sound sources and scatterers) comprising the environment and perhaps to form an image of the auditory scene. Detecting sound scatterers requires an acoustic background that may just be absent in the deep sea. So with the loss (for whatever reason) of fish communication sounds, hearing may be significantly less useful in the deep than it is in acoustically rich surface waters.

C. Mechanosensory Lateral Line

The mechanosensory lateral line has been reviewed by Bleckmann (1993) and Montgomery *et al.* (1995). It consists of patches of sensory hair cells with associated cells and an overlying gelatinous cupula. The whole structure is called a neuromast. Single hair cells are morphologically/functionally polarized such that they are maximally sensitive to deflection along one particular axis. Along this axis, deflection in one direction excites the hair cell, whereas deflection in the opposite direction inhibits it. Within each neuromast there are two populations of hair cells, polarized along the same axis, but facing different directions. Neuromasts are organized as free standing on the surface of the skin, or sunk beneath the skin into canals that may be rigid or membranous. As with other hair cell systems, the essential stimulus to the hair cells is differential movement between the cupula and the epithelium, so the associated anatomy plays a crucial role in exactly what aspects of water movements in the environment are encoded. In general terms many aspects of the structure/function of mechanosensory lateral lines can be interpreted in terms of the mechanical filtering properties of peripheral structure: maximizing sensitivity to signals of interest, while minimizing the response to extraneous noise such as stimulation generated by the animal's own movements.

Free-standing neuromasts will be stimulated by water movement over the surface of the skin. Typically, in a free-standing neuromast, the axis of greatest sensitivity of the hair cells is along the long axis of the cupula. This means that the cupula is friction-coupled to the water. In other words, effective movement of the cupula is generated by the velocity of the water movement along the sides of the cupula (Fig. 4). For this reason superficial neuromasts are generally described as being velocity sensitive. Close to the surface of the skin of the fish there is a boundary layer, the thickness of which increases as the velocity of water flow decreases. For oscillating flows, there is also a thickening of the boundary layer with decreasing frequency. So increasing sensitivity to slow flows and to low frequencies can be achieved by increasing the height of the cupula, or raising the superficial neuromast onto a small papilla. Further sensitivity to slow flows and low frequencies can be obtained by changing the orientation of the cupula so that the long axis of the cupula is at right angles to the axis of sensitivity of the hair cells. In this configuration the cupula is directly coupled to water movement, so in effect becomes displacement sensitive. Winding up the sensitivity of the superficial neuromasts is all very well, but in addition to increasing their response to biologically important sources, it also increases their susceptibility to self-generated noise. Deep-sea fishes, including the ceratioid anglerfishes have some of the most extraordinary superficial neuromasts known. They are typically papillate (Fig. 5), and in some species, for example, *Neoceratias*, occur on long stalks (Marshall, 1979) with the direction of the cupula in the displacement-sensitive configuration (Marshall, 1996). For neuromasts of this sort, holding self-generated noise levels down to an

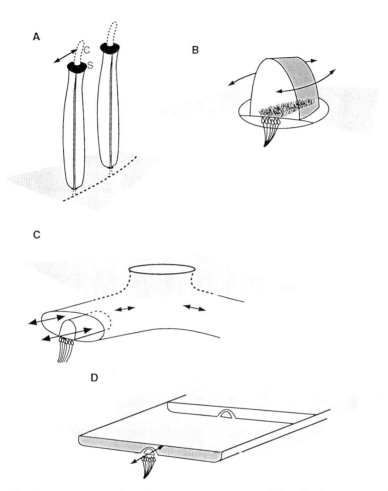

Fig. 4. Arrangements of mechanosensory neuromasts. (A) stalked neuromast of the deep-sea angler *Neoceratias*. The cupula (c, dotted appendage) is shown above the sensory epithelium (s). The double-headed arrow represents the water movements that will stimulate movements of the cupula (after Marshall, 1979). (B) Standard superficial neuromast found on the surface of the skin of fishes. The long axis of the cupula matches the axis of sensitivity of the hair cells. Friction of water movement past the sides of the cupula generates the cupula movement, which is the effective stimulus to the hair cells. (C) Neuromast embedded in a rigid canal, cupula movement is generated by water movements within the canal. (D) Membranous canals system of *Anaplogaster*. A broad shallow canal is covered by a thin membrane. The canal is divided into sections by bony partitions. Fluid movements in the canal sections stimulate the neuromasts, which are located in windows in the bony partitions (after Denton and Gray, 1988).

Fig. 5. Photograph of the deep-sea anglerfish *Phrynichthys wedli*. Note the prominent lateral-line system with all the neuromasts situated on dermal papillae. Photograph by J. Marshall and C. Diebel, Vision, Touch, and Hearing Research Center, University of Queensland, and Experimental Biology Research Group, University of Auckland.

acceptable level must place a huge premium on staying as still as possible. Buoyancy adaptations and low metabolic rate with corresponding reduced ventilatory demand (Denton and Marshall, 1958) can be seen as part of a suite of attributes that must reduce self-generated noise to the point at which papillate neuromasts can function effectively.

The recent description of a new form of superficial neuromast in two species of deep-sea fishes (Marshall, 1996) suggests an intriguing linkage between mechanosensory systems and olfaction. From first principles it makes sense for mechanosensory information to be utilized in the interpretation of olfactory signals. Many fish species have superficial neuromasts closely associated with the olfactory nares (J. C. Montgomery and A. G. Carton, unpublished observation) but the presence of a novel class of superficials arranged in a rosette around the olfactory nares of *Poromitra capito* is the best evidence we have of a possible functional linkage between lateral line and olfactory systems.

If lifting free-standing neuromasts above the boundary layer increases their responsiveness to low frequencies, placing them in rigid canals works in the other direction. The rigid canals so typical in shallow-water teleosts act as mechanical filters that attenuate lower frequencies (Montgomery *et al.*, 1994). Constrictions placed in the canals opposite the neuromast further attenuate low frequencies and mechanically amplify high frequencies (Denton and Gray, 1988). These morphologies are appropriate for the detection of high-frequency signals against a background of low-frequency noise, or while actively swimming, but these circumstances are not particularly appropriate to the situation facing most deep-sea fishes. These forms commonly have elaborate membranous canals. Marshall (1979) illustrates two benthopelagic species, a halosaur and a macrourid, both of which have extensive membranous canal systems over the surface of the head. Apparently it is not unusual for the area of the skin stretched over the canals to represent well over half the surface area of the head. Denton and Gray (1988) describe in some detail the membranous canal systems of *Poromitra* (family Melamphaidae) and *Anaplogaster* (family Anaplogastridae). Although these systems differ in anatomy, they appear to be roughly functionally equivalent. In *Poromitra* large flat neuromasts are found in the base of wide canals that are covered by a soft membrane perforated by small pores. In *Anaplogaster* the canals are broad but shallow with bony partitions along their length. The neuromasts are located in windows in the bony partitions. The whole system is covered by a thin soft membrane. Despite these anatomical differences, Denton and Gray (1988) predict that both will have a resonance in the region 5–10 Hz, providing as much as a 100-fold increase in sensitivity over this frequency range when compared with a shallow-water teleost such as a sprat.

Further discussion of the mechanical tuning of lateral-line canals runs into the problem that we know very little detail of the characteristics of real signals versus potential noise sources. Montgomery and Macdonald (1987) show that swimming plankton (including copepods) can produce appreciable water oscillations at frequencies from 4 or 5 Hz up to around 40 Hz. Bleckmann *et al.* (1991) show that swimming fish produce substantial hydromechanical energy at the fundamental frequency of their tail beat, anywhere from a few to 10 Hz, and significant frequencies up to 100 Hz. It is also likely that some deep-sea fishes produce communication signals for lateral-line detection such as those occurring in salmon courtship (Satou *et al.*, 1994). So the signals of interest generated by other animals could be anywhere in the range of 1 or 2 Hz to 100 Hz. Vortex sheets produced by swimming in a teleost fish with a standard homocercal tail persist for a considerable time after the fish has passed (Blickhan *et al.*, 1992), potentially providing a potent and useful stimulus to another fish that swims into the wake. The precise hydrodynamic wake generated by rat-tail fishes is not known, but intuitively rat-tails would produce much less of a "footprint" than a homocercal tail, and it is tempting to speculate that this morphology which is so common in the deep sea, acts as a lateral-line camouflage.

The largest source of noise will be self-generated noise, from fin and body movements and from ventilation. Recent studies of central lateral-line processing show that the first stage of sensory processing is a sophisticated adaptive filter that learns to cancel self-generated noise (Montgomery and Bodznick, 1994). Despite this, there will still be a premium on minimizing movement, not just to simplify the job for the central filter, but also to reduce the fish's hydrodynamic "visibility" to other animals.

Lateral-line detectors are typically distributed as a trunk lateral line and a series of lines on the head, above and below the eye and a preopercular mandibular line. The arrangement provides a system with a relatively high spatial acuity that can accurately determine the position and movement of objects close to the body, particularly around the head and close to the mouth. Mechanosense has been described as touch at a distance, or touch mediated via the intervening water movement. The lateral line has the added advantage that the other party can be felt without itself being touched. Despite the relatively short range of only a body length or so, mechanosense must provide a good adjunct to vision, or replacement of vision in many behavioral interactions in the deep sea.

D. Electrosense

Sharks, skates, rays, and chimaeras have a sense that is additional to those found in bony fishes. This is the electrosense, which is reviewed by Montgomery (1988a). Pore openings on the surface of the skin, particularly

---

Fig. 6. Photograph of the ventral surface of the skate (*Raja nasuta*). Although not a deep-sea species, this photograph serves to illustrate the arrangement of the mechanosensory and electrosensory lateral lines around the mouth in elasmobranchs. The canals of the mechanosensory lateral-line system have been injected with India ink. The injection site is evident on the right side of the photograph, and the canals are more extensively filled on this side. The pore openings of the electrosensory ampullae of Lorenzini are naturally pigmented in this species, so each of the black dots is a pore opening. In some cases, particularly the pores on the base of the pectoral fins, the jelly-filled canal leading away from the pore is evident. Note the concentration of both systems around the mouth, and particularly the rostrum in front of the mouth is heavily invested by the mechanosensory lateral-line system. Photograph by E. Skipworth, Experimental Biology Research Group, University of Auckland.

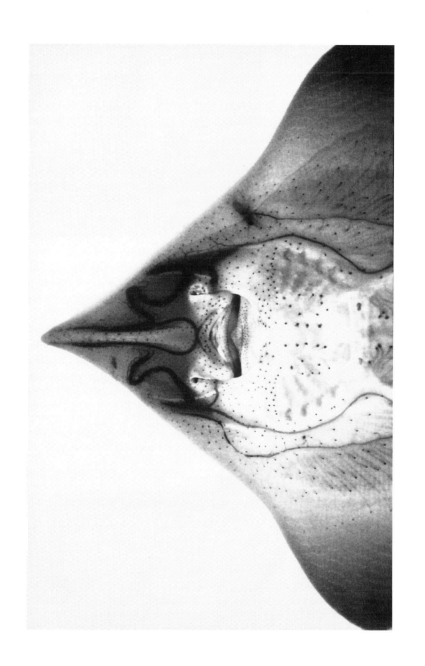

common around the mouth (Fig. 6), lead to jelly-filled canals that terminate in the ampullae of Lorenzini, in the walls of which are the electrosensory cells. The system is exquisitely sensitive to weak electric fields, with behavioral thresholds in the region of 5 nV/cm. One of the primary functions of the electrosense is prey detection, though it has also been shown to be involved in mate recognition (Tricas et al., 1995). Weak electric fields are produced by ion pumps used in osmoregulation and these quasi-dipole fields are modulated by body movements such as ventilation. Elasmobranchs can detect the fields up to a meter or so away, and can home in on the source. The high density of electrosensory organs around the mouth indicate a high spatial acuity in this region. Electrosense must provide a similar quality of information to mechanosensory lateral line, but mediated by electric fields produced by the other animal rather than hydrodynamic disturbances created by movement. The relatively short range of both electro- and mechanosense may be related to the extensive preoral surfaces seen particularly in deep-sea elasmobranchs, such as the long-nosed skates and chimaeras, but also in benthopelagic teleosts such as halosaurs and macrourids. Sensory information from preoral surfaces would clearly be useful in initiating and controlling predatory strikes (Montgomery, 1991).

In addition to the detection of other animals, electrosense has other potentialities. Movement through the earth's magnetic field produces electrical gradients within the sensitivity range of the elasmobranch electrosense. The intensity and direction of these gradients are related to the speed and direction of the movements that cause them, so elasmobranchs could use their electrosense in navigation (Paulin, 1995).

VI. GENERAL COMMENTS

A. Orientation and Navigation

Light imposes a very strong structuring influence on ocean inhabitants. The camouflage mechanisms of mesopelagic fishes demand that normal body posture (dorsal uppermost) is maintained at all times (Janssen et al., 1986), and to maximize use of available light, the visual axis is typically vertical, even in fish that swim at an oblique angle (Janssen et al., 1992). Below the level of natural light, these constraints are released. Gravity must be the strongest orientation cue, but there is no particular reason to maintain dorsal-up posture, and indeed it is not unusual to see fish at these depths swimming in unusual ways, such as bouncing along the bottom in a head-down posture (C. Diebel, personal communication, 1996).

Mesopelagic fish position themselves in the water column by undertaking vertical migrations to maintain themselves at particular light intensities.

To do this they must be capable of measuring something approximating absolute light levels (Denton, 1990). Below the level of natural light it is not easy to see what cues are available to fish to allow them to hold a particular level in the water column. For bathypelagic fishes neutral buoyancy is of importance (as discussed above) for minimizing movements, so they may well position themselves vertically along the density gradient to achieve this. However, is density set and the vertical position found, or is density regulated to achieve a particular vertical position? Pressure is clearly the variable we would use to determine depth, and pressure does have direct biological effects (Harper et al., 1987), but we currently do not know if fish, with or without swim bladders, have an absolute sense of pressure, though larval herring do respond to transient pressure changes (Colby et al., 1982). If the microstructure of the deep ocean continues into the bathypelagic zone, then small temperature changes, and even the current shear of adjacent layers, could both provide cues to the fish that it was moving vertically, and absolute temperature could be a reasonable proxy of depth. However, like pressure, there is little known of the ability fish have to detect absolute temperature levels, or the sensory mechanisms that might be employed. Behavioral experiments have shown thresholds to acute temperature change of as little as 0.03°C (Murray, 1971), and the suggestion has been made that bilateral input from the vestibular labyrinth could provide one source of temperature information (Montgomery, 1988b).

Similar problems must exist with movements in the horizontal plane. Fish must drift with the local currents, and some of these displacements would be counteracted by movement in different water masses at different stages of the life-history cycle. However, it is still likely that directed horizontal movements would be required. Active migrations are certainly indicated in the spawning aggregations of some mesopelagic fishes, which in examples such as orange roughy (*Hoplostethus atlanticus*) are site specific (Pankhurst, 1988). Metcalf et al. (1993) have shown that plaice can maintain consistent headings in midwater at night, in the apparent absence of visual and tactile clues. Their results suggest that the fish are using an external geophysical reference. The sensory basis of this in teleost fish is unknown, but could be by detection of the electrical field generated by the flow of sea water through the geomagnetic field, or detection of the earth's magnetic field (Walker, 1984). For elasmobranch fishes with specialized electroreceptors, the former mechanism is most likely (Paulin, 1995).

B. Comparisons with Antarctic Fishes

There are many similarities between antarctic seas and the deep ocean, the principal ones being that both are cold and dark. Antarctic high-latitude basins are effectively dark for the duration of the antarctic winter, and even during summer the light levels are extensively reduced by ice cover, and the continental shelf is deeper than for other continents. With respect to the fish faunas, perhaps the major difference is that the antarctic fauna is dominated by the family Nototheniidae, which belongs to the most numerous order of teleost fishes, the Perciformes, whereas the deep-sea fauna is overwhelmingly composed of nonperciform fishes. The antarctic fishes have also had a relatively short evolutionary period under these conditions compared with the deep-sea fishes. Despite these differences, antarctic fishes provide useful models of sensory function in a low-light environment. They appear to lack any particular visual adaptations for feeding under low light (Eastman, 1988; Montgomery et al., 1989) and rely on mechanosensory and tactile cues for feeding (Janssen et al., 1990, 1993; Janssen, 1992). The open cephalic lateral-line organs of the antarctic fish *Pleurogramma* are reminiscent of some deep-sea lateral lines, and the ice fishes have membranous canals (Montgomery et al., 1994), but nowhere near as well developed as those found in the deep-sea fauna. There is a striking parallel between the ice fish *Chionodraco hamatus* and the tripod fishes. Tripod fishes (*Bathypterois* spp.) sit motionless, supported up off the bottom by three extraordinary stiff elongate fin rays (Marshall, 1979), two modified pelvic fin rays, and one elongate ray from the ventral caudal fin. They face upstream, taking zooplankton brought to them by the current. *Chionodraco*, which is piscivorous, could be described as a bipod fish, because it has been observed sitting well up off the bottom on its elongate pelvic fins (Fig. 7). Marshall (1979) drew attention to the similarities between tripod fish and one of the ice fish, *Pagetopsis*, which was described by Robilliard and Dayton (1969) as perching on a sponge. The stance of *Chionodraco* is an even more striking parallel. Ice fishes and tripod fishes seem to have converged on a similar strategy to sit motionless above the substrate with the attendant benefits that motionlessness brings to nonvisual, particularly mechanosensory, function.

C. Deep-Sea Sensory Biology

The deep-sea fish fauna is an interesting phylogenetically diverse group that inhabits an unusual sensory world. There is a wide range of solutions to the ubiquitous demands of finding food and recognizing and locating mates. Nocturnal fishes, cave dwellers, and antarctic fishes can provide us with insights into some of the potential solutions and the potentialities and limitations of sensory systems working under similar constraints. But it does seem that deep-sea fishes have a more extreme development of their sensory systems than do fishes in any of the more accessible ecosystems.

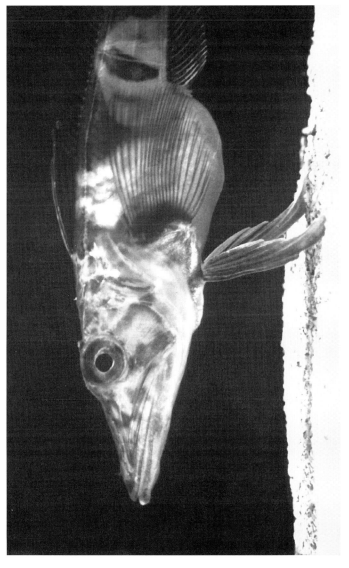

Fig. 7. Lateral view of an aquarium specimen of the ice fish *Chionodraco hamatus*. Note the stance, with the fish perched on its elongate modified pelvic fins.

Nowhere else do we see quite the development of function-specific olfactory and visual systems, elaborate tactile appendages, membranous lateral-line canals, and stalked superficial neuromasts. A diverse phylogeny, a long evolution, and the peculiar nature of the deep-sea environment have combined to produce a fascinating sensory physiology in the deep-sea fishes.

ACKNOWLEDGMENTS

We particularly wish to thank Justin Marshall and Carol Diebel for permission to use the photograph of the anglerfish.

REFERENCES

Aho, A.-C., Donner, K., Hydén, C., Larsen, L. O., and Reuter, T. (1988). Low retinal noise in animals with low body temperature allows high visual sensitivity. *Nature (London)* **334,** 348–350.

Atema, J., Fay, R. R., Popper, A. N., and Tavolga, W. N. (1988). "Sensory Biology of Aquatic Animals." Springer-Verlag, New York.

Best, A. C. G., and Nicol, J. A. C. (1980). Eyeshine in fishes. A review of ocular reflectors. *Can. J. Zool.* **58,**, 945–956.

Bleckmann, H. (1993). Role of the lateral line in fish behaviour. *In* "Behaviour of Teleost Fishes" (T. J. Pitcher, ed.), 2nd ed. pp. 177–202. Chapman & Hall, London.

Bleckmann, H., Breithaupt, T., Blickhan, R., and Tautz, J. (1991). The time course and frequency content of hydrodynamic events caused by moving fish and crustaceans. *J. Comp. Physiol. A* **168**, 749–757.

Blickhan, R., Krick, C., Zehren, D., Nachtigall, W., and Breithaupt, T. (1992). Generation of a vortex chain in the wake of a subundulatory swimmer. *Naturwissenschaften* **79,**, 220–221.

Bowmaker, J. K., Govardovskii, V. I., Shukolyukov, S. A., Zueva, L. V., Hunt, D. M., Sideleva, V. G., and Smirnova, O. G. (1994). Visual pigments and the photic environment: The Cottid fish of Lake Baikal. *Vision Res.* **34,**, 591–605.

Bullock, T. H., and Heiligenberg, W. (1986). "Electroreception." Wiley, New York.

Carter, H. J., and Musick, J. A. (1985). Sexual dimorphism in the deep-sea fish *Barathrodemus manatinus,* ophidiidae. *Copeia,* 69–72.

Caprio, J. (1988). Peripheral filters and chemoreceptor cells in fishes. *In* "Sensory Biology of Aquatic Animals" (J. Atema, R. R. Fay, A. N. Popper, and W. N. Tavolga, eds.), pp. 313–338. Springer-Verlag, New York.

Colby, D. R., Hoss, D. E., and Blaxter, J. H. S. (1982). *Fish. Bull.* **80,** 567–574.

Coombs, S., Gorner, P., and Münz, H. (1989). "The Mechanosensory Lateral Line: Neurobiology and Evolution." Springer-Verlag, New York.

Crescitelli, F. (1989). The visual pigments of a deep-water malacosteid fish. *J. Mar. Biol. Assoc. U.K.* **69,** 43–51.

Denton, E. J., (1990). Light and vision at depths greater than 200 metres. *In* "Light and Life in the Sea" (P. J. Herring, A. K. Campbell, M. Whitfield, and L. Maddock, eds.), pp. 127–148. Cambridge Univ. Press, Cambridge.

Denton, E. J., and Gray, J. A. B. (1988). Mechanical factors in the excitation of the lateral lines of fishes. *In* "Sensory Biology of Aquatic Animals" (J. Atema, R. R. Fay, A. N. Popper, and W. N. Tavolga, eds.), pp. 595–618. Springer-Verlag, New York.

Denton, E. J., and Locket, N. A. (1989). Possible wavelength discrimination by multibank retinae in deep-sea fishes. *J. Mar. Biol. Assoc. U.K.* **69** 409–435.

Denton, E. J., and Marshall, N. B. (1958). The buoyancy of bathypelagic fishes without a gas-filled swimbladder. *J. Mar. Biol. Assoc. U.K.* **37,**, 753–767.

Denton, E. J., Gilpin-Brown, J. B., and Wright, P. G. (1972). The angular distribution of light produced by some mesopelagic fish in relation to their camouflage. *Proc. R. Soc. London B* **182,**, 145–158.

Denton, E. J., Herring, P. J., Widder, E. A., Latz, M. F., and Case, J. F. (1985). The role of filters in the photophores of oceanic animals and their relation to vision in the oceanic environment. *Proc R. Soc. London B* **225,** 63–97.

Douglas, R. H., and Thorpe, A. (1992). Short-wave absorbing pigments in the ocular lenses of deep-sea teleosts. *J. Mar. Biol. Assoc. U.K.* **72,**, 93–112.

Douglas, R. H., Partridge, J. C., and Hope, A. J. (1995). Visual and lenticular pigments in the eyes of demersal deep-sea fishes. *J. Comp. Physiol. A* **177,** 111–122.

Eastman, J. T. (1988). Ocular morphology in antarctic notothenioid fishes. *J. Morphol.* **196,** 283–306.

Fine, M. L., Horn, M. H., and Cox, B. (1987). *Acanthonus armatus,* a deep-sea teleost with a minute brain and large ears. *Proc. R. Soc. London B* **230,** 257–265.

Harper, A. A., Macdonald, A. G., Wardle, C. S., and Pennec, J. P. (1987). The pressure tolerance of deep sea fish axons results of *Challenger* cruise 6B–85. *Comp. Biochem. Physiol. A* **88,** 647–654.

Herring, P. J. (1983). The spectral characteristics of luminous marine organisms. *Proc. R. Soc. London B* **220,** 183–217.

Howland, H. C., Murphy, C. J., and McCosker, J. E. (1992). Detection of eyeshine by flashlight fishes of the family Anomalopidae. *Vision Res.* **32,** 765–769.

Janssen, J. (1992). Responses of antarctic fishes to tactile stimuli. *Antartic. J. U.S.* **27,** 142–143.

Janssen, J., Harbison, G. R., and Craddock, J. E. (1986). Hatchetfishes hold horizontal attitudes during diagonal descents. *J. Mar. Biol. Assoc. U.K.* **66,** 825–833.

Janssen, J., Coombs, S., Montgomery, J. C., and Sideleva, V. (1990). Comparisons in the use of the lateral line for detecting prey by Notothenioids and sculpins. *Antarctic J. U.S.* **25,** 214–215.

Janssen, J., Pankhurst, N. W., and Harbison, G.R. (1992). Swimming and body orientation of *Notolepis rissoi* in relation to lateral line and visual function. *J. Mar. Biol. Assoc. U.K.* **72,** 877–886.

Janssen, J., Slattery, M., and Jones, W. (1993). Feeding responses to mechanical stimulation of the barbel in *Histidraco velifer* (Artedidracondidae). *Copeia,* 885–889.

Jumper, G. Y., and Baird, R. C. (1991). Location by olfaction: a model and application to the mating problem in the deep-sea hatchetfish *Argyropelecus hemigymnus. Am. Nat.* **138,** 1431–1458.

Locket, N. A. (1977). Adaptations to the deep-sea environment. *In* "The Visual System in Vertebrates" (F. Crescitelli, ed.), Handbook of Sensory Physiology, Vol. 7/5, pp. 67–192. Springer-Verlag, Berlin.

Locket, N. A. (1980). Variation of architecture with size in the multiple-bank retina of a deep sea teleost, *Chauliodus sloani. Proc. R. Soc. London B* **208,** 223–242.

Locket, N. A. (1985). The multiple bank fovea of *Bajacalifornia drakei,* an alepocephalid deep-sea teleost. *Proc. R. Soc. London B* **224,** 7–22.

Marshall, N. B. (1979). "Developments in Deep-Sea Biology," pp. 305–409. Blandford, Poole, U.K.
Marshall, N J. (1996). The lateral line systems of three deep-sea fish. *J. Fish Biol.* **49**(Suppl. A), 239–258.
Mensinger, A. F., and Case, J. F. (1990). Luminescent properties of deep sea fish. *J. Exp. Mar. Biol. Ecol.* **144,** 1–16.
Metcalf, J. D., Holford, B. H., and Arnold, G. P. (1993). Orientation of plaice (*Pleuronectes platessa*) in the open sea: Evidence for the use of external directional clues. *Mar. Biol.* **117,** 559–566.
Montgomery, J. C. (1988a). Sensory Physiology. *In* "Physiology of Elasmobranch Fishes" (T. Shuttleworth, ed.), pp. 79–98. Springer-Verlag, Berlin.
Montgomery, J. C. (1988b). Temperature compensation in the vestibulo-ocular reflex: a novel hypothesis of cerebellar function. *J. Theor. Biol.* **132,** 163–170.
Montgomery, J. C. (1991). "Seeing" with nonvisual senses: mechanosensory and electrosensory systems of fish. *News Physiol. Sci.* **6,** 73–77
Montgomery, J. C., and Bodznick, D. (1994). An adaptive filter cancels self-induced noise in the electrosensory and lateral line mechanosensory systems of fish. *Neurosci. Lett.* **174,** 145–148.
Montgomery, J. C., and Macdonald, J. A. (1987). Sensory tuning of lateral line receptors in antarctic fish to the movements of planktonic prey. *Science* **235,** 195–196.
Montgomery, J. C., Pankhurst, N. W., and Foster, B. A. (1989). Limitations on visual feeding in the planktivorous antarctic fish *Pagothenia borchgrevinki*. *Experientia* **45,** 395–397.
Montgomery, J. C., Coombs, S., and Janssen, J. (1994). Aspects of structure and function in the anterior lateral line of six species of antarctic fish of the suborder Notothenioidei. *Brain, Behav. Evol.* **44,** 299–306.
Montgomery, J. C., Coombs, S., and Halstead, M. B. D. (1995). Biology of the mechanosensory lateral line in fishes. *Rev. Fish Biol. Fish.* **5,** 399–416.
Munk, O. (1966). Ocular anatomy of some deep-sea teleosts. *Dana Report* **70,** 1–71.
Muntz, W. R. A. (1976). On yellow lenses in mesopelagic animals. *J. Mar. Biol. Assoc. U.K.* **56,** 963–976.
Murray, R. W (1971). Temperature receptors. *In* "Fish Physiology"(W. S. Hoar and D. J. Randall, eds.), Vol. 6, pp. 121–133. Academic Press, New York.
Nicol, J. A. C. (1978). Bioluminescence and vision. *In* "Bioluminescence in Action" (P. J. Herring, ed.), pp. 367–398. Academic Press, New York.
O'Day, W. T., and Fernandez, H. R. (1974). *Aristostomias scintillans* (Malocosteidae): A deep-sea fish with visual pigments apparently adapted to its own bioluminescence. *Vision Res.* **14,** 545–550.
Pankhurst, N. W. (1987). Intra- and interspecific changes in retinal morphology among mesopelagic and demersal teleosts from the slope waters of New Zealand. *Environ. Biol. Fish.* **19,** 269–280.
Pankhurst, N. W. (1988). Spawning dynamics of orange roughy, *Hoplostethus atlanticus*, in mid-slope waters of New Zealand. *Environ. Biol. Fish.* **21,** 101–116
Partridge, J. C., Archer, S. N., and Lythgoe, J. N. (1988). Visual pigments in the individual rods of deep-sea fishes. *J. Comp. Physiol. A* **162,** 543–550.
Partridge, J. C., Shand, J., Archer, S. N., Lythgoe, J. N., and van Groningen-Luyben, W. A. H. M. (1989). Interspecific variation in the visual pigments of deep-sea fishes. *J. Comp. Physiol. A* **164,** 513–529.
Paulin, C. D., Stewart, A. L., Roberts, C. D., and McMillan, P. J. (1989). New Zealand fish: A compete guide. *National Museum of New Zealand Miscellaneous Series,* No. 19.

Paulin, M. G. (1995). Electroreception and the compass sense of sharks. *J. Theor. Biol.* **174,** 325–339.
Popper, A. N., and Fay, R. R. (1993). Sound detection and processing by fish: A critical review and major research questions. *Brain Behav. Evol.* **41,** 14–38.
Robilliard, G. A., and Dayton, P. K. (1969). Notes on the biology of the chaenichthyid fish *Pagetopsis macropterus* from McMurdo Sound, Antarctica. *Antarctic J. U.S.* **4,** 304–306.
Satou, M., Takeuchi, H. A., Takei, K, Hasegawa, T., Matsushima, T., and Okumoto, N. (1994). Characterization of vibrational and visual signals which elicit spawning behavior in the male hime salmon (landlocked red salmon, *Oncorhynchus nerka*). *J. Comp. Physiol. A* **174,** 527–537.
Tavolga, W. N., Popper, A. N., and Fay, R. R. (1981). "Hearing and Sound Communication in Fishes." Springer-Verlag, New York.
Tricas, T. C., Michael, S. W., and Sisneros, J. A. (1995). Electrosensory optimization to conspecific phasic signals for mating. *Neurosci. Lett.* **202,** 129–132.
Walker, M. M. (1984). Learned magnetic field discrimination in yellowfin tuna *Thunnus albacares. J. Comp. Physiol. A* **155,** 673–679.
Westerberg, H. (1984). The orientation of fish and the vertical stratification at fine- and microstructure scales. *In* "Mechanisms of Migration in Fishes" (J. D. McCleave, G. P. Arnold, J. J. Dodson, and W.H. Neill, eds.), pp. 179–204. Plenum, New York.
Yamamoto, M. (1982). Comparative morphology of the peripheral olfactory organ in teleosts. *In* "Chemoreception in Fishes" (T. J. Hara, ed.), pp. 39–59. Elsevier, New York.

9

LABORATORY AND *IN SITU* METHODS FOR STUDYING DEEP-SEA FISHES

KENNETH L. SMITH, Jr., AND ROBERTA J. BALDWIN

I. Introduction
II. Laboratory Studies
   A. Animal Collections
   B. Animal Maintenance
III. *In Situ* Studies
   A. Animal Collection/Measurements
   B. Behavioral Observations
IV. Future Directions
   References

I. INTRODUCTION

The ocean deeper than 1000 m covers approximately 62% of the earth's surface (Gage and Tyler, 1991). This large habitat is devoid of solar illumination and is characterized by high hydrostatic pressure, low temperature, and low food supply (see Chapter 1, this volume). A wide diversity of fishes (Chapter 2, this volume), ranging from the gonostomatid genus, *Cyclothone*, which abundantly populates the midwater regions, to macrourids such as *Coryphaenoides*, which are more commonly associated with the benthic boundary layer, occupy this extreme environment. Studies of the physiology of these deep-sea fishes are difficult to conduct because of stresses inherent to changes in these environmental parameters. Two basic approaches have been used to collect and conduct physiological experiments on deep-sea fishes: laboratory and *in situ* studies. In this chapter we emphasize the collection and maintenance of live animals, both of which require specialized equipment and techniques. Classical equipment for collection of deep-sea fishes such as trawls, dredges, and baited traps are not equipped for the recovery of specimens in good physiological condition and are not considered here.

II. LABORATORY STUDIES

Laboratory studies of living, deep-sea fishes require their capture, recovery, and maintenance. Intrinsic to such procedures are a number of limitations that must be resolved and/or acknowledged.

1. The capture process, no matter how gentle, stresses the animal. This point is evident from initial increases in oxygen consumption noted immediately after *in situ* entrapment at bathypelagic depths of a variety of fishes, including the thornyback, *Sebastolobus altivelis* (Smith and Brown, 1983); the gonostomatid, *Cyclothone acclinidens* (Smith and Laver, 1981); and the macrourids, *Coryphaenoides acrolepis* (Smith and Hessler, 1974) and *Coryphaenoides armatus* (Smith, 1978).

2. Fishes captured at depth and brought to the surface with no insulation from ambient conditions undergo extensive increases in temperature (except in polar regions with isothermal water columns) as well as decompression. To alleviate these problems, traps and trawl cod ends can be thermally insulated and/or modified to retain pressure. Such devices are described below.

3. Solar or artificial light sources can adversely affect visual pigments of deep-sea fishes accustomed to low ambient light levels over a narrow spectrum of wavelengths (e.g., Douglas et al., 1995; Fernandez, 1978; O'Day and Fernandez, 1976).

4. Fishes are generally returned to a surface ship after collection where they are exposed to abnormal motions (yawing, pitching, rolling) and vibrations.

5. Confinement of fishes in containers during laboratory maintenance creates physical and biological stresses (e.g., Robison, 1973).

6. Once on the surface, fishes are generally held in surface seawater, ignoring possible water quality differences between *in situ* and surface water conditions.

Attempts to minimize the impact of these limitations on laboratory studies of deep-sea fishes are discussed below.

A. Animal Collections

Two primary factors influencing the physiological condition of deep-sea fishes collected for laboratory studies are increasing temperature and

9. METHODS FOR STUDYING DEEP-SEA FISHES

decompression. Hence, collections of fishes from bathypelagic depths require temperature insulation in geographic areas without an isothermal water column, and ambient pressure retention for species with pressure-sensitive biochemical reactions or tissues (e.g., swim bladder). The opaque traps and trawl cod ends used to maintain *in situ* temperature and pressure also serve to protect the visual pigments of the animals.

1. TEMPERATURE INSULATION

Thermally insulated cod ends have been successfully developed and used on a variety of opening–closing midwater trawls and epibenthic sleds to collect and recover living bathypelagic animals for metabolic studies in shipboard laboratories (Childress *et al.*, 1978; Childress, 1983). A typical cod-end device consists of a polyvinylchloride (PVC) or polypropylene tube with a mesh liner bag and with guillotine valves at each end (see Childress *et al.*, 1978). The valves are held open during deployment with pins attached by a lanyard to the closing bar of the trawl mouth. On closure of the trawl mouth, the sliding plates of each valve are released and pulled across the cod-end aperture by extension springs (surgical tubing), thus sealing the catch. Replacement of the guillotine closure with ball valves has been very effective in increasing the robustness of the device and its successful closure (see Childress *et al.*, 1978). With a tube wall thickness of 1.25 cm and a large enclosed volume of water (20–30 liters), thermal insulation can be maintained within $\sim 5°C$ of the *in situ* collection temperature. This opaque cod end also allows bathypelagic animals to be collected without exposure to surface light. Additional modifications have included a hydraulic activator to open and close the ball valves when this cod end is used on an epibenthic sled and trawl to collect benthopelagic animals (Childress, 1983).

Temperature-insulated traps also have been employed for collecting scavenging deep-sea fishes. Such traps have been either attached to long pull lines extending to depth from the surface or configured as free vehicles. As a pull line, the release of the ballast eases noticeable tension, signaling an observer to undertake a retrieval process, usually with a hydraulically driven line puller of sufficient power to haul these large-volume traps through a davit or A-frame on board a ship or small boat. An alternative method is to use free-vehicle systems, defined here as autonomous instrument packages that are deployed with attached flotation and disposable ballast and have the capability of releasing the ballast at depth, allowing the instrument to become positively buoyant and return to the surface for recovery (Isaacs and Schick, 1960; Phleger and Soutar, 1971; Smith *et al.*, 1979). A free-vehicle system generally consists of a mast assembly, flotation, mooring line, instrument package, ballast release, and disposable ballast

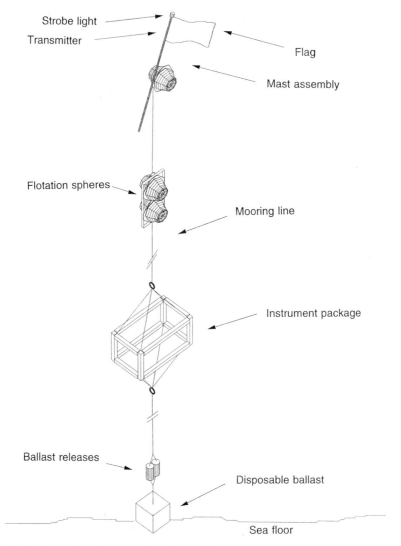

Fig. 1. Illustration of a generalized free-vehicle system with the main components identified.

(Fig. 1). The mast assembly attached to the top of the mooring is equipped with a submersible radio transmitter and strobe light, each of which has a pressure-activated on/off switch. Positively buoyant hollow glass spheres or containers of low-density petroleum products are attached to the mast

9. METHODS FOR STUDYING DEEP-SEA FISHES

and along the mooring to provide sufficient buoyancy to float the free vehicle when the disposable ballast is released. A mooring line of synthetic braided rope (nylon or polypropylene) or wire cable is used to attach the various components into one integrated autonomous system (e.g., Berteaux, 1991). Instrument packages can range from baited traps to more elaborate systems involving cameras, current meters, and acoustic transponders (described below). The mooring is usually anchored to the bottom with disposable ballast (e.g., scrap metal) of sufficient weight to overcome the positive buoyancy of the flotation and provide a secure anchor against local current activity. This ballast is attached to a release mechanism controlled either by a preset internal clock and firing circuit (e.g., Sessions and Marshall, 1971) or an acoustically activated mechanism (e.g., Berteaux, 1991) remotely triggered from the surface ship through a transducer. Simpler releases involving corrosive links (e.g., dissolution of a magnesium rod in seawater) (Isaacs and Schick, 1960) are also quite effective and in many cases are used as backups to the more sophisticated electronic releases.

Thermally insulated traps have been used to catch scavenging fishes at bathyal depths. PVC tubes up to 2 m in length and 30 cm in diameter have been deployed on mooring lines in the bathyal basins off southern California (Brown, 1975; R. McConnaughey, personal communication, 1980). These tube traps have a sealed end and an open end equipped with an internally hinged door. A barbless fish hook is attached to an elastic cord anchored near the fixed end and is held in a positioning hole in front of the trap. When a fish seizes the baited hook, the elastic cord pulls the hooked animal into the trap and the hinged door closes securely against a mating flange surface on the interior of the open end of the trap. The large volume of ambient water enclosed in the trap combined with the insulation provided by the PVC walls ensures temperature insulation during recovery to within 2–3°C. The trap closure has been rigged to a plier release (Phleger and Soutar, 1971) so that the ballast weight is released on capture of a fish and the instrument can then be recovered.

2. Pressure and Temperature Insulation

Incorporating an earlier design of a shallow-water fish trap (Brown, 1975), Phleger and co-workers (1979) developed the first pressure-retaining trap for the collection of deep-sea fishes. This trap took advantage of the scavenging behavior of common benthopelagic fishes in the bathyal basins off the southern California coast. The trap consisted of an aluminum tube with one fixed end and an internal hinged door with shock cords on the other end (see Phleger *et al.*, 1979). A baited hook protruded through the door opening of the trap and was attached to a spring motor mounted inside the tube near the fixed end. The baited hook also was rigged to a plier release for the ballast weight. When a fish seized the baited hook, the animal was rapidly drawn into the trap by the spring, triggering the release of the internal hinged door, which sealed against an O-ring on the inside collar of the tube. The closure of the trap door also triggered the release of the ballast weight. As the pressure external to the trap decreased on ascent, the pressure differential firmly sealed the closure of the door and the internal hydrostatic pressure up to 2250 psi was maintained. A gauge was attached to the fixed end of the trap to provide an analog measure of the hydrostatic pressure within the trap. A plexiglass window also was built into the fixed end of the trap to permit viewing of the contents.

A more elaborate hyperbaric trap was developed by Wilson and Smith (1985) using principals similar to those described by Phleger *et al.* (1979). Their hyperbaric trap/aquarium included temperature insulation, a more effective door closure mechanism, and a gas accumulation system for maintaining *in situ* pressure during recovery and subsequent maintenance in the laboratory (Fig. 2). The cylindrical aluminum trap was wrapped with an insulating tape to provide temperature insulation during recovery. The closure mechanism of the trap consisted of a wedge-shaped guillotine door mounted above the trap entrance, allowing an unrestricted opening to the interior. When a fish seized the barbless hook extending from the entrance of the trap, the animal was drawn completely inside the trap by a spring motor, which also activated the closure of the guillotine door. This hyperbaric trap was equipped with a burnwire timed release (Smith and Baldwin, 1983) and a backup corrosive (magnesium) link release (Isaacs and Schick, 1960). A gas accumulation system similar to that described by Yayanos (1978) was plumbed to the trap through the fixed end plate to restore internal pressure when the trap volume increased due to thermal expansion of the aluminum housing in warm surface waters during recovery (Wilson and Smith, 1985). The hyperbaric trap/aquarium also was equipped with an internal light source that could be energized through external electrical penetrators to permit viewing of the contents of the trap through a central plexiglass window in the fixed end plate. This trap was successfully used in the live recovery of five grenadier fish, *Coryphaenoides acrolepis*, from bathyal depths to 1314 m (Table I), with maintenance of internal temperature to within 3°C of ambient bottom temperatures.

B. Animal Maintenance

A critical aspect of laboratory studies on living deep-sea fishes is their maintenance under simulated *in situ* conditions. We have discussed collection procedures to ensure temperature and pressure insulation while avoiding surface light. Maintaining fishes at *in situ* temperature in the lab under

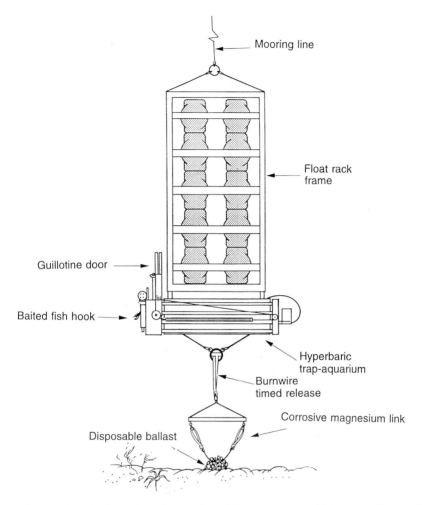

Fig. 2. A hyperbaric trap/aquarium configured for use as a free-vehicle system. Reprinted from *Deep-Sea Res.* **32,** Wilson, R. R., Jr., and Smith, K. L., Jr., Live capture, maintenance and partial decompression of a deep-sea grenadier fish (*Coryphaenoides acrolepsis*) in a hyperbaric trap-aquarium, 1571–1582. Copyright 1985, with kind permission from Elsevier Science Ltd, The Boulevard, Langford Lane, Kidlington OX5 1GB, UK.

controlled light conditions is routinely achieved using standard equipment available either on board ship or in shore-based facilities. One critical factor to consider in maintaining bathypelagic fishes is their strong avoidance–escape response when contacting surfaces (Robison, 1973). This response can be amplified by their exposure to light or vibration. To minimize these

Table I
Data from Five Successful Deployments of the Hyperbaric Trap/Aquarium[a]

Date	Location	Depth (m)	Bottom temperature (°C)	Bottom hydrostatic pressure (bars)	Internal pressure at surface (bars)	Drop in pressure (%)	Catch	Condition on recovery	Period of maintenance (h)
9/24/1984	San Clemente Basin	1241	3.2	125.7	117.1	7	*Coryphaenoides acrolepis*	Alive, upright	<1
9/27/1984	San Diego Trough	1155	3.5	117.0	79.2	32	*Coryphaenoides acrolepis*	Alive, on side	<1
11/24/1984	San Nicolas Basin	1115	3.5	112.9	72.3	36	*Coryphaenoides acrolepis*	Alive, on side	5
11/29/1984	San Nicolas Basin	1314	3.1	133.1	17.2	87	*Coryphaenoides acrolepis*[b]	Alive, ventral side up	30
01/30/1985	San Diego Trough	1152	3.5	116.7	96.4	17	*Coryphaenoides acrolepis*[b]	Alive slightly on side	41

[a] From Wilson and Smith (1985).
[b] Full pressure was restored for these specimens, with the trap inside the ship's refrigerated van.

effects on midwater fishes, Robison (1973) developed a darkened spherical maintenance system with incurrent water jets to provide an equatorial flow similar to that used in planktonkreisels (Greve, 1968). Two bathypelagic fishes, *Anoplogaster cornuta* and *Melanocetus johnsonii*, have been successfully maintained in flow-through aquaria for periods of weeks under *in situ* temperature with red light illumination to reduce adverse affects on visual pigments (Childress, 1973; B. Robison and K. Reisenbichler, personal communication, 1996). However, the planktonkreisel, which has proved an effective design for maintaining a variety of bathypelagic zooplankton and micronekton, is not suitable for faster swimming species with darting behavior (Hamner, 1990), a typical behavior of many deep-water fishes.

Few attempts have been made to maintain bathypelagic fishes in the laboratory under *in situ* temperature and pressure. Wilson and Smith (1985) developed a hyperbaric trap/aquarium for the collection of the grenadier fish, *Coryphaenoides acrolepis*, and maintenance at *in situ* temperature and pressure on board ship and in the laboratory ashore (Fig. 2). On recovery, this instrument was placed in a cold room and connected to a system created to maintain a constant flow of fresh chilled seawater through the trap without compromising the hydrostatic pressure. Five *C. acrolepis* were collected at depths greater than 1000 m and brought to the surface alive. Three of these animals were kept alive for longer periods, up to 41 h. The internal pressure of the aquarium in all instances dropped, ranging from 7 to 87% of the original *in situ* pressure (Table I). Attempts were made to decompress these fishes, which have a physoclistous swim bladder, through a slow, controlled reduction in hydrostatic pressure. Wilson and Smith (1985) hypothesized that a slow reduction in pressure would permit a full resorption of swim bladder gases and that these fishes could subsequently be maintained at atmospheric pressure, eliminating the continuous requirement for hyperbaric maintenance. However, no fish survived full decompression. The hope that such decompression would ultimately permit maintenance of these fish at atmospheric pressure and alleviate the complexities involved in maintenance at *in situ* pressures is yet to be realized.

Some bathypelagic fishes without swim bladders, such as the sablefish, *Anoplopoma fimbria*, can withstand rapid decompression and temperature changes during collection. These fish can be maintained in the laboratory in good physiological condition for long periods of time (months) at atmospheric pressure but near *in situ* temperatures in flowing aquarium systems (Sullivan, 1982; Sullivan and Smith, 1982).

III. *IN SITU* STUDIES

In situ methodology has been used to capture deep-sea fishes and measure physiological and behavioral parameters at depth, as well as to make behavioral observations on unrestrained fishes in their natural environment. However, the limitations to these approaches must be considered: (1) The capture process, with its associated stress to the fishes, is a problem relevant to *in situ* work as well as to the laboratory approach discussed previously (Smith and Hessler, 1974; Smith, 1978; Smith and Laver, 1981; Smith and Brown, 1983). (2) When submersibles and remotely operated vehicles (ROVs) are used for *in situ* collections and manipulations or observations, artificial lighting can overwhelm the visual pigments of deep-sea fishes, which normally experience attenuated light levels over a much narrower spectrum of wavelengths (e.g., O'Day and Fernandez, 1976; Fernandez, 1978; Douglas *et al.*, 1995). (3) There are severe limitations on the number of animals that reasonably can be captured and/or manipulated and on the complexity of the measurements and experiments that can be performed (Smith and Baldwin, 1983). (4) Containment of fishes in either flow-through or closed chambers creates artificial boundaries.

Given all the problems associated with both laboratory and *in situ* measurements, we feel that the *in situ* approach offers a closer approximation to the natural conditions experienced by deep-sea fish. Two basic *in situ* procedures have been used effectively to study bathypelagic fishes: containment in traps or other vessels and behavioral observations of unrestrained fishes.

A. Animal Collection/Measurements

Collection methods have involved the capture and containment of fishes through either pumping for nonscavenging species or baited traps for scavenging species.

1. PUMPING

Plankton/Nekton Respirometer. A plankton/nekton respirometer was developed to collect animals in midwater environments using a pumping mechanism for *in situ* incubations. This instrument, a slurp gun, was developed for use with a manned submersible having the manipulative and visual capabilities required for the instrument's effective use in gently collecting individual fishes. This system was described by Smith and Baldwin (1983) and contains three modules aligned horizontally on an aluminum frame (Fig. 3). Each module consists of an acrylic tube with right-angle slider valves at either end, which serve as the intake and outlet ports. A common manifold sequentially engages each respirometer module, supplying an intake hose through one right-angle valve and an outlet hose to the other valve and centrifugal pump. In operation, fishes are selected visually and one of the submersible's manipulators is used to position the intake hose

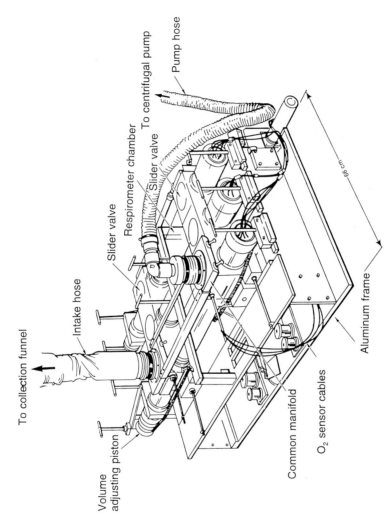

Fig. 3. The plankton/nekton respirometer for the collection of individual fishes with a manned submersible. Adapted from Smith and Baldwin (1983).

with collection funnel to gently suck the animal into the respirometer chamber. The animal is retained in the respirometer chamber by a coarse-mesh nylon filter covering the outlet valve. Each right-angle valve is then manually closed by the manipulator and a volume-adjusting piston is turned into the chamber through the intake valve assembly to alter the volume (200 to 2460 ml) based on the requirements of the metabolic measurements and the size of the fish. Each chamber is equipped with a polarographic oxygen sensor to measure respiration rates of individual animals continuously throughout the incubation period (Smith and Baldwin, 1983). A syringe system is actuated by a preset timer to withdraw water samples, for excretory product analysis, from each chamber during the incubation. Once each chamber of the plankton/nekton respirometer is filled and closed, the entire instrument is released from the submersible and tethered to a free-vehicle mooring line at the same depth as the fish were collected. This decoupling from the submersible permits respiration measurements to be made *in situ* without the temporal constraints imposed by the normal dive time of the submersible (usually <12 h) (Smith and Baldwin, 1983).

The plankton/nekton respirometer was successfully used with the submersible Alvin for individual collection of three adult females of the gonostomatid fish, *Cyclothone acclinidens,* at 1300 m depth in the Santa Catalina Basin off southern California, and to measure their oxygen consumption for a period of 28 h at the depth of collection (Smith and Laver, 1981). This instrumentation allowed the first live collection of this ubiquitous bathypelagic animal, which is very fragile and does not survive the rigors of trawl collections. These *in situ* measurements of oxygen consumption also provided the first evidence for a daily pattern in respiration, with nocturnal rates exceeding diurnal rates (Smith and Laver, 1981).

The plankton/nekton respirometer was also used to collect and measure the oxygen consumption of two pelagic juveniles of the bathyal thornyhead, *Sebastolobus altivelis.* Over a 48-h measurement period, the nighttime respiration was substantially higher than respiration during the day. Artificial lights from the submersible were used only when necessary during the collection process, but their use probably altered the physiological responses of these fishes.

The principle of the plankton/nekton respirometer has been used very effectively to develop an *in situ* respirometer for use with the Johnson Sea-Link submersibles (Bailey *et al.,* 1994, 1995). This respirometer consists of eight acrylic chambers and has been used to collect even the most fragile zooplankton and micronekton in excellent physiological condition. Sensors mounted in each chamber are used to measure oxygen consumption of the enclosed animals either while the respirometer unit is attached to the submersible or after it is tethered to a free-vehicle mooring line for longer

9. METHODS FOR STUDYING DEEP-SEA FISHES

incubations. This system was used to collect five specimens of the midwater eel, *Serrivomer beani,* in good physiological condition (Bailey *et al.* 1995), and in its present configuration could be used to collect and then measure respiration of bathyal fishes *in situ* (T. Bailey, personal communication, 1996).

2. Trapping

Baited Trap Respirometer. A trap respirometer was built as a free-vehicle system to collect and measure the respiration rates of fishes attracted to a bait source (Smith and Baldwin, 1983). A prototype of this respirometer was first built and used with a ROV and then with a submersible. The trap consisted of an acrylic box with a hinged door on one end and a bait source mounted inside to attract scavenging fishes. When a fish entered the trap, as observed with real-time video from the ROV or directly from the submersible, a manipulator arm closed the door of the respirometer. A control trap without a fish was used to assess any oxygen consumption due to the bait and enclosed water. This respirometer was first used with an ROV, the remote underwater manipulator (RUM) system, at a depth of 1230 m in the San Diego Trough. Two benthopelagic fishes, the macrourid, *Coryphaenoides acrolepis,* and the myxiniid, *Eptatretus deani,* were collected and their oxygen consumption measured. These were the first attempts to measure the metabolic activity of deep-sea fishes and revealed rates significantly ($p < 0.05$) lower than respiration rates in comparable shallow-water species (Smith and Hessler, 1974). Similar trap respirometers were used with the submersible Alvin to measure the respiration rates of three individual *Coryphaenoides armatus* from depths of 2753 and 3650 m in the western North Atlantic (Smith, 1978). These few measurements supported the physiological axiom that respiration increases as a fractional power of body weight in these deep-sea fishes. These reduced metabolic rates, when compared to shallower living fishes, suggested adaptation to a food-limited environment (Smith, 1978).

This baited trap respirometer required the use of an ROV or submersible to activate the closure mechanism. To eliminate this dependence and make the system an autonomous free vehicle, we developed a sensing system that detects the presence of a fish in the trap and closes the trap door via a burnwire release (Smith and Baldwin, 1983). This sensing system consists of a series of acoustic emitters on one wall of the trap and an aligned set of sensors on the other side (Fig. 4). A continuous disruption of the transmission between the paired emitters and sensors on opposing walls of the trap indicates the presence of an animal. Once this disruption is detected by the paired sensors next to the bait source (farthest from the trap door), the trap door is released and closed, sealing the fish in the

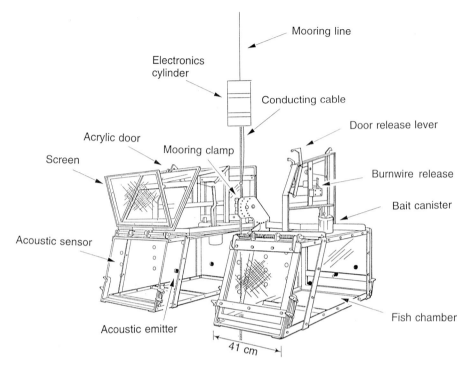

Fig. 4. Baited trap respirometers shown in tandem on a free-vehicle mooring with one trap in an open position and one in a closed position. Adapted from Smith and Baldwin (1983).

respirometer. This same electronic triggering device simultaneously withdraws the mesh-lined bait canister from the respirometer to minimize effects of the bait on the dissolved oxygen and nutrient content of the enclosed water and on the behavior of the animal (Smith and Baldwin, 1983). Two baited trap respirometers have been deployed in tandem on a standard free-vehicle mooring with the electronics cylinder (central controller and battery) secured to the mooring line above the traps (Fig. 4). We have used these trap respirometers to measure the respiration of the grenadiers *C. armatus* and *C. yaquinae* within the bottom 100 m of the water column at depths between 3600 and 6000 m in the eastern North Pacific (K. Smith, unpublished data, 1980).

B. Behavioral Observations

Behavioral observations can be used effectively to evaluate foraging strategies, swimming speeds, and dietary preferences, all of which are closely related to the physiology of deep-sea fishes. Bathypelagic fishes have been observed directly from submersibles and with remote camera systems.

1. SUBMERSIBLES

Submersibles and atmospheric diving suits have been used to observe the behavior of midwater fishes in the upper 1000 m of the water column (e.g., Barham, 1966; Robison, 1983; Auster *et al.*, 1992). At bathypelagic depths, Barham and associates (1967) made a series of dives with the bathyscaphe, Trieste, in the San Diego Trough, to depths as great as 1280 m. They observed the swimming behavior of the sablefish, *Anoplopoma fimbria*, and noted occasional gulping of sediments, suggesting feeding incidents. The swimming of the flatnose codling, *Antimora rostrata*, was also observed in association with the seafloor. Similar observations were made to evaluate the swimming performance of *A. rostrata* using the submersible Alvin at 2400 m depth in the western North Atlantic (Cohen, 1977). Photographic transects and observations from the submersible Alvin were conducted to a depth of 1800 m in the western North Atlantic, and the behavior of demersal and benthopelagic fishes belonging to the Macrouridae, Synaphobranchidae, Moridae, Halosauridae, and Alepocephalidae was recorded by Grassle and associates (1975). Their behavioral observations from the submersible confirmed the earlier contention of Barham and associates (1967) that these fishes appeared unaffected by the presence of the submarine and its lights, although they did feel that some attraction of these fishes to the sediment disturbance created by the submarine could not be discounted. In contrast, large midwater and bottom trawls rigged with lights caught more bathyal fishes than did trawls without lights, inferring a positive phototactic response in some fishes such as squalids, alepocephalids, and notacanthids (Pascoe, 1990).

2. FREE-VEHICLE SYSTEMS

a. Time-Lapse Cameras. Remote camera systems, either tethered to a ship by wire rope or configured as autonomous free vehicles, have been employed to observe the behavior of deep-sea fishes. Free-vehicle systems have been the most effective. Time-lapse camera systems have been deployed to study the behavior of fishes in the deep sea. The first such deployments were conducted by Isaacs and co-workers using motion picture and still camera systems deployed with bait to lure scavenging species within the field of view (e.g., Isaacs and Schwartzlose, 1975) (see Table II). These systems consisted of a free-vehicle apparatus (Fig. 1) with a camera and strobe unit mounted in a rigid frame to ensure the proper inclination and lighting within the desired field of view (oblique or vertical orientation) (e.g., Smith *et al.*, 1993). A bait package was placed below and in the field

Table II
Time-Lapse Camera Deployments to Examine the Behavior of Deep-Sea Fishes at Depths >1000 m

Location	Type of area	Depth (m)	Camera type[a]	Area photographed (m²)	Duration of measurements (days)	Number of deployments	Other attached instruments	Emphasis of study	Fish species observed	Reference
NE Pacific	Soft substrate	1200–1400	s/m	na[b]	na	na	None	Scavenging behavior, disturbance (bait)	*Coryphaenoides* sp., *Anoplopoma fimbria*	Isaacs and Schwartzlose (1975)
E Pacific	Hard substrate (vent nonvent)	2500	s	na	?	9	Baited trap	Scavenging behavior, disturbance (bait)	Skate, *Conocara* (?), *Antimora* sp., *Coryphaenoides bulbiceps*, *Coryphaenoides anguliceps*, *Acanthonus* sp., *Bassozetus* sp., *Spectrunculus* sp., aphyonid, zoarcid (large), zoarcid (small)	Cohen and Haedrich (1983)
NE Atlantic	Soft substrate	3852–4009	s	2	1	2	Baited trap, current meter	Scavenging behavior, disturbance (bait)	*Paraliparis bathybius*	Lampitt et al. (1983)
NE Pacific	Soft substrate	3800–5800	v	4	≤0.6	9	None	Scavenging behavior, disturbance (bait)	*Coryphaenoides* sp., Ophidiidae	Wilson and Smith (1984)

Location	Substrate	Depth			n	Equipment	Topic	Species	Reference
NE Pacific	Soft substrate	3790	v	4	<0.6	1	None	Attraction to light and sound	Coryphaenoides sp.	Wilson and Smith (1984)
NE Pacific	Soft substrate	1310	s	0.9	4–57	3	None	Scavenging behavior, disturbance (bait)	Eptatretus deani, Anoplopoma fimbria, Sebastolobus altivelis, Coryphaenoides acrolepis	Smith (1985)
NE Pacific	Soft substrate	5704–5763	v	4	<0.9	6	Acoustic tracking (ingestible transmitters)	Scavenging behavior, disturbance (bait)	Coryphaenoides yaquinae	Priede and Smith (1986)
NE Pacific	Soft substrate	4400–5900	v	8.9*	≤0.9	29	Acoustic tracking (ingestible transmitters)	Scavenging behavior, disturbance (bait)	Coryphaenoides armatus, Coryphaenoides yaquinae	Priede et al. (1990)
NE Pacific	Soft substrate	5900	v	8.9*	≤0.9	10	Acoustic tracking (ingestible transmitters)	Scavenging behavior, disturbance (bait)	Coryphaenoides yaquinae	Armstrong et al. (1991)
NE Atlantic	Soft substrate	4800–4900	s	8.7	na	19	Current meter, acoustic tracking (ingestible transmitters)	Scavenging behavior, disturbance (bait)	Coryphaenoides armatus, Synaphobranchus bathybius, Spectrunculus grandis, Barathrites sp.	Armstrong et al. (1992a)
NE Pacific	Soft substrate	4100	s	8.7	<1	11	Current meter, acoustic tracking (ingestible transmitters)	Scavenging behavior, disturbance (bait)	Coryphaenoides armatus	Priede et al. (1994a)

(*continues*)

Table II *Continued*

Location	Type of area	Depth (m)	Camera type[a]	Area photographed (m^2)	Duration of measurements (days)	Number of deployments	Other attached instruments	Emphasis of study	Fish species observed	Reference
NE Atlantic	Soft substrate	1517–4050	s	3.84–6.42	≤1	8	Current meter, acoustic tracking (ingestible transmitters)	Scavenging behavior, disturbance (bait)	*Hexanchus griseus*, zoarcid (?), *Paraliparus* spp. (?), *Centroscymnus coelolepis*, *Lepidion eques* (?), *Hydrolagus affinis*, *Synaphobranchus kaupi*, *Synaphobranchus bathybius*, *Antimora rostrata*, *Halosauropris macrochir* (?), *Spectrunculus grandis*, *Coryphaenoides armatus*	Priede *et al.* (1994b)
NE Pacific	Soft substrate	4100	s	20	120	4	None	Swimming/foraging behavior (no bait)	*Coryphaenoides armatus*, *Coryphaenoides yaquinae*	K. L. Smith (unpublished data, 1996)

[a] s, Still camera; m, motion picture system; v, video camera.
[b] na, Not available.
[c] Published as 20 m^2 and later corrected to 8.9 m^2 (I. G. Priede, personal communication, 1992).

of view of the camera, where it generally was released along with the disposable ballast. Other instruments have been attached to these camera systems to provide complementary information on individual movements of fishes using acoustic tracking devices and on local current velocities using current meters (Table II).

A wide range of studies using free-vehicle camera systems have been conducted in the Atlantic and Pacific oceans to observe the scavenging behavior of fishes at depths up to 5900 m (Table II). Most of these studies have been conducted on species in close proximity to the seafloor and in areas with soft substrates (Table II). Only one study that we are aware of reports behavioral observations of fishes over hard-substrate environments (Cohen and Haedrich, 1983); this work included baited trap and camera deployments in both hydrothermal vent and adjacent nonvent areas in the eastern Pacific. Fish species identified in these camera studies over both soft and hard substrates were dominated by macrourids, zoarcids, ophidiids, and synaphobranchids (Table II).

Most of the time-lapse camera deployments have been conducted with baited systems. However, one study was conducted with a video camera system without bait to examine the attraction of benthopelagic fishes to the physical presence of this system, with its associated light and noise (Wilson and Smith, 1984). Their results with both baited and unbaited deployments of a free-vehicle video system indicate that macrourids (*Coryphaenoides* spp.) are neither attracted to nor repelled by light. Attraction or repulsion by the sound of the camera system either hitting the bottom or its operation during the deployment was not adequately tested. Smith and associates (1993) developed an unbaited time-lapse camera system capable of photographing ~20 m^2 of the seafloor for periods up to 4 months at hourly intervals (Table II). This camera system has been used to monitor the presence of grenadiers at an abyssal time-series station in the eastern North Pacific during four 4-month deployments and one 1-month deployment. This monitoring has substantiated earlier findings that there is no consistent pattern in presence or absence of these fishes that could be correlated with the arrival of the camera system on the seafloor or with the hourly firing of the strobe light.

b. Acoustic Tagging/Tracking. One of the most sophisticated instruments to study the scavenging behavior and movements of bathyal fish has incorporated a camera and an acoustic tracking system. This free-vehicle system, referred to as AUDOS (Aberdeen University Deep Ocean Submersible), consists of a downward-looking camera and flash system mounted in a rigid aluminum tubular frame (Fig. 5). Also mounted on the frame is a scanning sonar tracking system, an electronic cylinder with microprocessor

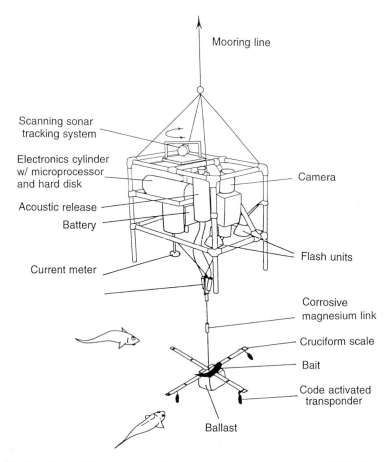

Fig. 5. A free-vehicle acoustic tracking system and time-lapse camera (AUDOS, Aberdeen University Deep Ocean Submersible). Illustration courtesy of I. G. Priede, University of Aberdeen.

and hard disk, and a battery to power the entire system (Armstrong et al., 1992a,b). Two ballast releases are incorporated into the system, one under acoustic command and the other a corrosive magnesium link. An electromagnetic current meter is also included in the instrument to provide current speed and direction during each deployment. Small ingestible acoustic transmitters are wrapped in minced mackerel flesh, stuffed into nylon fine-mesh bags, and attached in the field of view of the camera (Priede and Smith, 1986). The transmitters currently being used are code-activated transponders operating at a frequency of 77 kHz and are 13 mm in diameter and 45 mm long (Bagley, 1992). These ingestible transponders are attached below the camera on a cruciform scale used to estimate the size of fish in the field of view of the camera. Each transponder has a different activation code and can be individually interrogated by the directional scanning sonar on the instrument. Macrourid fishes frequently ingest these transponders and their movements have been tracked within 500 m of the camera system (Armstrong *et al.*, 1992b), although this system has a theoretical range of >1000 m (Bagley, 1992). Retention of these tags in the stomach of macrourids is at least 21 h (Armstrong and Baldwin, 1990), at which time the fish are usually out of the tracking range (Armstrong *et al.* 1992b). This retention time may actually be several weeks given the inverse relationship between digestion rate and water temperature (Armstrong *et al.*, 1992b). The camera system serves to identify a fish with the time of ingestion of specific tags. This methodology has been used very effectively to track the dispersal rate and direction of macrourid fishes in the north Pacific and north Atlantic oceans to depths of 5900 m (Table II). The current meter on AUDOS records current speed and direction which can then be related to the movements of the fish tracked with the sonar system.

c. Acoustic Monitoring. Free-vehicle acoustic monitoring systems have been developed to study the abundance, movements, and behavior of bathypelagic animals using noninvasive active sonar. The first system developed consisted of a split-beam line array with a beam pattern narrow in the vertical and omnidirectional in the horizontal to detect individual animals (acoustic targets) ≥2 cm in size within an insonified radius of 100 m around the moored array (Richter *et al.*, 1985; Smith *et al.*, 1989). This acoustic array operated at a frequency of 72 kHz and was used to detect individual pelagic animals, measure their acoustic target strengths, and then track their movements across specific narrow depth boundaries where the free vehicle was moored above the seafloor. This active sonar system had considerable operational flexibility so that the acoustic pinging sequence could be programmed to transmit groups of closely spaced pings to enable the tracking of individual targets and reconstruct their trajectories through the insonified field of the array. These arrays were deployed in the central north Pacific at altitudes of 100 and 600 m above the bottom (water depth of ~5800 m) for sampling periods of up to 52 h (Smith *et al.*, 1992). A total of 14 deployments of these arrays yielded 26 identifiable targets. Although these targets could not be positively identified from the acoustic records, trawling and baited trap collections used to "ground truth" the acoustic array data at these same depths suggest that the grenadier *Coryphaenoides yaquinae*, and the deep-sea eel, *Monognathus rosenblatti*, were probable suspects. We evaluated the possible influence of this active sonar system in eliciting attraction or avoidance responses of animals and noted an increased frequency of targets detected during the first 6 h of the sampling period at 100 m above bottom. However, the targets were randomly distributed around the array, which does not support the premise of attraction (Smith et al., 1992).

To better examine the vertical movements of bathypelagic animals, especially those associated with the benthic boundary layer, we have designed a split-beam acoustic array with upward- and downward-looking transducers. This array consists of a structural frame with transducers, central controller, and batteries attached to a free-vehicle mooring. No acoustically reflective components are placed above the array and the mooring below is acoustically dampened. Each of these components is described in Table III and illustrated in Fig. 6. The structural frame supports the upward- and downward-looking transducers, which are mounted at sufficient distances from the glass flotation spheres to avoid any acoustic reflection. This acoustic system operates at a frequency of 150 kHz and can resolve acoustic targets ≥1 cm (target strength, −83 dB) (e.g., Wiebe et al., 1990), to a distance of 200 m from each transducer, and insonify a conical volume of water of ~16,000 m^3 in both the upward and downward directions. If the downward-looking transducer is moored at 200 m above the seafloor, approximately 240 m^2 of the bottom is insonified. This volume can be sampled with a vertical resolution of ~50 cm throughout the 200-m segment of the water column insonified by each upward- or downward-looking array.

Prototypes of these vertically profiling acoustic arrays with single transducers have now been successfully used to study nekton over a seamount and under the Antarctic pack ice. A downward-looking acoustic array was deployed at mesopelagic depths over Fieberling Guyot in the eastern North Pacific to monitor the diel vertical movements of benthopelagic animals including the macrourids, *Malacocephalus laevis* and *Nezumian* sp. (K. Smith, unpublished data, 1992). In addition, an upward-looking acoustic array system has been successfully used to study the effects of seasonal pack ice on the distribution and movements of epipelagic nekton in the Weddell Sea (Kaufmann et al., 1995). Development of combined upward- and downward-looking acoustic arrays is now planned to examine the movements of abyssopelagic species at a long time-series station in the eastern North Pacific off central California (water depth 4100 m). Such noninvasive acoustic techniques show great promise in examining the behavior of bathypelagic fishes and addressing questions concerning migrations on a wide variety of temporal scales.

Table III
Description of the Components of a Vertically Profiling Acoustic Array

Component	Description
Instrument package (single upward/downward-looking unit)	
Structural frame	Constructed of titanium angle, reinforced epoxy, and fiberglass to minimize weight and corrosion in seawater. Vane provides stability and orientation of instrument into current
Transducer	Circular upward- and downward-looking transducers, each operating at 150 kHz. Gimbal mountings on titanium brackets elevated above and below surrounding frame and flotation
Central controller	Located in electronics cylinder. Controls battery power distribution, ping generation, T/R switch, and A/D converter. Final data storage on four 1.2-gigabyte hard drives. Transmitted pulse programmable regarding pulse length and ping sequence
Batteries	Alkaline battery pack in titanium cylinder
Mooring	
Flotation	Eight evacuated glass spheres (43.2 cm o.d.) (Benthos Inc.) with a total of 200 kg positive buoyancy hard fastened on titanium angle supports around periphery of structural frame but below upper transducer and above lower transducer
Mooring line	Acoustically transparent braided nylon line (Samson 1.9-cm diameter), 200 m long
Mooring release	Tandem acoustic releases (Benthos Model 865A) with stainless-steel cases
Disposable ballast	Steel train wheels with penetrometer base attached to releases. Penetrometer base places over 95% of the ballast at or below the sediment surface in fine-grained sediments, minimizing acoustic interference

IV. FUTURE DIRECTIONS

With the advent of electronic miniaturization and new technology, it is now possible to expand greatly the number of measurements and experiments that can be reliably conducted *in situ*. For example, fiber optic chemical sensors are now being developed to measure physiologically important parameters such as dissolved oxygen, carbon dioxide, and ammonia (Tokar *et al.*, 1990; Wolfbeis, 1991; Klimant *et al.*, 1995). These sensors are very stable over long periods of time, are not affected by electromagnetic noise,

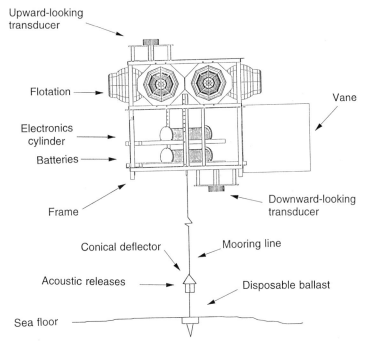

Fig. 6. Illustration of a vertically profiling acoustic array on a free-vehicle mooring positioned 200 m above the seafloor.

and could be effectively incorporated into pumping or trapping systems for *in situ* metabolic studies of deep-sea fishes.

The ingestible baited transmitters developed by Priede and co-workers (e.g., Priede *et al.*, 1990, 1994a) provide a means of getting sensing systems into scavenging deep-sea fishes. The acoustic telemetry is recorded by a central receiver, which can then transmit the data directly to a surface ship or buoy via an acoustic modem (Catipovic *et al.*, 1990; Merriam and Porta, 1993). Possible sensors that could be incorporated into the ingestible transmitters include an acoustic doppler flow probe to detect movement of organs or tissues within the fish, permitting monitoring of parameters such as blood flow, heart beat, and gill ventilation. An accelerometer could be used to record swimming movements. Fiber optic chemical sensors introduced into the stomach via the ingestible transmitters could monitor the surrounding chemical environment. Such new technologies applied to *in situ* studies of deep-sea fish physiology can provide unique insights that cannot be achieved through the study of captive or dead specimens. Biologists are often reluctant to embrace new technology, but many of the tools required to advance our science are available. All that is required is the scientific rationale, some innovation, and the perseverance to obtain the necessary funding to realize our goals.

ACKNOWLEDGMENTS

This paper benefitted from discussions with Tom Bailey, I.G. (Monty) Priede, Kim Reisenbichler, and Ray Wilson. Stace Beaulieu, Jeff Drazen, Rob Glatts, Ron Kaufmann, and Lynn Lauerman provided good constructive comments on various drafts of this paper. Fred Uhlman, the master of autocad, patiently worked with us to develop some of the figures. This work was supported by National Science Foundation Grant NSFOCE92-17334.

REFERENCES

Armstrong, J. D., and Baldwin, R. J. (1990). A method for testing retention of transmitters swallowed by deep-sea fish. *J. Fish. Biol.* **36,** 273–274.

Armstrong, J. D., Priede, I. G., and Smith, K. L., Jr. (1991). Temporal change in foraging behaviour of the fish *Coryphaenoides* (*Nematonurus*) *yaquinae* in the central North Pacific. *Mar. Ecol. Prog. Ser.* **76,** 195–199.

Armstrong, J. D., Bagley, P. M., and Priede, I. G. (1992a). Photographic and acoustic tracking observations of the behaviour of the grenadier *Coryphaenoides* (*Nematonurus*) *armatus*, the eel *Synaphobranchus bathybius*, and other abyssal demersal fish in the North Atlantic Ocean. *Mar. Biol.* **112,** 535–544.

Armstrong, J. D., Bagley, P. M. and Priede, I. G. (1992b). Tracking deep-sea fish using ingestible transmitters and an autonomous sea-floor instrument package. *In* "Wildlife Telemetry: Remote Monitoring and Tracking of Animals" (I. G. Priede and S. M. Swift, eds.), pp. 376–386. Ellis Horwood, New York.

Auster, P. J., Griswold, C. A., Youngbluth, M. J., and Bailey, T. G. (1992). Aggregations of myctophid fishes with other pelagic fauna. *Environ. Biol. Fishes* **35,** 133–139.

Bagley, P. M. (1992). A code-activated transponder for the individual identification and tracking of deep-sea fish. *In* "Wildlife Telemetry: Remote Monitoring and Tracking of Animals" (I. G. Priede and S. M. Swift, eds.), pp. 111–119. Ellis Horwood, New York.

Bailey, T. G., Torres, J. J., Youngbluth, M. J., and Owen, G. P. (1994). Effect of decompression on mesopelagic gelatinous zooplankton: A comparison of *in situ* and shipboard measurements of metabolism. *Mar. Ecol. Prog. Ser.* **113,** 13–27.

Bailey, T. G., Youngbluth, M. J., and Owen, G. P. (1995). Chemical composition and metabolic rates of gelatinous zooplankton from midwater and benthic boundary layer environments off Cape Hatteras, North Carolina. *Mar. Ecol. Prog. Ser.* **122,** 121–134.

Barham, E. G. (1966). Deep scattering layer migration and composition: Observations from a diving saucer. *Science* **151,** 1399–1403.

Barham, E. G., Ayer, N. J., and Boyce, R. E. (1967). Macrobenthos of the San Diego Trough: Photographic census and observations from bathyscaphe, Trieste. *Deep-Sea Res.* **14,** 773–784.

Berteaux, H. O. (1991). "Coastal and Oceanic Buoy Engineering." H. O. Berteaux, Woods Hole, Massachusetts.

Brown, D. M. (1975). Four biological samplers: Opening–Closing midwater trawl, closing vertical tow net, pressure fish trap, free vehicle drop camera. *Deep-Sea Res.* **22,** 565–567.

Catipovic, J., Frye, D. F., and Porta, D. (1990). Compact digital signal processing enhances acoustic data telemetry. *Sea Technol.* **31,** 10–15.

Childress, J. J. (1983). Capture and live recovery of deep-sea crustaceans. *Natl. Geogr. Soc. Res. Rep.* **21,** 67–69.

Childress, J. J. (1973). Observations on the feeding behavior of a mesopelagic fish (*Anoplogaster cornuta:* Beryciformes). *Copeia,* 602–603.

Childress, J. J., Barnes, A. T., Quetin, L. B., and Robison, B. H. (1978). Thermally protecting cod ends for the recovery of living deep-sea animals. *Deep-Sea Res.* **25,** 419–422.

Cohen, D. M. (1977). Swimming performance of the gadoid fish *Antimora rostrata* at 2400 meters. *Deep-Sea Res.* **24,** 275–277.

Cohen, D. M., and Haedrich, R. L. (1983). The fish fauna of the Galapagos thermal vent region. *Deep-Sea Res.* **30,** 371–379.

Douglas, R. H., Partridge, J. C., and Hope, A. J. (1995). Visual and lenticular pigments in the eyes of demersal deep-sea fishes. *J. Comp. Physiol. A* 177, 111–122.

Fernandez, H. R. C. (1978). Visual pigments of bioluminescent and nonbioluminescent deep-sea fishes. *Vision Res.* **19,** 589–592.

Gage, J. D. and Tyler, P. A. (1991). "Deep-Sea Biology: A Natural History of Organisms at the Deep-Sea Floor." Cambridge Univ. Press, Cambridge.

Grassle, J. F., Sanders, H. L., Hessler, R. R., Rowe, G. T., and McLellan, T. (1975). Pattern and zonation: A study of the bathyal megafauna using the research submersible Alvin. *Deep-Sea Res.* **22,** 457–481.

Greve, W. (1968). The "Planktonkreisel," a new device for culturing zooplankton. *Mar. Biol.* **1,** 201–203.

Hamner, W. M. (1990). Design developments in the planktonkreisel, a plankton aquarium for ships at sea. *J. Plankton Res.* **12,** 397–402.

Isaacs, J. D., and Schick, G. B. (1960). Deep-sea free instrument vehicle. *Deep-Sea Res.* 7, 61–67.

Isaacs, J. D. and Schwartzlose, R. A. (1975). Active animals of the deep-sea floor. *Sci. Am.* **233,** 85–91.

Kaufmann, R. S., Smith, K. L., Jr., Baldwin, R. J., Glatts, R. C., Robison, B. H., and Reisenbichler, K. R. (1995). Effects of seasonal pack ice on the distribution of macrozooplankton and micronekton in the northwestern Weddell Sea. *Mar. Biol.* **124,** 387–397.

Klimant, I., Meyer, V., and Kuhl, M. (1995). Fiber-optic oxygen microsensors, a new tool in aquatic biology. *Limnol. Oceanogr.* **40,** 1159–1165.

Lampitt, R. S., Merrett, N. R., and Thurston, M. H. (1983). Inter-relations of necrophagous amphipods, a fish predator, and tidal currents in the deep sea. *Mar. Biol.* **74,** 73–78.

Merriam, S., and Porta, D. (1993). DSP-based acoustic telemetry modems. *Sea Technol.* **34,** 24–30.

O'Day, W. T., and Fernandez, H. R. (1976). Vision in the lanternfish *Stenobrachius leucopsarus* (Myctophidae). *Mar. Biol.* **37,** 187–195.

Pascoe, P. L. (1990). Light and the capture of marine animals. In "Light and Life in the Sea" (P. J. Herring, A. K. Campbell, M. Whitfield, and L. Maddock, eds.), pp. 229–244. Cambridge Univ. Press, New York, NY.

Phleger, C. F., and Soutar, A. (1971). Free vehicles and deep-sea biology. *Am. Zool.* **11,** 409–418.

Phleger, C. F., McConnaughey, R. R., and Crill, P. (1979). Hyperbaric fish trap operation and deployment in the deep sea. *Deep-Sea Res.* **26,** 1405–1409.

Priede, I. G., and Smith, K. L., Jr. (1986). Behaviour of the abyssal grenadier, *Coryphaenoides yaquinae*, monitored using ingestible acoustic transmitters in the Pacific Ocean. *J. Fish. Biol.* **29**(Suppl. A), 199—206.

Priede, I. G., Smith, K. L., Jr., and Armstrong, J. D. (1990). Foraging behavior of abyssal grenadier fish: Inferences from acoustic tagging and tracking in the North Pacific Ocean. *Deep-Sea Res.* **37**, 81–101.

Priede, I. G., Bagley, P. M., and Smith, K. L., Jr. (1994a). Seasonal change in activity of abyssal demersal scavenging grenadiers *Coryphaenoides* (*Nematonurus*) *armatus* in the eastern North Pacific Ocean. *Limnol. Oceanogr.* **39**, 279–285.

Priede, I. G., Bagley, P. M., Smith, A., Creasey, S., and Merrett, N. R. (1994b). Scavenging deep demersal fishes of the Porcupine Seabight, north-east Atlantic: Observations by baited camera, trap and trawl. *J. Mar. Biol. Assoc. U.K.* **74**, 481–498.

Richter, K. E., Bennett, J. C., and Smith, K. L., Jr. (1985). Bottom-moored acoustic array to monitor density and vertical movement of deep-sea benthopelagic animals. *IEEE J. Oceanogr. Eng.* **10**, 32–37.

Robison, B. H. (1973). A system for maintaining midwater fishes in captivity. *J. Fish. Res. Bd. Can.* **30**, 126–128.

Robison, B. H. (1983). Midwater biological research with the WASP ADS. *Mar. Tech. Soc. J.* **17**, 21–27.

Sessions, M. H., and Marshall, P. M. (1971). A precision deep-sea time release. SIO Reference Ser. 71-5. University of California, Scripps Institution of Oceanography, San Diego.

Smith, C. R. (1985). Food for the deep sea: Utilization, dispersal, and flux of nekton falls at the Santa Catalina Basin floor. *Deep-Sea Res.* **32**, 417–442.

Smith, K. L., Jr. (1978). Metabolism of the abyssopelagic rattail *Coryphaenoides armatus* measured *in situ*. *Nature* (*London*) **274**, 362–364.

Smith, K. L., Jr., and Baldwin, R. J. (1983). Deep-sea respirometry: *In situ* techniques. *In* "Polarographic Oxygen Sensors: Aquatic and Physiological Applications" (E. Gnaiger and H. Forstner, eds.), pp. 298–319. Springer-Verlag. New York.

Smith, K. L., Jr., and Brown, N. O. (1983). Oxygen consumption of pelagic juveniles and demersal adults of the deep-sea fish *Sebastolobus altivelis*, measured at depth. *Mar. Biol.* **76**, 325–332.

Smith, K. L., Jr., and Hessler, R. R. (1974). Respiration of benthopelagic fishes: *In situ* measurements at 1230 meters. *Science* **184**, 72–73.

Smith, K. L., Jr., and Laver, M. B. (1981). Respiration of the bathypelagic fish *Cyclothone acclinidens*. *Mar. Biol.* **61**, 261–266.

Smith, K. L., Jr., White, G. A., Laver, M. B., McConnaughey, R. R., and Meador, J. P. (1979). Free vehicle capture of abyssopelagic animals. *Deep-Sea Res.* **26A**, 57–64.

Smith, K. L., Jr., Alexandrou, D., and Edelman, J.L. (1989). Acoustic detection and tracking of abyssopelagic animals: A description of an autonomous split-beam acoustic array. *Deep-Sea Res.* **36**, 1427–1441.

Smith, K. L., Jr., Kaufmann, R. S., Edelman, J. L., and Baldwin, R. J. (1992). Abyssopelagic fauna in the central North Pacific: Comparison of acoustic detection and trawl and baited trap collections to 5800 m. *Deep-Sea Res.* **39**, 659–685.

Smith, K. L., Jr., Kaufmann, R. S., and Wakefield, W. W. (1993). Mobile megafaunal activity monitored with a time-lapse camera in the abyssal North Pacific. *Deep-Sea Res.* **40**, 2307–2324.

Sullivan, K. M. (1982). The bioenergetics of the sablefish *Anoplopoma fimbria* occurring off southern California and energy allocation during low-frequency feeding in deep-living benthopelagic fishes. Ph.D. Thesis, University of California, San Diego.

Sullivan, K. M., and Smith, K. L., Jr. (1982). Energetics of sablefish, *Anoplopoma fimbria*, under laboratory conditions. *Can. J. Fish. Aquat. Sci.* **39**, 1012–1020.

Tokar, J. M., Woodward, W. F., and Goswami, K. (1990). Fiber optic chemical sensors: Exploring the light fantastic. *Sea Technol.* **31**, 45–49.

Wiebe, P. H., Greene, C. H., Stanton, T. K., and Burczynski, J. (1990). Sound scattering by live zooplankton and micronekton: Empirical studies with a dual-beam acoustical system. *J. Acoust. Soc. Am.* **88**, 2346–2360.

Wilson, R. R., Jr., and Smith, K. L., Jr. (1984). Effect of near-bottom currents on detection of bait by the abyssal grenadier fishes *Coryphaenoides* spp., recorded *in situ* with a video camera on a free vehicle. *Mar. Biol.* **84**, 83–91.

Wilson, R. R., Jr., and Smith, K. L., Jr. (1985). Live capture, maintenance and partial decompression of a deep-sea grenadier fish (*Coryphaenoides acrolepis*) in a hyperbaric trap-aquarium. *Deep-Sea Res.* **32**, 1571–1582.

Wolfbeis, O. S. (1991). "Fiber Optic Chemical Sensors and Biosensors." CRC Press, Boca Raton, Florida.

Yayanos, A. A. (1978). Recovery and maintenance of live amphipods at a pressure of 580 bars from an ocean depth of 5700 meters. *Science* **200**, 1056–1059.

INDEX

A

A_1 adenosine receptor 257–260
  A_1 adenosine receptor–G
    protein–adenylyl cyclase pathway,
    257, 262
Abyssal, 14–16
  depths, 1, 7, 15–16, 23, 126–127, 169
  plains, 14–15, 23–25, 88
  species 16, 35, 125, 160, 163
Acclimatization or acclimation, 254,
    267–268, 299–307
  effects on metabolism, 301–303
Actin, 245–247, 263, 266, 269, 312
Actinopterygii (ray-finned fishes),
    53–74
Adaptation, 43, 46, 80–81, 102–104, 106,
    196, 297, 306, 329, 344
  capacity, 245, 263–268
  tolerance, 244–263
Albuliformes, 53–54
Alepisauridae, 61, 152, 154, 176
Alepocephalidae, 58, 83, 135, 138, 150, 147,
    157,177, 365
Ampullae of Lorenzini, 342
Anaerobic metabolism, 205, 247, 295, 306
Anaesthesia, 295–296
Ancient deep-water fish, 81
Anguillidae, 54
Anguilliformes, 54–55
Annual production, 31–33, 91–92
Antarctic fishes, 217, 221, 223, 225, 333,
    343–344
Aphakic space, see Vision
Aphyonids, 63, 140, 149
Argentiniformes, 56–58
Artedidraconidae, 333
Aulopiformes, 60, 134

B

Back-diffusion, 209–210
Barbourisidae, 83
Bathyal, 14–16, 37, 178, 321, 355
Bathylagidae, 121–122, 157, 165, 167
Bathymetric profiles, 13–18
Behavior, 4, 9, 15, 28–29, 118, 132–159,
    212, 240, 253, 284–286, 299–301, 355,
    359–360, 364–368
Benthic, 13–15, 17, 24, 34–35, 79, 117
  boundary layer, 4, 15, 136, 351, 372
Benthopelagic, 14, 79, 117, 160, 174–175,
    330, 334, 339, 353, 355, 365, 369, 372
Biodiversity, 18, 34, 84
Bioluminescence, see Vision
Blood, 6, 203, 205–213, 292–293
  plasma contents, 223–226, 288
Brain, 292
  electroencephalogram (EEG), 286, 295
Bregmacerotidae, 154

C

Camera, 35, 87, 355, 365
  free-vehicle system, 369–387
  time-lapse, 365–369
Camouflage, 9, 325, 328, 332
Capacity adaptations, see Adaptation
Carangidae, 82
Carbonic anhydrase, 206–208, 233
Carcharhinidae, 50
Carcharhiniformes, 49
Catecholamines
  brain, 287, 292
  plasma, 288
Cell membrane, 208, 297–299
  composition, 254, 298, 304–306

379

fluidity, 253–257, 298, 304–306, 312
  homeoviscous adaptation, 214, 253–254, 305, 312
  phospholipid, 214, 253–254, 298, 304–306
  protein, 253–257, 298
Ceratiodei, 83, 117, 133, 145, 152–154
Chauliodontidae, 83
Chemoreception, *see* Olfaction
Chiamodontidae, 152
Chimaeriformes, 48, 49, 144
Chloride cells, 307
Chlorophthalmus, 61
Cholesterol, 214–217, 220
Chondrichthyes (cartilagenous fishes), 44–53, 83
Circulation, 286–289
Cladistics, 45
Cod-end, 352
  device, 353
  thermally insulated, 353
Communities, 8, 15–18, 24, 26, 30, 34–35, 88–90, 95, 175
Compression, 213
  effect, 290
  rate, 289–290
Congridae 55
Continental
  shelf, 1, 2, 15, 21, 24, 96, 117, 268
  slope, 36, 96, 120, 268
Copharyngid, 153
Coryphaenoides, 35, 64 , 86, 95, 99, 119, 120, 124, 127, 132, 133, 136, 138, 146, 160, 163, 177, 214, 245, 246, 255, 258, 282, 351, 352, 356, 359, 363, 364, 369, 371
Countercurrent, 202, 205, 209–211
Crustaceans, 9, 91, 121, 137, 139, 144, 154–157, 263–266
Cupula, 336
Cyclic adenosine monophosphate (cAMP), 259

D

Demersal, 79, 81, 83, 88, 117–121, 123–127, 129–150, 161–163, 169–170, 217, 222, 224, 227
Demersal deep-sea species, 35, 82, 86, 93, 90, 118–119

Derichthyidae, 54–55, 154, 156
Detritivores, 13, 31–32, 131, 149–150
Development, 104–106, 161, 202, 224
Diet, 9, 35, 100–102, 118–120, 124, 129–172, 216, 220
  analyses, 100, 123
  derived from scavenging, 127
  seasonal changes, 169
Diretmoides, 67, 332

E

Eggs, 104–106
Elasmobranchii, 49–53, 217–219, 223, 227, 230
Elopiformes, 53
Elopocephala, 53–74
Elopomorpha, 53–56
Energy, 7, 96, 105
  content, 176–177
  expenditure, 196–202, 212, 217, 220, 222, 225, 228
  lipid content, 177
  protein content, 177, 304, 306, 310
  Q_c definition, 178
Energy flow, 101, 169
  sinking rates, 173
  vertical coupling, 172
Energy transport, 174
Enzyme
  activity, 177, 205, 246, 309, 311
  citrate synthase, 309
  creatine phosphokinase, 309
  cytochrome *c* oxidase (COX), 295, 298, 302, 309
  glyceraldehyde 3-phosphate dehydrogenase, 249
  isocitrate dehydrogenase, 303
  lactate dehydrogenase (LDH), 246–251, 303, 309
  malate dehydrogenase (MDH), 249–251, 303, 309
  Na^+-K^+ ATPase, 255–257, 263, 267, 269, 291, 303, 307, 309
  pyruvate kinase, 266, 295, 302, 309
  V_{max}, 248–249, 261, 309
  K_m, 248–250, 259, 261–263, 309
Enzyme–substrate binding, 241, 247–253
Epibenthic, 139–142, 160
  sleds, 353

INDEX   **381**

Erythrocyte, 292
Euphausiids, 100, 121, 137, 149, 162, 166
Euryphagous, 130–132, 137
Eurypharyngidae, 56, 117, 159
Euteleostei, 56–74
Evermannellidae, 61, 152–154
Exocoetidae, 82

F

Fatty acids, 214–217, 219–220, 251, 254, 294
Faunal boundaries, 34, 87
Fecundity, 105
Feeding, 91, 99–102, 225
    activities, 127, 170
    aggregations, 138
    apparatus, 125
    bathygadine, 163
    behavior, 116, 125, 128–129, 150, 227
    chronology, 116
    deposit-feeding, 124
    diel patterns, 169–172, 167
    energy transfer, 179
    generalists, 121, 123, 131, 158–159
    habits, 116, 118–129
    macrourine, 163
    net-feeding, 118
    opportunist, 118, 123, 129, 131, 136, 144, 165–166
    pelagic, 120, 128, 154
    seasonal patterns, 167, 169–172
    sporadic, 134
    swallowers, 128
    "vacuum cleaner," 150
Florescence polarization, 242
Fluorescent probe, 304
Food availability, 16, 93, 102–104, 127, 130, 176–177, 220, 225, 264, 266
Foodfalls, 172–174
Food webs, 30–32, 34, 99–100
Foragers
    active, 127, 135, 153–154, 174, 264
    macronekton, 131, 137–139
Foraging, 87, 106, 120–121
    benthopelagic, 135
    optimal, 127

Free-vehicle system, 353–354, 365–373

G

Gadidae (codfishes, haddocks, and cuskfishes), 65
Gadiformes, 86, 131
Gas density, 6–7, 196–197, 211
Gas gland cells, 203–208, 210, 213
Gas law, 197, 201
Gempylidae, 73, 152, 154
Gibbs free energy, 242, 244
Giganturidae, 62, 152–154
Gill, 211
Glyceralderhyde 3-phosphate dehydrogenase, *see* Enzymes
Gonostomatidae, 83, 117, 127, 154, 167–169, 172, 175, 351–352, 362
Guanosine diphosphate (GDP), 261
Guanosine triphosphate (GTP), 259, 262
Gut evacuation rates, 116, 166–168

H

Hadal, 1, 14, 24
Hair cell receptors, 334, 336
Halocline, 10
Halosaur, 125, 130, 133, 139, 143, 145, 162, 163, 327, 339, 342
Halosauridae, 53, 54, 117, 133, 365
Hearing, *see* Vestibular system
Heart, 290
Heincke's Law, 95
Helium, 296
Hemoglobin, 208, 212–213, 251–252, 292
Heterodontiformes, 49
Hexanchiformes, 50
Holocephali, 48–49
Hyaline, 102–103
Hydrodynamic lift, 227–229
Hydrostatic pressure, 5, 7, 196–197, 239–240, 252, 262, 282–284, 351, 356, 359, 369
Hydrothermal vents, 13, 23, 249, 251, 255–267, 271
Hyperbaric, 205, 212, 280
    chambers 240, 264

I

Inert gases, 208, 209, 211, 213, 281, 296
  partial pressure, 281
Infrared spectroscopy, 242, 270

L

Lactate dehydrogenase, see Enzymes
Lake Baikal, 70, 330
Lamniformes, 50
Lateralis systems, 333–342
  electroreception, 333, 340–342
  hearing, 334–335
  mechanosensory lateral line, 335–336, 344
  neuromast, 336–338
Le Chatelier's principle, 242
Light, 7–11, 30, 325, 328, 352–354, 356–357, 360, 362, 365, 369
  intensity, 7–9, 171, 177, 329, 342
Lipid–protein interactions, 241, 253–257

M

Macrohamphsoides platycheilus, 74
Macroplanktonivores, 146–147
Macrouridae, 83, 117, 119, 129, 130, 131, 160–161, 327, 365
Macrouroides, 64
Macrourids, 118–120, 124–127, 130–132, 135–141, 146, 160–163, 169, 170, 173, 174, 177, 245, 246, 335, 339, 342, 351, 352, 363, 369, 371, 372
Malate dehydrogenase, see Enzymes
Marine snow, 172–174
Melamphaidae, 129, 154, 157
Melanocetidae, 66, 152
Membrane fluidity, see Cell membrane
Metabolic rate ($\dot{M}_{O_2}$), 117, 139–142, 151–154, 171, 200, 263–268, 284, 289–290, 295, 296, 299–301, 308, 311–312, 352, 362, 363
Micronektonivores, 121, 151
Migration
  diel, 15, 16, 32, 98, 100, 128, 151, 169–171, 174, 342
  ontogenetic, 29, 98, 170
  vertical, 16, 29, 32, 120–121, 128, 151, 153, 154, 169, 171, 174, 197, 212, 213, 229, 246, 254, 265, 267
Moridae, 83, 118, 119, 126, 135, 136, 138, 139, 141, 143, 259, 327, 365
Mucous cells, 307
Muscle
  biochemistry, 293–295, 308–310
  NAD/NADH, 247, 248, 294
  structural changes, 221–222, 226–227, 306
Myctophidae, 83, 106, 117, 121–123, 127, 133, 151, 154, 158, 165–170 172–175, 179, 212, 213
Myoglobin, 72, 251–253

N

NAD/NADH 294, see also Muscle
NADP/NADPH, 294
Necrophagivores, 131, 147–149
Nektonic organisms, 123
Nektonivores, see Predator
Nemichthyidae, 55, 154–155
Nepheloid layer, 173
Neritic, 14, 175
Nervous system, 240, 257, 286–289
Nettastomatidae, 55
Neurotransmitter, 292, 303
Nitrogen, 4, 7, 177, 197, 208, 296
Notacanthids, 54, 144, 145
Notacanthiformes, 83, 150, 365
Nuclear magnetic resonance (NMR), 242, 270
Nucleotides, 294
  ATP, ADP, AMP, 259–260, 262, 294, 301
  energy charge, 294

O

Ocean
  basins, 18–26
  biophysics, 32
  circulation, 4, 5, 12, 18, 22, 23, 26–29
  eddies, 29, 198, 227
  upwelling, 29
Olfaction, 127, 140, 174, 326–328
  mate location and recognition, 327
  pheromonal communication, 327

INDEX

Olfactory lamellae, 326
Omnivore, 123
Omosudidae, 153
Ophidiidae, 63, 119, 124, 135, 138, 139, 141–143, 145, 146, 176, 177, 226, 369
Ophidioidei, 83
Opisthoproctidae, 57, 122, 157
Opportunistic predators, *see* Predation
Orectolobiformes, 49
Orientation and navigation, 342–343
Osmolarity, 218, 223, 224, 226
Osteoglossiformes, 53
Otoliths, 102–103, 334, 171
Oxygen consumption, *see* Metabolic rate

P

Paralepididae, 61, 152–154
Pelagic, 14, 15, 28, 29, 34, 36, 79, 81–83, 87–88, 116, 117, 121–123, 127–129, 131, 132, 150–160, 163–170, 196, 198, 200, 217, 218, 224, 227, 228
Pentose phosphate shunt, 206, 210
Perciformes, 69, 86, 344
Pheromone, 327–328
Phosichthyidae, 154
Photic zone, 326
Photophores, 9, 328, 332
Phylogenetic, 103, 105, 269–271, 344
  context, 269
  information, 269
  limitations, 269
Phylogeny, 44–45, 204
Physoclist, 202, 204, 259
Physostome, 202, 204
Piscivores, 121, 132–137
Platytroctidae, 57–58, 147
Pleuronectiformes, 83, 86, 133, 143
Pneumatic duct, 202
Predation, 9, 91 105, 327, 332
  ambush, 125, 128, 136, 153
  benthivorous, 125, 142–144
  nonselective, 118, 130
  opportunistic, 123
Predator, 131
  ambush, 132, 153, 156
  ambush nectivores, 153

generalists, 121, 123, 131, 139–142, 158–159
nektonivores, 28, 121
sit-and-wait ambush, 153, 254
zooplanktivores, 121, 127, 154, 158
Pressure, 5
  effects, 279–314, 343
    shallow-water versus deep-water fishes, 310–313
    short-term 284–299
  hydrostatic, *see* Hydrostatic pressure
  pressure–temperature interaction, 23, 297
  thresholds, 284, 291
Prey selectivity, 141, 165
Prey size, 91, 129–131, 166
Productivity, 13, 22, 32–34, 160, 176, 177, 265
  primary, 28, 30, 84, 88, 91–93, 142, 265
Protein
  denaturation, 243, 244
  polymerization, 244, 260
  protein–protein interactions, 245–247, 353
  structure, 244, 309, 310
Pseudoceanic, 117, 175

R

Radioactive isotopes, 13
Rajidae, 53, 119
Rajiformes, 52, 86
Regnard, 284
Remote underwater manipulator (RUM), 363
Remotely operated vehicles (ROVs), 181, 360
Reproduction, 103–106, 343
  K-strategists, 104
  R-strategists, 104
  livebearers, 104
  oviphagous, 104
  hermaphroiditism, 104
Respirometer, 308
  baited trap, 363–364
  plankton/nekton, 360–363
Rete mirabile, 202–205, 209–211, 213
Root effect, 207, 208, 210, 212–213, 252

INDEX

S

Saccopharyngiiformes, 55–56
Saccopharyngidae, 83, 117, 152, 153
Salting out effect, 207–210
Sampling
　gear, 85
　techniques, 85
Scavenging, 120, 125–127, 136, 144, 147, 148, 163, 327, 355
Sciaenid, 160, 225
Scopelarchidae, 61, 62, 152–154
Scorpaenids, 69, 133–134, 146, 247
Scorpaeniformes, 69, 86
Scyliorhinidae, 49, 135
Seamounts, 120, 174, 372
Serrivomeridae, 55
Skin, 150, 197, 198, 222, 227, 336
Sound, 9–10, 212
　production, 335
　transmission, 10
Species
　biomass, 91–93, 97, 101
　commercial, 86
　diversity, 17, 86–87, 93, 97
　numbers, 45, 81, 82, 85, 86, 93
　patchiness, 28
　richness, 18, 24
　size spectrum, 31
　spatial distribution, 31, 87–91
Squalene, 201, 202, 215–219, 221
Squalids, 51–52, 104, 119, 136, 365
Squaliformes, 51, 86
Sternoptychidae, 59, 83, 117, 127, 154, 168, 169
Stomach contents, 99, 118–120, 124, 165
Stomiatoidei, 83
Stomiidae, 59, 117, 123, 152–155, 158, 165, 168–170
Submersible, 360, 365
　Aberdeen University Deep Ocean (AUDOS), 369–371
　bathyscaphe, 365
　Johnson Sea-Link, 362
　manned, 360, 361
Survival time, 284–286
Swim bladder, 118, 150, 151, 176, 177, 196, 200–214, 220, 221, 227, 229, 230, 252, 335, 343, 359
　lipid-fillled, 213–214
　oval, 202, 211

Swimming hydrodynamics, 339–340
Synaphobranchidae, 54, 117, 126, 135, 136, 143, 327, 365, 369
Systematics, 43–77

T

Temperature, 4–6, 96, 196, 197, 201, 216, 217, 222, 247, 249–251, 253–255, 257, 262–265, 268, 296–297, 343, 353–356
　acclimatization, 305–306
Thermally insulated cod ends, *see* Cod ends
Thermocline, 10
Thunnidae, 82
Tissue composition, 303–304
Torpenoidei, 52
Touch, 333
Tracking, acoustic, 369–371
Transducers, split beam acoustic, 372
Transponder, ingestible, 371
Traps
　baited, 127–128, 144, 148, 352, 355–356
　hyperbaric, 356–358
　pressure-retaining, 355–356
　shallow-water, 355–356
　temperature-insulated, 353
Triakidae, 50

V

Ventilation, 286–289
Vestibular system, 334–335
Video system, 363, 369
　free-vehicle, 369
Vision, 136, 140, 142, 149, 153, 328–333
　aphakic space, 147, 332
　bioluminescence, 9, 138, 153, 329–331, 333
　emission spectra, 329
　intraspecific communication, 333
　multiple banks of rods, 332
　pure rod retinas, 331
　tapeta lucida, 332
　tubular eyes, 62, 153, 331
　visual pigments, 330, 352, 353, 359

W

Water
　column stability, 11

density, 4–6, 196–199, 211, 214, 219, 223, 225, 227
dissolved carbon dioxide, 6, 373
dissolved oxygen, 7, 11, 97, 252, 364, 373
heat capacity, 2
nutrients, 13, 364
pH, 3, 6
properties, 2
salinity, 3, 97, 196, 223, 225
Wax ester, 213, 215–222

X

X-ray crystallography, 242

Z

Zeiformes, 83
Zoarcidae, 83, 117, 128, 134, 142, 176, 249, 250, 256, 260, 369
Zonations, 8, 15, 16
  abyssopelagic, 14, 15
  bathypelagic, 14, 82
  epipelagic, 14, 82, 151, 172
  euphotic zone, 10
  mesopelagic, 14, 82, 152, 169, 330, 342
Zooplanktivores, *see* Predator

OTHER VOLUMES IN THE FISH PHYSIOLOGY SERIES

VOLUME 1	Excretion, Ionic Regulation, and Metabolism
	Edited by W. S. Hoar and D. J. Randall
VOLUME 2	The Endocrine System
	Edited by W. S. Hoar and D. J. Randall
VOLUME 3	Reproduction and Growth: Bioluminescence, Pigments, and Poisons
	Edited by W. S. Hoar and D. J. Randall
VOLUME 4	The Nervous System, Circulation, and Respiration
	Edited by W. S. Hoar and D. J. Randall
VOLUME 5	Sensory Systems and Electric Organs
	Edited by W. S. Hoar and D. J. Randall
VOLUME 6	Environmental Relations and Behavior
	Edited by W. S. Hoar and D. J. Randall
VOLUME 7	Locomotion
	Edited by W. S. Hoar and D. J. Randall
VOLUME 8	Bioenergetics and Growth
	Edited by W. S. Hoar, D. J. Randall, and J. R. Brett
VOLUME 9A	Reproduction: Endocrine Tissues and Hormones
	Edited by W. S. Hoar, D. J. Randall, and E. M. Donaldson
VOLUME 9B	Reproduction: Behavior and Fertility Control
	Edited by W. S. Hoar, D. J. Randall, and E. M. Donaldson
VOLUME 10A	Gills: Anatomy, Gas Transfer, and Acid–Base Regulation
	Edited by W. S. Hoar and D. J. Randall
VOLUME 10B	Gills: Ion and Water Transfer
	Edited by W. S. Hoar and D. J. Randall
VOLUME 11A	The Physiology of Developing Fish: Eggs and Larvae
	Edited by W. S. Hoar and D. J. Randall

VOLUME 11B	The Physiology of Developing Fish: Viviparity and Posthatching Juveniles *Edited by W. S. Hoar and D. J. Randall*
VOLUME 12A	The Cardiovascular System *Edited by W. S. Hoar, D. J. Randall, and A. P. Farrell*
VOLUME 12B	The Cardiovascular System *Edited by W. S. Hoar, D. J. Randall, and A. P. Farrell*
VOLUME 13	Molecular Endocrinology of Fish *Edited by N. M. Sherwood and C. L. Hew*
VOLUME 14	Cellular and Molecular Approaches to Fish Ionic Regulation *Edited by Chris M. Wood and Trevor J. Shuttleworth*
VOLUME 15	The Fish Immune System: Organism, Pathogen, and Environment *Edited by George Iwama and Teruyuki Nakanishi*